STOCHASTIC CALCULUS

and

STOCHASTIC MODELS

Probability and Mathematical Statistics

A Series of Monographs and Textbooks

---

1. Thomas Ferguson. Mathematical Statistics: A Decision Theoretic Approach. 1967
2. Howard Tucker. A Graduate Course in Probability. 1967
3. K. R. Parthasarathy. Probability Measures on Metric Spaces. 1967
4. P. Révész. The Laws of Large Numbers. 1968
5. H. P. McKean, Jr. Stochastic Integrals. 1969
6. B. V. Gnedenko, Yu. K. Belyayev, and A. D. Solovyev. Mathematical Methods of Reliability Theory. 1969
7. Demetrios A. Kappos. Probability Algebras and Stochastic Spaces. 1969
8. Ivan N. Pesin. Classical and Modern Integration Theories. 1970
9. S. Vajda. Probabilistic Programming. 1972
10. Sheldon M. Ross. Introduction to Probability Models. 1972
11. Robert B. Ash. Real Analysis and Probability. 1972
12. V. V. Fedorov. Theory of Optimal Experiments. 1972
13. K. V. Mardia. Statistics of Directional Data. 1972
14. H. Dym and H. P. McKean. Fourier Series and Integrals. 1972
15. Tatsuo Kawata. Fourier Analysis in Probability Theory. 1972
16. Fritz Oberhettinger. Fourier Transforms of Distributions and Their Inverses: A Collection of Tables. 1973
17. Paul Erdös and Joel Spencer. Probabilistic Methods in Combinatorics. 1973
18. K. Sarkadi and I. Vincze. Mathematical Methods of Statistical Quality Control. 1973
19. Michael R. Anderberg. Cluster Analysis for Applications. 1973
20. W. Hengartner and R. Theodorescu. Concentration Functions. 1973
21. Kai Lai Chung. A Course in Probability Theory, Second Edition. 1974
22. L. H. Koopmans. The Spectral Analysis of Time Series. 1974
23. L. E. Maistrov. Probability Theory: A Historical Sketch. 1974
24. William F. Stout. Almost Sure Convergence. 1974
25. E. J. McShane. Stochastic Calculus and Stochastic Models. 1974

In Preparation

Z. Govindarajulu. Sequential Statistical Procedures

Roger Cuppens. Decomposition of Multivariate Probabilities

STOCHASTIC CALCULUS and *STOCHASTIC MODELS*

E. J. McShane

Department of Mathematics
University of Virginia
Charlottesville, Virginia

1974

Academic Press New York San Francisco London

A Subsidiary of Harcourt Brace Jovanovich, Publishers

ACADEMIC PRESS, INC.
111 Fifth Avenue, New York, New York 10003

United Kingdom Edition published by
ACADEMIC PRESS, INC. (LONDON) LTD.
24/28 Oval Road, London NW1

Library of Congress Cataloging in Publication Data

McShane, Edward James, Date
Stochastic calculus and stochastic models.

(Probability and mathematical statistics series,
Bibliography: p.
1. Stochastic integrals. 2. Stochastic differential equations. I. Title.
QA274.22.M32 519.2 74-1640
ISBN 0-12-486250-0

PRINTED IN THE UNITED STATES OF AMERICA

CONTENTS

PREFACE

The study of systems subjected to random disturbances requires a calculus of stochastic processes and a procedure for using such a calculus to construct mathematical models of these systems. There are many published researches on this subject, most of them making use of the Itô integral. But, as A. H. Jazwinski remarks in the preface to his excellent book, *Stochastic Processes and Filtering Theory*, "available literature on the continuous nonlinear theory is quite esoteric and controversial." The purpose of this book is to present a stochastic calculus and a method of constructing mathematical models that is accessible to scientists who have neither the time nor the inclination to study measure theory deeply, and that avoids the controversial choice between one formulation and another by having a unified theory that applies equally well when the input noise is determinate and smooth and when it is "white" or "colored."

In this book we define two kinds of stochastic integral. The first is clearly a close relative of the Riemann integral; we call it the "belated" integral. The second has a definition that resembles that of the belated integral, but is not quite as close a replica of Riemann's definition. In spite of its apparent resemblance to the Riemann integral, in deterministic situations it generalizes the Lebesgue integral, and in stochastic situations it generalizes the Itô integral. We call it the "Itô-belated" integral. But in spite of the generality of the Itô-belated integral, we have chosen to assign the principal role in this book to the belated integral because of a belief that it is especially easily comprehended. At the ends of Chapters II, III, and IV there are sections extending the theory to the Itô-belated integral. But the theory of stochastic differential equations is at best rather intricate, and the added generality obtained by using the Itô-belated integral does not seem to produce enough profit in the form of interesting applications to justify the added complexity, at least at present.

While we have been willing to accept the continuity restrictions on integrands that accompany a Riemann type of integral, we have regarded it as of primary importance to make as few assumptions as possible about the noise processes z. It is undesirable to base a modeling procedure for physical systems on a calculus that is restricted to martingales z, which are not physically realizable. Our stochastic calculus puts only physically reasonable restrictions on the noise processes involved. This makes it easier to compare the solutions of equations with different input noise processes. In particular, when the classical model is extended to allow a large class of noises (*not* replaced by a different equation to be used only for martingale noises!) according to the procedure in Section 3 of Chapter II, the sources of controversy dry up. For example, under suitable hypotheses, if $z_1, z_2, \ldots$ is a sequence of random functions whose sample functions tend uniformly to those of a process z_0, and the solution of our model equation with noise-input z_n is x_n, then the sample functions of x_n almost surely converge uniformly to those of x_0.

ACKNOWLEDGMENTS

I owe thanks to Professor A. V. Balakrishnan for having encouraged me, five years ago, to undertake the writing of this book, and also for having called my attention to the desirability of proving several theorems that I had not considered then, but that are now included in the stochastic calculus. I also am grateful to Professor Eugene Wong for stimulating conversations, and in particular for convincing me that an approximation theorem of mine needed strengthening. My colleagues Professors James Murphy and James Rovnyak have been most cooperative and helpful, and their suggestions have produced several improvements in the manuscript. Professor Daniel Stroock did me the favor of pointing out the need of proving a continuity theorem of the kind that is in Section 5 of Chapter VI. Joseph Ball and Boyd Pearson called my attention to the fact that some results in a paper by W. M. Wonham did not follow from my first version of Section 5 of Chapter VI, although they could be deduced from preceding theorems; this impelled me to develop the present version of that section. When the manuscript was in what I thought was final form, it was read by Professors Albert Schwarzkopf and James Wendel, each of whom made numerous valuable comments about content and style. As a result I have revised the text extensively, and I believe that it is significantly more readable and usable than it was in the earlier version. I am happy to express my gratitude for their cooperation.

My colleague, Professor Hui-Hsiung Kuo, did me the great favor of reading the proof sheets of the whole book. This resulted in a considerable reduction in the number of uncorrected typographical errors. More importantly, Professor Kuo pointed out several places in which statements, as I had written them, were simply wrong but could be corrected by a small change in wording.

Finally I am much indebted to Susan Sabin, Elizabeth Jordan, Marie Brown, and Georgia Murphy who typed the manuscript with skill, dispatch, and unfailing good nature.

Part of the research contained in this book was supported by the Army Research Office, under Grants ARO-D-G182 and ARO-D-G1005.

STOCHASTIC CALCULUS and *STOCHASTIC MODELS*

I Introduction

0 *Motivation, and a forward look*

The differential calculus is the art of setting up and solving differential equations; and Newton's prime incentive in its development was to use differential equations to construct mathematical models of physical systems. If, for example, the state of a system is characterized by a number x, and the system is subjected to an external disturbance whose amount during a time interval $[s, t]$ is characterized by the increment $z(t) - z(s)$ of some continuously differentiable function z, it often happens that experimental evidence indicates that the state $x(t)$ of the system at time t satisfies an equation

$$(0\text{-}1) \qquad x(t) = x(0) + \int_0^t f(s, x(s))\, ds + \int_0^t g(s, x(s))\, dz(s),$$

the last integral being a Riemann–Stieltjes integral. (Extensions with vectors x and z and matrices f and g are obvious.) When z is known and smooth, the situation has been mastered.

However, in recent years it has become important to consider the time development of such systems when the disturbance z is a random function, and we are interested in the distribution of the random variable $x(t)$ rather than in the particular value of $x(t)$ that corresponds to some particular sample of the random function z. Furthermore, for valid reasons engineers have been particularly interested in the idealization, possessing especially desirable mathematical properties but never encountered experimentally, in which the disturbance is "Gaussian white noise," i.e., z is a Wiener process. In this case the last integral in our equation cannot be interpreted as a Riemann–Stieltjes or Lebesgue–Stieltjes integral. We need to extend the calculus so as to be able to set up and solve differential equations involving random functions of fairly general types; and we need a method of using such differential equations to construct mathematical models of randomly disturbed systems that will accord adequately well with experimental observations. The purpose of this book is to develop such a calculus and such a method of constructing models.

The Itô integral $\int g\,dz$, with z a martingale, leads to a rich theory, and to important applications, for example in the study of Markov processes. McKean has devoted the whole of an interesting book, *Stochastic Integrals* [9], to the study of the Itô integral and related topics. In view of this success, it is natural to try to extend the calculus by including the Itô integral, and to model systems affected by "white noise" by keeping the same form of differential equation (0-1) as for smooth noises z and interpreting $\int g\,dz$ as an Itô integral. This has frequently been done; and when $g(t, x)$ is sufficiently simple (e.g., independent of x) the model is successful. But for less simple g, the situation is less satisfactory.

The Itô integral $\int g\,dz$ exists under weak hypotheses on g, but only for martingales z. So the calculus becomes a patchwork, with one definition of the integral for smooth z, another for martingale z, and none at all for some z for which a symbol like $\int g\,dz$ has been used with the overt or tacit idea that it is an Itô integral, although it is not. Besides this lack of unity, the modeling procedure of the preceding paragraph has two further defects. First, if $z_0, z_1, z_2, \ldots$ are processes such that the sample functions of z_n converge uniformly to those of z_0 (even if the z_n are polygons more and more finely inscribed in a Wiener process z_0), the corresponding solutions x_n need not converge in any sense to x_0. Second, if x is a vector in R^n, and we change to a new coordinate system y, by standard calculus procedures Eq. (0-1) is transformed into a new equation. When z is smooth, the solutions of this new equation are the transforms into y coordinates of the solutions of the original equation (0-1). But when z is a Wiener process and the integral is an Itô integral, this is not necessarily the case.

In Chapter II we shall give a definition of $\int g\,dz$ for a class of processes

z large enough to include all the types we have mentioned. The definition is very close to that of the Riemann integral. We also state a definition of the integral that includes the Lebesgue integral, but for easier accessibility we give the leading role to the Riemann-type integral.

However, for constructing models the integrals $\int g\,dz$ prove insufficient. We also need "second-order" integrals $\int h\,dz^1\,dz^2$, which are limits of Riemann sums of a form suggested by the symbol for the integral. These, and also the less important integrals of third and higher order, are defined in Chapter II. Theorems on the existence and the properties of these integrals are the subject of Chapters II–IV. This is the "stochastic calculus" of the title. There are books that contain sections entitled "stochastic calculus"; but those sections could more appropriately be entitled "martingale calculus," since they are based on the Itô integral and permit only martingale z. By contrast, the present theory could be called "unified calculus," since it includes the ordinary Riemann–Stieltjes integral. But this is unnecessary; if z denotes one single function of bounded variation, we can regard this as a process that assigns probability 1 to that particular z. So a stochastic calculus should properly include ordinary calculus as a special case.

Since our stochastic calculus includes the ordinary calculus, we are no longer faced with the dilemma of having two different sets of computation rules, one for ordinary calculus and one for martingale calculus. In ordinary calculus, if some well-behaved F depends on t and some z that are continuously differentiable, or more generally Lipschitzian [i.e., have bounded increment ratios $(z(t') - z(t''))/(t' - t'')$], we can replace dF by an expression formed by the computation rule:

(0-2) Expand ΔF formally as a Taylor's series in $\Delta t, \Delta z^1, \ldots, \Delta z^r$. Count each Δz^ρ as having one factor Δt. Discard all terms having a factor $(\Delta t)^c$ with $c > 1$; in the rest replace $\Delta t, \Delta z^\rho$ by dt, dz^ρ, respectively.

To extend this to cover our stochastic calculus, with processes z^ρ satisfying the weak hypotheses at the beginning of Chapter IV, we need only this small change:

(0-2′) *Expand ΔF formally as a Taylor's series in $\Delta t, \Delta z^1, \ldots, \Delta z^q$. Count Δz^ρ as having a factor Δt if the sample functions of z^ρ are known to be Lipschitzian and as having a factor $(\Delta t)^{1/2}$ otherwise. Discard all terms having a factor $(\Delta t)^c$ with $c > 1$; in the rest replace $\Delta t, \Delta z^\rho$ by dt, dz^ρ, respectively.*

This rule reduces to (0-2) in the ordinary calculus situation of smooth z. For more general z, it gives us the correct form of the conclusion of the chain rule of differentiation [IV(2-4)] and the substitution theory of integrals (Chapter IV, Section 4).

In Chapter V we prove a theorem on the existence of solutions of differential equations of the form

$$x(t) = x(0) + \int_0^t f(s, x(s))\, ds + \int_0^t g(s, x(s))\, dz(s)$$

$$+ \int_0^t h(s, x(s))\, [dz(s)]^2,$$

with extension to vector-valued z and x, and we show that with slight strengthening of hypotheses the solutions are limits of approximations formed by an adaptation of the classical polygons of Cauchy.

When z is Lipschitzian, the second-order integral in the last equation vanishes, so it is equivalent to the first equation no matter how we choose h. But for other z the second-order integral need not vanish. We propose that the second equation, with a certain h defined in terms of g, be regarded as the canonical form of the stochastic differential equation for modeling systems affected by random disturbances. Chapter VI is devoted to showing that such models are free of the infelicities that are associated with the customary Itô-integral model.

It is painfully obvious that some of the proofs in this book are tediously long. I hope that it will also be noticed that the definition of the integral is simple and that the computation rules are easily remembered extensions of those of ordinary calculus.

1 Random variables

This book is written in the hope that it will on the one hand provide a useful mathematical tool for scientists working with systems in which random noises have to be considered, and will on the other hand constitute the beginning of the study of a part of the theory of stochastic processes that mathematicians may wish to study more deeply. With such an assorted set of potential readers, it seems advisable to begin with some general remarks about probability theory and random functions. Most readers will find that they can pass rapidly over many parts of this chapter, consisting of long-familiar material. But readers with different backgrounds will find different parts omissible, and we have preferred to err on the side of overinclusiveness.

In order to make our statements more easily read by those who are not specialists in probability theory, we shall use a rather less compact notation than is customary, usually avoiding such abbreviations as the omission of the variable in naming a function or random variable.

Today many schemes are used for indicating functions. Consider for example the function that assigns to each real number t the "functional value" t^3. Many people still find it convenient to denote the function itself by the symbol t^3, although this symbol has already been assigned a different meaning, namely the cube of the real number t. More explicitly the function (as a whole) can be denoted by $(t^3 : t \in R)$, or more compactly by $(\cdot)^3$, or by $t \mapsto t^3$. We shall use a compromise between the two notations t^3 and $(\cdot)^3$; we replace the dot by a bold-faced letter. Thus $\mathbf{t}$ is not a real number, but marks a place in which any real number can be inserted. The function under discussion is then $\mathbf{t}^3$. (If any one objects that this is excessive reliance on typography, he should think of $\mathbf{t}$ as the name of the identity function on the reals; then $t \mapsto t^3$ is $\mathbf{t}^3$.) In the same way, if $\mathbf{k}$ is a "dot" replaceable by any point in another copy of the real number system (or is the identity function on that other copy), the Fourier transform of the function $f(\mathbf{t})$ is the function $\hat{f}(\mathbf{k})$ whose value at k is

$$\hat{f}(k) = (1/2\pi)^{1/2} \int_{-\infty}^{\infty} f(t) e^{-ikt}\, dt.$$

The mathematical distinction between a function and a value of that function is not mere hairsplitting; but any one who considers it so has the privilege of ignoring the fact that some letters in some formulas are bold-faced.

Many of our functions will be defined on a set called Ω. We shall always use $\boldsymbol{\omega}$ for the identity function on Ω, so that a function $f: \omega \mapsto f(\omega)$ can be written as $f(\boldsymbol{\omega})$. Also, sets of the type $\{\omega \in \Omega : f(\omega) \in A\}$ are often met, A being some set. We shall, in accordance with custom, abbreviate this to $\{f \in A\}$. Likewise, if f is real valued and c is a real number, $\{f < c\}$ is the set $\{\omega \in \Omega : f(\omega) < c\}$, and so on.

If X is any set, a (nonempty) collection $\mathcal{A}$ of its subsets is an **algebra** of sets if whenever A_1 and A_2 are in $\mathcal{A}$, so are $X \backslash A_1$ and $A_1 \cup A_2$ (and therefore $A_1 \cap A_2$ also). The algebra A is a **σ-algebra**, or **Borel field**, of sets if it is also true that whenever $A_1, A_2, A_3, \ldots$ is a sequence of sets all belonging to $\mathcal{A}$, their union $\bigcup_{n=1}^{\infty} A_n$ also belongs to $\mathcal{A}$.

Given any collection $\mathcal{K}$ of subsets of X, there are always σ-algebras that contain all the members of $\mathcal{K}$; for example, the collection of all subsets of X is such a σ-algebra. There are also subsets B of X with the property that whenever $\mathcal{A}$ is a σ-algebra of subsets of X and all members of $\mathcal{K}$ belong to $\mathcal{A}$, the set B also belongs to $\mathcal{A}$; for example, every set in the collection $\mathcal{K}$ is such a set B. The family of all such sets B is itself a σ-algebra; it is called the σ-algebra **generated** by the collection $\mathcal{K}$. For example, let $\mathcal{K}$ be the class of all sets that consist of exactly two positive integers. Then the family of all sets of real numbers is a σ-algebra containing all sets belonging to $\mathcal{K}$. So is the family of all sets of rational numbers, and so also is the family of all sets of positive integers. But every set consisting of a single integer n must belong to every σ-algebra containing $\mathcal{K}$, since it is $\{n, n + 1\} \cap \{n, n + 2\}$; and so must every finite or countable union of such singletons. So the family of all sets of positive integers is the σ-algebra generated by $\mathcal{K}$.

We are primarily interested in real-valued functions and in functions whose values lie in Euclidean n-space R^n for some positive integer n. However, it is often convenient to allow infinite values. By $\bar{R}$ we shall denote the extended real number system, consisting of the set R of real numbers with two elements ∞ and $-\infty$ adjoined; of course $-\infty < r < \infty$ for all r in R, and the neighborhoods of ∞ will be the sets of the form $\{x \in \bar{R}: x > a\}$ for all a in R, and analogously for $-\infty$. A function with values in $\bar{R}$ will be called an "extended-real-valued" function.

If X is R^n or $\bar{R}^n$, the σ-algebra generated by the open sets is the class of Borel sets. It is easy to see that it is also the σ-algebra generated by the closed sets, or by the right-open intervals

$$\{x \in X: a^1 \leqq x^1 < b^1, \ldots, a^n \leqq x^n < b^n\} \qquad (a^1, \ldots, b^n \in \bar{R}),$$

etc. (Here and henceforth we use superscripts attached directly to letters as enumerative indices; the coordinates of the point x are $x^1, \ldots, x^n$.)

If $\mathcal{A}$ is a σ-algebra of sets in X, an extended-real-valued function f on X is $\mathcal{A}$-measurable if for every c in $\bar{R}$, the set $\{x \in X: f(x) \leqq c\}$ belongs to $\mathcal{A}$. The family of sets S in $\bar{R}$ such that $\{x \in X: f(x) \in S\}$ belongs to $\mathcal{A}$ is a σ-algebra, because $\mathcal{A}$ is. It contains all half lines $(-\infty, c]$, and therefore contains the smallest σ-algebra that contains all half lines, which is the class of Borel sets. So $\{x \in X: f(x) \in B\}$ belongs to $\mathcal{A}$ whenever B is a Borel set. Likewise, if $f(x) = (f^1(x), \ldots, f^n(x))$ is an n-vector-valued function, f is $\mathcal{A}$-measurable if the set $\{x \in X: f^1(x) \leqq c^1, \ldots, f^n(x) \leqq c^n\}$ belongs to $\mathcal{A}$ whenever $c^1, \ldots, c^n$ are in $\bar{R}$. Equivalently, f is $\mathcal{A}$-measurable if and only if the set $\{x \in X: f(x) \in B\}$ belongs to $\mathcal{A}$ whenever B is a Borel set in $\bar{R}^n$. In particular, if X is R^n and $\mathcal{A}$ is the σ-algebra of Borel sets in X, an $\mathcal{A}$-measurable function is said to be **Borel-measurable**.

If Φ is any collection of extended-real-valued functions on a set X, the smallest σ-algebra of subsets of X that contains all the sets $\{\phi \leqq c\}$ with ϕ in Φ and c in $\bar{R}$ is called the σ-algebra **generated** by the collection Φ, and is often denoted by $\mathcal{B}(\Phi)$. In particular, if Φ contains finitely many members $\phi_1, \ldots, \phi_n$, it is easy to verify that $\mathcal{B}(\Phi)$ consists of the family of all sets of the form

$$\{x \in X: (\phi_1(x), \ldots, \phi_n(x)) \in B\}$$

in which B is a Borel set in $\bar{R}^n$.

A **measure** m is a nonnegative function defined on a σ-algebra $\mathcal{A}$ of sets that has the properties

(i) $m\emptyset = 0$,
(ii) if $A_1, A_2, \ldots$ is a sequence of disjoint members of $\mathcal{A}$, $m(\bigcup_i A_i) = \sum_i mA_i$.

Although this definition permits mA to be ∞, we shall make use only of finite-valued measures. For these, (i) is a consequence of (ii).

In particular, a **probability measure** on a set Ω is a measure P defined on a σ-algebra $\mathcal{A}$ of subsets of Ω and satisfying

$$P(\Omega) = 1.$$

The set Ω, the σ-algebra $\mathcal{A}$, and the probability measure P taken together constitute the **probability triple** $(\Omega, \mathcal{A}, P)$. The probability measure of A will be denoted by $P(A)$, but sometimes we omit the parentheses and write PA for $P(A)$.

If m is a measure defined on a σ-algebra $\mathcal{A}$ of sets, a **negligible** (or more specifically a ***dm*-negligible**) **set** is a set contained in a set A_0 of $\mathcal{A}$ with $mA_0 = 0$. We can always extend such a measure m by adjoining to $\mathcal{A}$ all sets A^* such that for some A in $\mathcal{A}$, $(A - A^*) \cup (A^* - A)$ is a negligible set, and defining $mA^* = mA$. The measure thus extended is a **complete** measure, in the sense that if A is any set with $mA = 0$, then $mA^* = 0$ whenever $A^* \subset A$. Merely for convenience, we shall always understand that P denotes a complete measure. Therefore a set A is negligible if and only if $P(A) = 0$.

In measure theory it is customary to say that a statement concerning a point x of X (in which a measure m is defined) is true "almost everywhere" if the set of points for which the statement is false is a negligible set. When the measure is a probability measure, it is customary to use the expression "almost surely" (henceforth abbreviated a.s.) with the same meaning. Equivalently, a statement is true a.s. if and only if it is true on the complement of a set of P-measure 0.

An n-vector-valued function on Ω is a **random variable** (henceforth abbreviated r.v.) if it is $\mathcal{A}$-measurable. It is also called a **random *n*-vector**.

If f is a random n-vector, and to each Borel set B in $\bar{R}^n$ we assign the number

$$F(B) = P\{\omega \in \Omega: f(\omega) \in B\},$$

we see readily that F is also a measure; thus $(\bar{R}^n, \mathcal{B}, F)$ is a probability triple, $\mathcal{B}$ being the class of Borel sets in $\bar{R}^n$. The measure F is called the **distribution** of f.

The distribution of $f(\omega)$ has the important virtue that it represents all the knowledge about $f(\omega)$ that is experimentally available to us. We estimate the relative frequency $p(I)$ with which a r.v. falls in an interval I, and use the resulting function of intervals as an estimate for the distribution of the r.v. The probability triples $(\Omega, \mathcal{A}, P)$ are mathematical artifacts; a given distribution on $\bar{R}^n$ will correspond to infinitely many r.v.'s, on different triples. However, the artifacts are mathematically convenient. If, for instance, we consider Ω to be a set of labels for all possible "states of the world" (e.g., the set of all possible wave functions of the universe), we avoid having to stop in the middle of a discussion and start afresh because of discovering that our n functions $f^1(\omega), \ldots, f^n(\omega)$ were merely the first n components of an $(n+m)$-vector that we are obliged to consider.

If $(\Omega, \mathcal{A}, P)$ is a probability triple and $f(\omega)$ is a real r.v., the integral

$$\int_\Omega f(\omega)\, P(d\omega)$$

can be defined by any of the usual procedures for defining the (Lebesgue, or Lebesgue–Stieltjes, or Radon–Lebesgue) integral. It is customary to call this integral the **expectation** of f, and to denote it by $E(f)$.

For all $p \geqq 1$, the set of all those r.v.'s $f(\omega)$ for which $E(|f|^p)$ is finite is a linear space; if f and g are in it, and c is a real number, cf and $f + g$ are in it. If we define

$$\|f\|_p = [E(|f|^p)]^{1/p},$$

this is a pseudo-norm on the space; that is, the conditions

$$\|cf\|_p = |c|\,\|f\|_p \qquad (c \text{ real}),$$

$$\|f+g\|_p \leqq \|f\|_p + \|g\|_p,$$

$$\|f\|_p = 0 \qquad \text{if} \quad f(\omega) = 0 \text{ for all } \omega \text{ in } \Omega$$

are satisfied. For $\|f\|_p$ to be a norm of f, the last condition would have to be "if and only if." But it is not; $\|f\|_p = 0$ if and only if the set $\{f \neq 0\}$ is negligible.

Although $\|f\|_p$ is merely a pseudo-norm, we shall miscall it the "L_p-norm of f." We shall have much more use for $\|f\|_2$ than for $\|f\|_p$ with $p \neq 2$, so we shall usually drop the subscript p when $p = 2$ and define

$$\|f\| = \|f\|_2.$$

The space of r.v.'s $f(\omega)$ with finite $\| f \|_p$ is complete; that is to say, if $f_1(\omega), f_2(\omega), f_3(\omega), \ldots$ is a sequence of r.v.'s in that space, there exists an f_0 in the space such that $\| f_n - f_0 \|_p \to 0$ as n increases if and only if the sequence satisfies the "Cauchy condition": for each positive ϵ, there is an n_ϵ such that $\| f_m - f_n \|_p < \epsilon$ whenever $m, n > n_\epsilon$. For a proof, see, e.g., Loève [7, p. 161].

By the Cauchy–Buniakowski–Schwarz inequality,

$$(1\text{-}1)\qquad \| f \|_1 = \int | f(\omega) | \, P(d\omega) \leqq \left\{ \int | f(\omega) |^2 \, P(d\omega) \right\}^{1/2} \left\{ \int 1 \, P(d\omega) \right\}^{1/2} = \| f \|_2.$$

Three kinds of convergence are often used in probability theory, and to these we shall add a fourth kind. We state their definitions for the case of sequences; other processes, such as convergence of Riemann sums as the partitioning becomes finer, will require only obvious modifications. Consider then a sequence $f_0, f_1, f_2, \ldots$ of a.s. finite extended-real r.v.'s.

(1-2) $f_n \to f_0$ a.s. if there exists a negligible subset N of Ω such that if $\omega \in \Omega \backslash N$, the numbers $f_1(\omega), f_2(\omega), \ldots$ converge to $f_0(\omega)$.

In other words, whatever the "state of the world" (indicated by ω) may be, unless it happens to be one of a set of states with chance 0 of happening, $\lim f_n(\omega) = f_0(\omega)$. To different ω may correspond different speeds of convergence; but convergence takes place, slow or fast, unless a misfortune of probability 0 has happened.

(1-3) $f_n \to f_0$ **in quadratic mean,** or **in L_2-distance,** if

$$\lim_{n \to \infty} \| f_n - f_0 \| = 0.$$

In other words, to each positive ϵ there corresponds an $n(\epsilon)$ such that if we first choose an $n > n(\epsilon)$, and then repeatedly determine by experiment the values of f_0 and f_n that are caused by the various "states of the world" we meet, and estimate the quadratic mean difference between them, we find that it is less than ϵ. This has an obvious generalization to convergence in L_p-norm.

(1-4) $f_n \to f_0$ **in probability** if for each positive ϵ,

$$\lim_{n \to \infty} P\{ | f_n - f_0 | \geqq \epsilon \} = 0.$$

In other words, for large n it is "highly probable" that the difference between $f_n(\omega)$ and $f_0(\omega)$ is "small." More precisely, if ϵ and δ are positive numbers, there is a certain $n(\epsilon, \delta)$ such that for all $n > n(\epsilon, \delta)$, the event "$|f_n(\omega) - f_0(\omega)| < \epsilon$" occurs with probability greater than $1 - \delta$.

(1-5) $f_n \to f_0$ **in near-L_p-norm** if for each positive ϵ there exists a set A in $\mathcal{A}$ with $P(A) > 1 - \epsilon$ such that

$$\lim_{n\to\infty} \int_A |f_n(\omega) - f_0(\omega)|^p P(d\omega) = 0;$$

that is, if 1_A is the indicator function of A, with value 1 on A and 0 elsewhere, then $\| 1_A(f_n - f_0) \|_p \to 0$.

If we regard the f_n as a sequence of estimates for f_0, convergence in probability is well suited to the type of experimental situation in which one wishes for an estimate that "very probably" is "very near" to the correct value. However, the other three types of convergence have some important mathematical advantages. It is well known that a.s. convergence implies convergence in probability. Obviously convergence in L_p-distance implies convergence in near-L_p-norm. If $f_n \to f_0$ in near-L_p-norm, and $\epsilon > 0$, there is a set A with $P(A) > 1 - \epsilon/2$ such that for all sufficiently large n

$$\int_A |f_n(\omega) - f_0(\omega)|^p P(d\omega) < \epsilon^{p+1}/2.$$

For the last to hold it must be true that if we define

$$B = \{\omega \in A : |f_n(\omega) - f_0(\omega)| \geqq \epsilon\},$$

we have $P(B) < \epsilon/2$. So $|f_n(\omega) - f_0(\omega)| < \epsilon$ unless ω is in $B \cup [\Omega - A]$, which has P-measure less than ϵ; and $f_n \to f_0$ in probability.

A sequence $f_1, f_2, f_3, \ldots$ of r.v.'s may converge in probability to a r.v. f_0 without converging in any of the other ways. However, it is always possible to extract a subsequence f_{n_i} $(i = 1, 2, 3, \ldots)$ that converges a.s. to f_0. For a proof see, e.g., Loève [7, p. 116].

A function $\rho(x, y)$ defined for pairs x, y of members of a set X is called a **metric**, or **distance function**, on X if for all x_1, x_2, x_3 in X it is true that

$$\rho(x_1, x_2) = \rho(x_2, x_1) \geq 0,$$

$$\rho(x_1, x_2) = 0 \qquad \text{if and only if} \qquad x_1 = x_2,$$

$$\rho(x_1, x_2) + \rho(x_2, x_3) \geqq \rho(x_1, x_3).$$

It is a **pseudo-metric** if the second condition is weakened to "$\rho(x_1, x_1) = 0.$"

There is no norm or pseudo-norm that defines convergence in probability, but there is a pseudo-metric, which we call $\bar{\rho}$ and define as follows.

(1-6) *If f_1 and f_2 are a.s. finite-valued r.v.'s, $\bar{\rho}(f_1, f_2)$ is the infimum of numbers ϵ such that $|f_1(\omega) - f_2(\omega)| < \epsilon$ except on a set of measure less than ϵ:*

$$\bar{\rho}(f_1, f_2) = \inf\{\epsilon: P(|f_1 - f_2| \geqq \epsilon) < \epsilon\}.$$

It is easy to prove that $\bar{\rho}$ is a pseudo-metric, and that $f_n \to f_0$ in probability if and only if $\bar{\rho}(f_n, f_0) \to 0$. Moreover, the space of r.v.'s is **complete,** with metric $\bar{\rho}$, in the usual sense; if $f_1, f_2, f_3, \ldots$ is a sequence of r.v.'s, there exists a r.v. f_0 such that $\bar{\rho}(f_n, f_0) \to 0$ if and only if the "Cauchy condition" [for each positive ϵ there is an n_ϵ such that $\bar{\rho}(f_n, f_m) < \epsilon$ whenever $m, n > n_\epsilon$] is satisfied. For a proof see, e.g., Loève [7, p. 117].

If a^2 is any number greater than $\|f_1 - f_2\|_1$, and $a > 0$, the set on which $|f_1(\omega) - f_2(\omega)| \geqq a$ cannot have measure as great as a, so

$$P(|f_1 - f_2| \geqq a) < a.$$

By definition, $\bar{\rho}(f_1, f_2) \leqq a$, and since this holds for all a greater than $(\|f_1 - f_2\|_1)^{1/2}$, it must be true that

$$\bar{\rho}(f_1, f_2) \leqq (\|f_1 - f_2\|_1)^{1/2}.$$

2 *Conditional expectations*

Suppose that f is an extended-real r.v. with finite $E(f^2)$. The concept of "best estimate" for f can be variously interpreted. If we understand it to mean the constant estimate that comes closest to f in quadratic mean, it is the number c that minimizes $E(|f(\omega) - c|^2)$. Since this last is merely a quadratic in c, we readily find that its minimum is reached by setting $c = E(f)$. This estimate $E(f)$ for f is "best" in another sense; when the "state of the world" is ω, the error in using $E(f)$ as an estimate for f is $E(f) - f(\omega)$, so the average error is $E([E(f) - f(\omega)]) = 0$.

It often happens, however, that before we have to estimate $f(\omega)$ we know the values of some other random variables $g_1, \ldots, g_n$. In this case we can make an estimate for $f(\omega)$ that depends on the known numbers $g_1(\omega), \ldots, g_n(\omega)$; ω is itself not usually known. So among all the Borel-measurable real-valued functions c, defined on n-dimensional space R^n, we seek one that gives the smallest value to the expectation of the square of the error:

$$E([c(g_1(\omega), \ldots, g_n(\omega)) - f(\omega)]^2) = \min. \tag{2-1}$$

It should be noted that after $g_1(\omega), \ldots, g_n(\omega)$ are known we can find the estimate $c(g_1(\omega), \ldots, g_n(\omega))$ corresponding to any estimator function c, but cannot in general find the value of $f(\omega)$ because many different ω can give the same values to the $g_j(\omega)$.

We now discuss an important mathematical reformulation of this problem. For notational simplicity we shall use g to denote the n-vector $(g_1, \ldots, g_n)$, and likewise shall use y to denote a point $(y^1, y^2, \ldots, y^n)$ of R^n. Also, if S is a set in R^n, we denote by $g^{-1}(S)$ the set $\{g(\omega) \in S\}$, and likewise for other functions.

The family of all sets $g^{-1}(B)$ with B a Borel set in R^n is contained in $\mathcal{A}$, and it is easily seen to be a σ-algebra. It is in fact the σ-algebra $\mathcal{B}(g)$ generated by g. If c is a real-valued function on R^n and we define $\phi(\omega) = c(g(\omega))$, it is clear that for every Borel set B in R^1

$$\phi^{-1}(B) = g^{-1}(c^{-1}(B)).$$

If c is Borel-measurable, $c^{-1}(B)$ is a Borel set, and $g^{-1}(c^{-1}(B))$ belongs to $\mathcal{B}(g)$; that is, ϕ is $\mathcal{B}(g)$-measurable. It is possible, but less elementary, to show that if $\phi(\omega)$ is a $\mathcal{B}(g)$-measurable r.v., there exists a Borel-measurable function c on R^n such that $\phi(\omega) = c(g(\omega))$ (cf. Doob [3, p. 603] or Neveu [11, p. 36]). Consequently our problem (2-1) can be rephrased as follows: Among all $\mathcal{B}(g)$-measurable r.v.'s $\phi(\omega)$, find one for which

$$E([\phi(\omega) - f(\omega)]^2) = \min. \tag{2-2}$$

In this form the problem looks more abstract, and the connection with the available data $g_1(\omega), \ldots, g_n(\omega)$ is no longer direct but takes place through the intermediation of the σ-algebra $\mathcal{B}(g)$. But in compensation the new formulation is convenient for mathematical discussion. In particular, there is no trouble at all in extending to the case in which there are infinitely many g, since $\mathcal{B}(g)$ has already been defined in this case.

The existence of a $\mathcal{B}(g)$-measurable function $\phi(\omega)$ satisfying (2-2) is an immediate consequence of the completeness of the space $L_2(\Omega, \mathcal{B}(g), P)$; we shall not present it here. From this point on we shall consider σ-subalgebras $\mathcal{F}$ of $\mathcal{A}$, dropping all mention of the functions $g(\omega)$ that might have been used to generate them. If $\mathcal{F}$ is a σ-subalgebra of $\mathcal{A}$, and f is a r.v. with finite $E(f^2)$, we define $E(f \mid \mathcal{F})$ (ambiguously!) to be any $\mathcal{F}$-measurable function $\phi(\omega)$ for which

$$E([\phi(\omega) - f(\omega)]^2) = \min. \tag{2-3}$$

Any such ϕ is called a **version** of the conditional expectation $E(f \mid \mathcal{F})$.

If A is any set in $\mathcal{A}$, the conditional probability of A conditioned by $\mathfrak{F}$ (or given $\mathfrak{F}$) is defined to be the conditional expectation of its indicator function 1_A,

$$P(A \mid \mathfrak{F}) = E(1_A \mid \mathfrak{F}), \tag{2-4}$$

just as $P(A)$ is equal to $E(1_A)$. If A is defined by a statement, such as $\{x(\omega) \in B\}$, we denote $P(A \mid \mathfrak{F})$ by the symbol $P(x \in B \mid \mathfrak{F})$ whenever convenient.

If ϕ is any version of $E(f \mid \mathfrak{F})$, and ψ is an $\mathfrak{F}$-measurable r.v. with finite $E(\psi^2)$, by (2-3) the function of t defined by

$$E([\phi(\omega) + t\psi(\omega) - f(\omega)]^2) \tag{2-5}$$

has its minimum value at $t = 0$. So, by differentiating and setting $t = 0$, we find

$$E(\psi(\omega)[\phi(\omega) - f(\omega)]) = 0. \tag{2-6}$$

In particular, if A is a set belonging to $\mathfrak{F}$ and $1_A(\omega)$ is its indicator function (with value 1 if $\omega \in A$ and 0 elsewhere), we take $\psi = 1_A$; then (2-6) takes the form

$$\int_A \phi(\omega)\, P(d\omega) = \int_A f(\omega)\, P(d\omega). \tag{2-7}$$

Conversely, if $\phi(\omega)$ is $\mathfrak{F}$-measurable and $E(\phi^2)$ is finite, and $\phi(\omega)$ satisfies (2-7) for all sets A in $\mathfrak{F}$, we see that (2-6) holds for all indicator functions of sets in $\mathfrak{F}$, hence for all finite linear combinations of such sets, hence (by a familiar limit process) for all $\psi(\omega)$ that are $\mathfrak{F}$-measurable and have $E(\psi^2) < \infty$. But then if $\phi_1(\omega)$ is $\mathfrak{F}$-measurable and $E(\phi_1^2) < \infty$, by taking $\psi = \phi_1 - \phi$ we obtain

$$\begin{aligned} E([\phi_1 - f]^2) &= E([\phi + \psi - f]^2) \\ &= E([\phi - f]^2) + 2E([\phi - f]\psi) + E(\psi^2). \end{aligned}$$

By (2-6) the middle term in the right member is 0, so $E([\phi_1 - f]^2) \geqq E([\phi - f]^2)$, and (2-3) is satisfied.

We can now extend our definition of conditional expectation to r.v.'s in which $E(f)$ exists, but not necessarily $E(f^2)$. For each such f, and for each σ-subalgebra $\mathfrak{F}$ of A, we define the conditional expectation $E(f \mid \mathfrak{F})$ to be any function $\phi(\omega)$ that is $\mathfrak{F}$-measurable and satisfies (2-7) whenever A is a set that belongs to $\mathfrak{F}$. If $E(f^2)$ happens to be finite, this new definition agrees with the previous one.

From (2-7) it is evident that if ϕ is a version of $E(f \mid \mathfrak{F})$, and ϕ_1 is $\mathfrak{F}$-measurable and $\phi_1(\omega) = \phi(\omega)$ a.s., then ϕ_1 is also a version of $E(f \mid \mathfrak{F})$.

On the other hand, if ϕ and ϕ_1 are both versions of $E(f \mid \mathfrak{F})$, by (2-7) we have

$$\int_A [\phi(\omega) - \phi_1(\omega)] P(d\omega) = 0$$

for all A in $\mathfrak{F}$, and since $\phi - \phi_1$ is $\mathfrak{F}$-measurable, this implies that it is 0 a.s. Thus $E(f \mid \mathfrak{F})$ is determined uniquely up to equivalence in P-measure.

A conditional expectation is a sort of Lebesgue integral, and we can extend to conditional expectations many of the useful theorems of integration theory, with practically the same proofs. We shall merely list some that we need, referring to Loève [7, pp. 347–351] for proofs. To save repetition, we assume that all f_i mentioned are r.v.'s for which $E(f_i)$ exists.

(2-8) *If* $f = c$ *(constant) a.s.,* $E(f \mid \mathfrak{F}) = c$ *a.s.*

(2-9) *If* $f_1 = f_2$ *a.s.,* $E(f_1 \mid \mathfrak{F}) = E(f_2 \mid \mathfrak{F})$ *a.s.*

(2-10) *If* c_1 *and* c_2 *are real numbers,*

$$E(c_1 f_1 + c_2 f_2 \mid \mathfrak{F}) = c_1 E(f_1 \mid \mathfrak{F}) + c_2 E(f_2 \mid \mathfrak{F}) \quad a.s.$$

(2-11) *If* $r \geqq 1$, $E(|f_1 + f_2|^r \mid \mathfrak{F}) \leqq 2^{r-1}\{E(|f_1|^r \mid \mathfrak{F}) + E(|f_2|^r \mid \mathfrak{F})\}$.

(2-12) *If* $E(f_1^2)$ *and* $E(f_2^2)$ *are finite,*

$$\{E(f_1 f_2 \mid \mathfrak{F})\}^2 \leqq E(f_1^2 \mid \mathfrak{F}) E(f_2^2 \mid \mathfrak{F}) \quad a.s.$$

More generally, the Hölder inequality holds: *if* $c_1, \ldots, c_n$ *are positive numbers and* $\sum c_i = 1$, *and* $E(f_j)$ *is finite for* $j = 1, \ldots, n$, *then a.s.*

$$E(|f_1|^{c_1} \cdots |f_n|^{c_n} \mid \mathfrak{F}) \leqq [E(|f_1| \mid \mathfrak{F})]^{c_1} \cdots [E(|f_n| \mid \mathfrak{F})]^{c_n}.$$

(2-13) *If* $E(f_1^2)$ *and* $E(f_2^2)$ *are finite,*

$$\{E([f_1 + f_2]^2 \mid \mathfrak{F})\}^{1/2} \leqq \{E(f_1^2 \mid \mathfrak{F})\}^{1/2} + \{E(f_2^2 \mid \mathfrak{F})\}^{1/2} \quad a.s.$$

If $\mathfrak{F}$ is a subalgebra of $\mathfrak{A}$ and $p \geqq 1$, as usual we define the space $L_p(\Omega, \mathfrak{F}, P)$ to consist of all $\mathfrak{F}$-measurable functions $f(\omega)$ with

$$\|f\|_p = \{E(|f|^p)\}^{1/p} < \infty,$$

and with f_1 and f_2 identified if $\|f_1 - f_2\|_p = 0$. In particular, where $p = 2$ we usually omit the subscript 2 from $\|\cdot\|_2$. The space $L_2(\Omega, \mathfrak{F}, P)$ is a

Hilbert space, and is a linear subspace of $L_2(\Omega, \mathcal{A}, P)$. In (2-3) we defined $E(f \mid \mathcal{F})$ to be the point of $L_2(\Omega, \mathcal{F}, P)$ closest to $f(\omega)$. By the elementary theory of Hilbert spaces this is equivalent to saying that $f(\omega) - E(f \mid \mathcal{F})$ is perpendicular to $L_2(\Omega, \mathcal{F}, P)$; and this in fact is the content of (2-6). So $E(f \mid \mathcal{F})$ is the perpendicular projection of $f(\omega)$ on $L_2(\Omega, \mathcal{F}, P)$. From this, again by the elementary geometry of Hilbert space, we obtain the following statement. Let $\mathcal{F}_1$ and $\mathcal{F}_2$ be σ-algebras such that $\mathcal{F}_1 \subseteq \mathcal{F}_2 \subseteq \mathcal{A}$. Then the (perpendicular) projection of $f(\omega)$ on $L_2(\Omega, \mathcal{F}_1, P)$ is the same as the projection on $L_2(\Omega, \mathcal{F}_1, P)$ of the projection $E(f \mid \mathcal{F}_2)$ of $f(\omega)$ on $L_2(\Omega, \mathcal{F}_2, P)$; that is,

(2-14) $$E(E(f(\omega) \mid \mathcal{F}_2) \mid \mathcal{F}_1) = E(f(\omega) \mid F_1).$$

This equation can be extended to all r.v.'s $f(\omega)$ with finite $E(f)$ by defining $f_n = \min\{n, \max\{-n, f\}\}$, applying (2-14) to each f_n and letting $n \to \infty$, or it can easily be established directly (Loève [7, p. 350]).

If $f(\omega)$ is a function a.s. equal to a constant, it is evident that $E(f) = f$ a.s. and $E(fg) = fE(g)$ a.s. whenever $E(g)$ exists. For conditional expectations, these properties of constant functions are shared by the $\mathcal{F}$-measurable r.v.'s, as follows.

(2-15) *If $f(\omega)$ is $\mathcal{F}$-measurable, $E(f(\omega) \mid \mathcal{F}) = f(\omega)$ a.s.*

For in (7) we can choose $f(\omega)$ for $\phi(\omega)$, so $f(\omega)$ is a version of $E(f(\omega) \mid \mathcal{F})$.

(2-16) *If $f_1(\omega)$ is $\mathcal{F}$-measurable and $E(f_1)$ exists, for all $f_2(\omega)$ with finite $E(f_2)$ we have*

$$E(f_1(\omega)f_2(\omega) \mid \mathcal{F}) = f_1(\omega)E(f_2(\omega) \mid \mathcal{F}) \quad a.s.$$

For the proof, we refer to Loève [7, p. 350].

Two events A, B (in $\mathcal{A}$) are independent if $P(A \cap B) = P(A) \cdot P(B)$; that is, if

$$E(1_A \cdot 1_B) = E(1_A)E(1_B).$$

Similarly, if $\mathcal{F}$ is a σ-subalgebra of $\mathcal{A}$, we say that A and B are **independent given** $\mathcal{F}$ if a.s.

(2-17) $$E(1_A 1_B \mid \mathcal{F}) = E(1_A \mid \mathcal{F})E(1_B \mid \mathcal{F}).$$

If $\mathcal{A}_1$, $\mathcal{A}_2$ are subsets of $\mathcal{A}$, they are independent given $\mathcal{F}$ if Eq. (2-17) holds whenever $A \in \mathcal{A}_1$ and $B \in \mathcal{A}_2$. Likewise, if f and g are r.v.s, they are independent given $\mathcal{F}$ if, whenever J_1 and J_2 are intervals of real numbers, (2-17) holds a.s. when $A = \{\omega : f(\omega) \in J_1\}$ and $B = \{\omega : g(\omega) \in J_2\}$.

Let F and G be step functions on R; say F has values $c_1', \ldots, c_k'$ on intervals $J_1', \ldots, J_k'$, and G has values $c_1'', \ldots, c_h''$ on intervals

$J_1'', \ldots, J_h''$. Let I_i' be the indicator function of $\{\omega: f(\omega) \in J_i'\}$ and I_j'' the indicator function of $\{\omega: g(\omega) \in J_j''\}$. If f, g are independent, a.s.

$$E(F(f) \cdot G(g) \mid \mathfrak{F}) = \sum_{i=1}^{k} \sum_{j=1}^{h} c_i' c_j'' E(I_i' I_j'' \mid \mathfrak{F})$$

$$= \sum_{i=1}^{k} \sum_{j=1}^{h} c_i' E(I_i' \mid \mathfrak{F}) c_j' E(I_j' \mid \mathfrak{F})$$

$$= E(F(f) \mid \mathfrak{F}) E(G(g) \mid \mathfrak{F}).$$

If F and G are nonnegative continuous functions on R, F is the limit of a rising sequence of nonnegative step functions $s_1', s_2', \ldots$, and G of a rising sequence $s_1'', s_2'', \ldots$. Since the preceding equation holds for each product s_i', s_i'', by the monotone convergence theorem we find that a.s.

$$E(F(f)G(g) \mid \mathfrak{F}) = E(F(f) \mid \mathfrak{F}) E(G(g) \mid \mathfrak{F}), \tag{2-18}$$

provided that the right members exist. The requirement that F and G be nonnegative is easily removed by applying (2-18) to $F^+(f)G^+(g)$, $F^+(f)G^-(g)$, $F^-(f)G^+(g)$, and $F^-(f)G^-(g)$ and combining. Equation (2-18) can easily be shown to hold if F and G are Borel-measurable functions for which the conditional expectations in the right member have meaning; but we do not need this, so we leave it as an exercise.

3 Stochastic processes

A point in $\bar{R}^n$ is represented by an n-tuple $(x^1, \ldots, x^n)$ of numbers in $\bar{R}$, which can equally well be thought of as a function $x^i = (x^i: i \in \{1, \ldots, n\})$ on the set $\{1, \ldots, n\}$. Thus $\bar{R}^n$ can be thought of as the set of all extended-real-valued functions x^i on $\{1, \ldots, n\}$. From this point of view, given any set T, it is natural to think of the extended-real-valued functions $x(\mathbf{t})$ on T as being the points of a space $\bar{R}^T$, having as many dimensions as there are points in T. Since an open interval in $\bar{R}^1$ is by definition a set having one of the forms

$$\{x \in \bar{R}^1 : a < x < b\}, \quad \{x \in \bar{R}^1 : a < x\}, \quad \{x \in \bar{R}^1 : x < b\}, \quad \bar{R}^1,$$

in which a and b belong to $\bar{R}^1$, it is also natural to attempt to define an open interval in $\bar{R}^T$ as the set of all x in $\bar{R}^T$ such that for each t in T, $x(t)$ lies in an open interval $I(t)$ in $\bar{R}^1$. However, this turns out to be too inclusive a definition. Instead, we shall define an "open interval" in $\bar{R}^T$ to be a set S defined thus:

(3-1) There is a finite subset $\{t_1, \ldots, t_n\}$ of T and there is a finite set $I_1, \ldots, I_n$ of open intervals in $\bar{R}^1$ such that S consists of all those points x in $\bar{R}^T$ that satisfy

$$x(t_i) \in I_i \qquad (i = 1, \ldots, n).$$

Thus only finitely many coordinates are at all restricted.

A right-open interval in $\bar{R}^1$ is the union of an open interval and its left end-point. To define a right-open interval in $\bar{R}^T$ we merely replace "open" by "right-open" in (3-1). Analogous statements hold for left-open and closed intervals.

Now the definition of vector-valued r.v. extends straightforwardly to r.v.'s with values in $\bar{R}^T$. Let $(\Omega, \mathcal{A}, P)$ be a probability triple, and let $x(\mathbf{t})(\boldsymbol{\omega})$, which we abbreviate to $x(\mathbf{t}, \boldsymbol{\omega})$, be a function on Ω with values in $\bar{R}^T$. This function is a r.v. if for every right-open interval I in $\bar{R}^T$ the set $\{x(\mathbf{t}, \omega) \in I\}$ belongs to $\mathcal{A}$. It would make no difference if we replaced *right-open* by *open* or by *closed*. Moreover, if as before we define the class of Borel sets in $\bar{R}^T$ to be the smallest σ-algebra of subsets of $\bar{R}^T$ that contains the open intervals, we can replace *right-open interval* by *Borel set*.

For brevity, a vector-valued r.v. is often called a "random vector." In the same way, when $(x(\mathbf{t}, \omega) : \omega \in \Omega)$ is a r.v. with values in $\bar{R}^T$, each value $x(\mathbf{t}, \omega)$ is itself a function on T, so such a r.v. is often called a "random function." On the other hand, by fixing t at any point in T we obtain a function $x(t, \boldsymbol{\omega})$ which is an extended-real-valued r.v. So $x(\mathbf{t}, \boldsymbol{\omega})$ can also be thought of as a r.v.-valued function on T.

The name "stochastic process" can be applied to any random function. However, it is customary to reserve it for random functions in which T is an infinite set. In later pages we shall be even more specific; a stochastic process will be a r.v. with values in $\bar{R}^T$, where T is a set of real numbers. The variable t will usually admit of being interpreted as time, and our language will be chosen in accordance with this interpretation.

A random function $x(\mathbf{t}, \boldsymbol{\omega})$ has a distribution defined on the σ-algebra $\mathcal{B}$ of Borel sets in $\bar{R}^T$; for each B in $\mathcal{B}$, let us use $F(B)$ to denote $P\{x(\mathbf{t}, \omega) \in B\}$. Suppose that Φ is a finite subset $\{t_1, \ldots, t_n\}$ of T. Given any set S_Φ contained in $\bar{R}^\Phi$, we can "pad it out" to obtain a set $S_\Phi{}^+$ in $\bar{R}^T$ by defining $x(\mathbf{t})$ to be in $S_\Phi{}^+$ if and only if

$$(x(t_1), \ldots, x(t_n)) \in S_\Phi.$$

The set of all S_Φ, such that the corresponding S_Φ^+ is in $\mathcal{B}$, is again a σ-algebra, which we call $\mathcal{B}_\Phi$. If we define $F_\Phi(S_\Phi) = F(S_\Phi^+)$ for all S_Φ in $\mathcal{B}_\Phi$, we obtain a distribution function; $(\bar{R}^\Phi, \mathcal{B}_\Phi, F_\Phi)$ is a probability triple. The distributions thus defined, corresponding to the "marginal distributions" of statistics, are the finite-dimensional distributions of the random function $x(\mathbf{t}, \boldsymbol{\omega})$. From an experimental point of view, they are the only characteristics of $x(\mathbf{t}, \boldsymbol{\omega})$ that are in our reach. This indicates the desirability of stating such concepts as "nearness" of two random functions in terms of their finite-dimensional distributions.

It also indicates the desirability of studying random functions beginning with their finite-dimensional distributions. We have seen that if a random function $x(\mathbf{t}, \boldsymbol{\omega})$ is given, we can define its finite-dimensional distributions. Suppose now that we are given a collection of distributions F_Φ, one for each finite subset Φ of T; can we find a random function $x(\mathbf{t}, \boldsymbol{\omega})$ whose finite-dimensional distributions are the given F_Φ? It is clear that there is a consistency condition that must be satisfied. Suppose that $\Phi = \{t_1, t_2, \ldots, t_n\}$, and that $\Psi = \{t_1, \ldots, t_{j-1}, t_{j+1}, \ldots, t_n\}$. Let S_Ψ be a set in $\mathcal{B}_\Psi$, and let S_Φ be the set in $\bar{R}^\Phi$ consisting of all points $(x(t_1), \ldots, x(t_n))$ such that

$$(x(t_1), \ldots, x(t_{j-1}), x(t_{j+1}), \ldots, x(t_n)) \in S_\Psi.$$

Then the set S_Φ^+ obtained by "padding out" S_Φ is the same as the set S_Ψ^+ obtained by "padding out" S_Ψ. So if F_Φ and F_Ψ are the finite-dimensional distributions corresponding to a random function $x(\mathbf{t}, \boldsymbol{\omega})$ with distribution F, we must have

$$F_\Phi(S_\Phi) = F(S_\Phi^+) = F(S_\Psi^+) = F_\Psi(S_\Psi).$$

That is, in order that the finite-dimensional distributions shall all correspond to one random function, it is necessary that (with the preceding notation) the consistency condition

$$F_\Phi(S_\Phi) = F_\Psi(S_\Psi) \tag{3-2}$$

be satisfied.

Kolmogorov has shown [3, p. 609; 11, p. 82] that if to each finite subset Φ of T there corresponds a distribution F_Φ on R^Φ, and these F_Φ satisfy the consistency condition (3-2), there is a probability triple $(\Omega, \mathcal{A}, P)$ and a random function $x(\mathbf{t}, \boldsymbol{\omega})$ such that the finite-dimensional distributions of $x(\mathbf{t}, \boldsymbol{\omega})$ are the F_Φ. In fact, we can choose $\Omega = R^T$.

The first example of a random function was constructed by Wiener [14] in 1921. This is the "Brownian-motion." It is historically important because it represents the first significant introduction of Lebesgue theory into probability theory, as well as the first really significant extension of measure theory beyond the finite-dimensional spaces. It continues to be important both in theory and in applications, being the most tractable and the most intensively studied of stochastic processes. Very soon after the invention of the microscope it was noticed that microscopically small particles in a fluid were in incessant motion. In the early nineteenth century this attracted the interest of a botanist, Robert Brown, who performed a number of experiments and published his findings in 1828 and 1829. Measurements indicate that each component of the displacement during a time interval (t_1, t_2) can be represented with adequate accuracy (if $t_2 - t_1$ is of the order of 1 sec or more) as a Gaussian r.v. with mean 0 and variance proportional to $t_2 - t_1$; for each set of disjoint intervals, the displacements during those intervals are independent. If the x coordinate of the position at time t ($\geqq 0$) is represented by $x(t)$, with $x(0) = 0$, and the time unit is suitably chosen, for each finite set $(t_1, \ldots, t_n)$ with $0 < t_1 < t_2 < \cdots < t_n$ the probability density for $(x(t_1), \ldots, x(t_n))$ at the place $(u_1, \ldots, u_n)$ is then

(3-3) $$(2\pi)^{-n/2}[t_1(t_2 - t_1) \cdots (t_n - t_{n-1})]^{-1/2} \times \exp\left[-\frac{u_1^2}{2t_1} - \frac{(u_2 - u_1)^2}{2(t_2 - t_1)} - \cdots - \frac{(u_n - u_{n-1})^2}{2(t_n - t_{n-1})}\right].$$

Wiener proved that it is indeed possible to find a σ-algebra $\mathcal{A}$ of subsets of the set $R^{[0,\infty)}$ of all real-valued functions on $[0, \infty)$, and a probability measure defined on $\mathcal{A}$, in such a way that the finite-dimensional distributions satisfy (3-3). Moreover, the subset of $R^{[0,\infty)}$ consisting of the continuous functions with value 0 at $t = 0$ has probability measure 1 in this model. The process thus defined is the "Brownian-motion" or "Wiener" process. For details of its definition and a study of many of its important properties we refer to McKean [9].

Corresponding to the four kinds of convergence of r.v.'s defined in (1-2)–(1-5), there are four kinds of continuity of a stochastic process. Let T be a set of real numbers, and let $x(\mathbf{t}, \boldsymbol{\omega}) = (x(t, \omega) : t \in T, \omega \in \Omega)$ be a stochastic process. Then if s is in T,

(3-4) $x(\mathbf{t}, \boldsymbol{\omega})$ is **a.s. continuous at** s if there exists a subset N of Ω with $P(N) = 0$ such that if $\omega \in \Omega \backslash N$,

$$\lim_{t \to s} x(t, \omega) = x(s, \omega);$$

(3-5) $x(\mathbf{t}, \boldsymbol{\omega})$ is **continuous in L_p-distance** at s if

$$\lim_{t \to s} \| x(t, \boldsymbol{\omega}) - x(s, \boldsymbol{\omega}) \|_p = 0;$$

(3-6) $x(\mathbf{t}, \omega)$ is **continuous in probability at** s if $x(s, \omega)$ is a.s. finite and for each positive ϵ,

$$\lim_{t \to s} P\{| x(t) - x(s) | \geqq \epsilon\} = 0;$$

(3-7) $x(\mathbf{t}, \omega)$ is **continuous in near-L_p-norm at** s if for each positive ϵ there is a set A in $\mathcal{A}$ with $P(A) > 1 - \epsilon$ such that

$$\lim_{t \to s} \int_A | x(t, \omega) - x(s, \omega) |^p P(d\omega) = 0.$$

If $x(\mathbf{t}, \omega)$ is continuous a.s., or in L_p-norm or in near-L_p-norm ($p \geqq 1$) at s, it is continuous in probability at s; but $x(\mathbf{t}, \omega)$ may be continuous in probability at s without being continuous in any of the other manners. The condition that for almost all ω the sample curve $x(\mathbf{t}, \omega)$ is continuous on T sounds much like the condition that $x(\mathbf{t}, \omega)$ be a.s. continuous at each point of T. But it is in fact much stronger. Consider for example the Poisson process, which for some positive λ satisfies the conditions

$$P\{x(0) = 0\} = 1,$$

$$P\{x(t) - x(s) = k\} = [\lambda^k (t - s)^k / k!] \exp(-\lambda[t - s]),$$

$$0 \leqq s < t, \quad k = 0, 1, 2, \ldots.$$

Then $x(\mathbf{t}, \omega)$ is a.s. continuous at each nonnegative s, but the set of ω such that $x(\mathbf{t}, \omega)$ is continuous on $[0, \infty)$ has P-measure 0.

Just as our only access to a r.v. is by way of its distributions, our only access to a stochastic process is by way of its finite-dimensional distributions. As a first result of this point of view, we consider two processes indistinguishable if they have the same finite-dimensional distributions. This is the case, in particular, if the two processes, say $x(\mathbf{t}, \omega)$ and $y(\mathbf{t}, \omega)$, are defined on the same probability triple and, for each t in T, the set $\{x(t) \neq y(t)\}$ has P-measure 0. If this last assertion holds, the processes $x(\mathbf{t}, \omega)$ and $y(\mathbf{t}, \omega)$ are called **equivalent**, and are said to be **versions** of each other.

When $f(\mathbf{t}, \omega)$ is a stochastic process on a Borel set T, in addition to the probability measure P on $\mathcal{A}$ we are usually given a measure m defined on the family $\mathcal{B}$ of Borel subsets of T. Often we encounter an iterated integral of the form

$$\int_T E(f(t, \omega))\, m(dt),$$

and we wish to replace it by

$$E\left(\int_T f(t, \boldsymbol{\omega})\, m(dt)\right).$$

This can be done by way of Fubini's theorem provided that f has a certain property which we now describe. As is usual in measure theory, we define a "measurable rectangle" in $T \times \Omega$ to be a set of the form $B \times A$ in which B is a member of $\mathcal{B}$ and A is a member of $\mathcal{A}$. Then the product-algebra $\mathcal{B} \times \mathcal{A}$ is defined to be the least σ-algebra of sets that contains all measurable rectangles. A process $f(\mathbf{t}, \boldsymbol{\omega})$ is said to be measurable if, considered as a function $(f(t, \omega) : t \in T, \omega \in \Omega)$ on $T \times \Omega$, it is $\mathcal{B} \times \mathcal{A}$-measurable.

If m is a measure defined on $\mathcal{B}$, to the measurable rectangle $B \times A$ we assign measure $(mB)(PA)$. This can be extended uniquely to a measure, which we call $dm\, dP$-measure, on the σ-algebra $\mathcal{B} \times \mathcal{A}$.

When f is a measurable process, we can use Fubini's theorem and Tonelli's theorem (cf. Neveu [11, p. 76]): if either of the integrals

$$E\left(\int_T |f(t, \boldsymbol{\omega})|\, m(dt)\right), \qquad \int_T E(|f(t, \boldsymbol{\omega})|)\, m(dt)$$

is finite, the integral

$$\int_{T\times\Omega} f(t, \omega)\ (dm\, dP)$$

exists and is finite; and if the last named integral exists and is finite, then $E(f(t, \boldsymbol{\omega}))$ exists for almost all t in T, and $\int_T f(t, \omega)\, m(dt)$ exists for almost all ω in Ω, and the three integrals

$$\int_{T\times\Omega} f(t, \omega)\ (dm\, dP), \qquad E\left(\int_T f(t, \boldsymbol{\omega})\, m(dt)\right), \qquad \int_T E(f(t, \boldsymbol{\omega}))\, m(dt)$$

all exist and are equal.

If $(\Omega, \mathcal{A}, P)$ is a probability triple and T is a set of real numbers, a family $\mathcal{F}_{\boldsymbol{\tau}} = \{\mathcal{F}_\tau : \tau \in T\}$ of σ-subalgebras of $\mathcal{A}$ is an **increasing** family if $\mathcal{F}_\tau \subseteq \mathcal{F}_{\tau'}$ whenever τ and τ' are in T and $\tau \leqq \tau'$. When typographically more convenient, we shall use the symbol $\mathcal{F}(\tau)$ to denote $\mathcal{F}_\tau$. A process $f(\boldsymbol{\tau}, \boldsymbol{\omega})$ on T is **adapted** to the family $\mathcal{F}_{\boldsymbol{\tau}}$ if $f(\tau, \boldsymbol{\omega})$ is $\mathcal{F}_\tau$-measurable for every τ in T.

There is a well-known theorem that if a real-valued function $f(\mathbf{t})$ is measurable and its square is Lebesgue integrable over an interval $[a, b]$, there are step functions arbitrarily close to f in L_2-distance. This theorem can be generalized to measurable processes on $[a, b]$. Let us define a **simple**

process $g(\mathbf{t}, \omega)$ on an interval $[a, b]$ to be one with the following property: there are finitely many numbers $t_1^* = a < t_2^* < \cdots < t_{h+1}^* = b$ and r.v.'s $v_1, v_2, \ldots, v_h$ such that for each j in the set $\{1, \ldots, h\}$, the r.v. v_j is $\mathfrak{F}(t_j^*)$-measurable, and

$$g(t, \omega) = v_j \qquad (t_j^* \leqq t < t_{j+1}^*).$$

[If we wish $g(b, \omega)$ to be defined, we can take it to be v_h.] Measurable processes with integrable $\| f \|^2$ can be approximated by simple functions in the sense of the following theorem.

(3-8) *THEOREM Let $f(\mathbf{t}, \omega)$ be a measurable process on an interval $[a, b]$, adapted to an increasing family $\mathfrak{F}_{\boldsymbol{\tau}}$, and having $\| f(t) \|^2$ Lebesgue integrable over $[a, b]$. Then for each positive ϵ there is a bounded simple process $g(\mathbf{t}, \omega)$ on $[a, b]$ that is adapted to $\mathfrak{F}_{\boldsymbol{\tau}}$ and satisfies*

$$\int_a^b \| f(t) - g(t) \|^2 \, dt < \epsilon.$$

For the proof we refer to Doob [3, pp. 440 and 441].

(3-9) *COROLLARY Let $f(\mathbf{t}, \omega)$ be a measurable process on an interval $[a, b]$, adapted to an increasing family $F_{\boldsymbol{\tau}}$, and having $\| f(t) \|_1$ integrable over $[a, b]$. Then for each positive ϵ there is a bounded simple process $g(\mathbf{t}, \omega)$ on $[a, b]$ that is adapted to $F_{\boldsymbol{\tau}}$ and satisfies*

$$\int_a^b \| f(t) - g(t) \|_1 \, dt < \epsilon.$$

For any three numbers a, b, c we use the familiar symbol $\operatorname{mid}\{a, b, c\}$ to denote the "second" one when they are arranged in nondecreasing order; that is,

$$\operatorname{mid}\{a, b, c\} = \min\{\max\{a, b\}, \max\{b, c\}, \max\{a, c\}\}.$$

By Tonelli's theorem $f(\mathbf{t}, \omega)$ is integrable over $[a, b] \times \Omega$ with respect to $dt\, P(d\omega)$ measure. So by the dominated convergence theorem we can find an N such that if $f_N(t, \omega) = \operatorname{mid}\{ f(t, \omega), N, -N\}$,

$$\int_a^b E(| f - f_N |) \, dt < \epsilon/2.$$

Clearly f_N is adapted to $\mathfrak{F}_{\boldsymbol{\tau}}$, is measurable and has $\| f_N(t) \|_2^2$ integrable. So by Theorem (3-8) there is a bounded simple process g such that

$$\int_a^b E([f_N(t) - g(t)]^2)\, dt < \epsilon^2/4(b-a).$$

By the Cauchy–Buniakowski–Schwarz inequality,

$$\int_a^b E(|f_N(t) - g(t)|)\, dt \leqq \left\{\int_a^b E([f_N(t) - g(t)]^2\, dt\right\}^{1/2} \left\{\int_a^b 1\, dt\right\}^{1/2}$$
$$< \epsilon/2.$$

This and the first inequality imply the conclusion.

Among stochastic processes two familiar ones have been especially thoroughly studied. The first family consists of the Markov processes; but it happens that we shall not refer to these. The second family consists of the martingales, submartingales, and supermartingales. We shall later have an important use for a theorem concerning submartingales, so we define them. As is customary, if f is any real-valued function we define

$$f^+ = \max\{f, 0\}.$$

(3-10) *DEFINITION A process $f(\boldsymbol{\tau}, \omega)$ on T is a* **submartingale** *[a* **martingale***] relative to the increasing family $\mathfrak{F}_{\boldsymbol{\tau}}$ of σ-algebras if it is adapted to $\mathfrak{F}_{\boldsymbol{\tau}}$, and for all τ in T*

$$E(f^+(\tau, \omega)) < \infty,$$

and for every s, t in T with $s \leqq t$ we have a.s.

$$E(f(t, \omega) \mid \mathfrak{F}_s) \geqq f(s, \omega),$$
$$[E(f(t, \omega) \mid \mathfrak{F}_s) = f(s, \omega)].$$

We shall need the following theorem which is a portion of a theorem due to Doob.

(3-11) *THEOREM If $X(\mathbf{t}, \omega)$ is a submartingale on an interval $[a, b]$, and D is a countable subset of $[a, b]$, then for every positive number N*

$$P(\sup\{X(t, \omega) : t \in D\} > N) \leqq E(X^+(b))/N.$$

The proof consists of the first two sentences of the proof of Theorem 3.2 of Doob [3, p. 353], or of the first part of the proof of Proposition IV.5.2 of Neveu [11, p. 133].

In many later theorems we encounter a process adapted to a family $\mathfrak{F}_{\boldsymbol{\tau}}$ of sets, and wish to replace it by an equivalent process with some desirable properties. In general this new process may fail to be adapted to $\mathfrak{F}_{\boldsymbol{\tau}}$, since for each τ it differs from the original on a negligible set in $\mathfrak{A}$ that may not belong to $\mathfrak{F}_\tau$. We can avoid verbosity by introducing new algebras $\bar{\mathfrak{F}}_\tau$ as follows.

(3-12) *DEFINITION If $\mathfrak{F}$ is a σ-subalgebra of $\mathfrak{A}$, $\bar{\mathfrak{F}}$ is the smallest σ-algebra that contains $\mathfrak{F}$ and also contains all negligible sets in $\mathfrak{A}$.*

It is easy to verify that $\bar{\mathfrak{F}}$ consists of the class of all sets of the form $(A\backslash N_1) \cup N_2$, where A is in $\mathfrak{F}$ and N_1 and N_2 are negligible sets belonging to $\mathfrak{A}$. It is also easy to verify that a r.v. $x(\omega)$ is $\bar{\mathfrak{F}}$-measurable if and only if it is equivalent to an $\mathfrak{F}$-measurable r.v.

Consequently, if $x(\boldsymbol{\tau}, \omega)$ is adapted to $\mathfrak{F}_{\boldsymbol{\tau}}$, every process equivalent to $x(\boldsymbol{\tau}, \omega)$ is adapted to $\bar{\mathfrak{F}}_{\boldsymbol{\tau}}$. We would in fact lose little generality if we restricted our attention to families $\mathfrak{F}_{\boldsymbol{\tau}}$ such that $\mathfrak{F}_\tau = \bar{\mathfrak{F}}_\tau$ for all τ.

The rest of this section is concerned with a property called separability. This is a property of considerable importance in the study of stochastic processes. But for those readers whose chief interest is not in the stochastic processes themselves, but in their use in stochastic differential equations and modeling procedures as presented in Chapters V and VI, this can be omitted; in Chapter IV, Section 1 we show that for the processes of greatest importance to us, the sample functions are a.s. continuous, and this is a stronger property than separability and allows us to get along without it. So those readers who are interested chiefly in stochastic differential equations and applications should ignore the rest of this section.

Since uncountable collections of measurable sets are intractable and uncountable collections of measurements are impossible, the processes which we can reasonably hope to investigate theoretically and use practically are those in which all the information we need can be deduced from those finite-dimensional distributions corresponding to subsets $\{t_1, \ldots, t_n\}$ of some countable subset of T. This idea takes precise form in the definition of the property of **separability**. Let $x(\mathbf{t}, \omega)$ be a random n-vector on a set T of real numbers. (The extension to processes with values in more general spaces is easy.) By $\bar{S}$ we mean the closure of S.

(3-13) *DEFINITION With the notation just described $x(\mathbf{t}, \omega)$ is* **separable** *if there is a countable subset S of T and a set N with $P(N) = 0$ such that if $\omega \in \Omega\backslash N$, for all t in T it is true that*

$$x(t, \omega) \in \overline{x(I \cap S, \omega)}$$

for every open interval I in R that contains t.

A countable set S with the property specified in Definition (3-13) is called a separating set, or separant.

We shall have need of the following well-known theorem on separability.

(3-14) *THEOREM* *Let* $X(\mathbf{t}, \boldsymbol{\omega})$ *be a real-valued process defined and continuous in probability on an interval* $[a, b]$. *Then there exists an extended-real-valued process* $\tilde{X}(\mathbf{t}, \boldsymbol{\omega})$ *equivalent to* $X(\mathbf{t}, \boldsymbol{\omega})$ *such that* $\tilde{X}$ *is separable and measurable and for each* ω *in* Ω *and each* t_0 *in* $[a, b]$, $\tilde{X}(t_0, \omega)$ *is the upper limit of* $X(t_n, \omega)$ *for a sequence* $t_1, t_2, \ldots$ *of points of* $[a, t_0]$ *converging to* t_0.

This is proved by Neveu [11, p. 92]. Although the property of $\tilde{X}$ mentioned last in Theorem (3-14) is not specifically stated by Neveu in his theorem, it is mentioned in the proof.

II Stochastic Integrals

1 Stochastic models, and properties they should possess

In any of the sciences we frequently meet systems whose state is specified, adequately for our needs, by an n-tuple of numbers $x^1, x^2, \ldots, x^n$, which in the absence of outside disturbances change in accordance with a system of differential equations

$$dx^i = f^i(t, x)\, dt \qquad (i = 1, \ldots, n).$$

However, in recent years situations have often been encountered in which outside disturbances, or "noises," have been too large to ignore. Suppose that there are r sources of noise, and that for the ρth noise the total (net) amount of input, measured in some appropriate units, from the start of the process to time t, has value $z^\rho(t)$ $(\rho = 1, \ldots, r)$. The input during time interval (s, t) is $z^\rho(t) - z^\rho(s)$. The rate of input at time t is then $\dot{z}^\rho(t)$ if this derivative exists. The sensitivity of the ith coordinate x^i to the ρth noise is a function $g_\rho{}^i(\mathbf{t}, \mathbf{x})$ of the state and the time. In many problems the underlying scientific theory permits us to conclude that at least for sufficiently smooth noises z^ρ the evolution of the system is governed by the differential equations

$$dx^i = f^i(t, x)\, dt + \sum_{\rho=1}^{r} g_\rho{}^i(t, x)\, dz^\rho. \tag{1-1}$$

Granted reasonable smoothness for the z^ρ, every such equation can be handled by long-established methods. But this is not usually what we want. The individual z^ρ are not known; instead, they are random functions out of some class, and we assume that our information about the noises consists of some knowledge about the joint distributions of the $z^\rho(t_j)$ at finite sets of times $\{t_1, \ldots, t_k\}$. So the appropriate model for the set of noises is an r-vector-valued random function $((z^1(t, \omega), \ldots, z^r(t, \omega)) : t \in T, \omega \in \Omega)$. Correspondingly, what we want is information about the finite-dimensional distributions of the $x^i(\mathbf{t}, \boldsymbol{\omega})$ that are the states of the system when such noises are present.

However, in choosing the types of stochastic processes that we shall use as models of the noises we meet a dilemma. On the one hand, there is no physical basis for considering any sample functions $z^\rho(\mathbf{t})$ except those of a rather simple structure. Properties such as nondifferentiability belong to the mathematical model rather than to the system itself, since they cannot be detected by experiment. In fact, the noise input $z^\rho(t) - z^\rho(s)$ is measured by some sort of indicator. If this is mechanical, it cannot move faster than the velocity of light. If it is electrical, it cannot support more than some limited current or voltage difference without destruction, and so on. Thus the whole theory of systems subject to noises should be adequately covered if we admit, at the most, noise processes $z^\rho(\mathbf{t}, \boldsymbol{\omega})$ in which all $z^\rho(\mathbf{t}, \omega)$ $(\rho = 1, \ldots, r;\ \omega \in \Omega)$ satisfy one and the same Lipschitz condition. [Recall that a function $z(t)$ is Lipschitzian, or satisfies a Lipschitz condition, on $[a, b]$, if there is a number L such that

$$| z(t') - z(t'') | \leqq L \,| t' - t'' |$$

whenever t' and t'' are both in $[a, b]$.]

On the other hand, if we are to learn anything from our differential equations, the processes z^ρ involved must be of a kind that we can manage mathematically. These, unfortunately, are rare. Beyond martingales and Markov processes we are in almost unexplored territory. In fact, usually only Brownian-motion processes ("white noise") and simple modifications of Brownian-motion processes have been used in studying systems affected by noise. Except in trivial cases, the sample curves $z^\rho(\mathbf{t}, \omega)$ of such processes are almost surely *not* Lipschitzian. Thus the physically reasonable (Lipschitzian) models for noise and the mathematically manageable models have nothing in common. Yet in studying noisy systems we must include both kinds, one to preserve contact with phenomena and the other to permit mathematical manipulations.

A mathematical model of a system is useful precisely because it simplifies and idealizes a complicated physical situation. But we must be careful not to use such a model except in situations in which the error resulting from the idealization is satisfactorily small. The model of a liquid as a homogeneous fluid with a continuous density is satisfactory for studying the flow of water in a pipe; it is misleading when used for phenomena on the molecular scale. Similarly the Wiener model for the Brownian motion is satisfactory for phenomena involving time intervals of the order of a second. When very small time intervals are involved, as is surely the case when stochastic differential equations are involved, we cannot feel confident a priori that a model involving the Wiener process will be satisfactorily accurate. If it differs excessively from the results corresponding to Lipschitzian noises, we cannot feel absolutely sure that the latter are right, but we can be practically certain that the results obtained by using the Wiener process are wrong. In this case we are not required to abandon classical physics, nor to abandon the use of the mathematically manageable martingales, "white noise," etc. All that we need to do is to try to use these last in some more appropriate way, so that in the new model the departure from the more trustworthy Lipschitzian models is demonstrably small.

We are therefore motivated to try to construct mathematical models for noisy systems that have three important properties, which we now describe. The first is an obvious consequence of the foregoing:

(1-2) *The property of inclusiveness* The model must apply to systems in which the permitted noises are processes belonging to some family large enough to include uniformly Lipschitzian processes, Brownian motion processes, and such modifications as have proved convenient in applications.

The second property is important because the Lipschitzian noises, even though they form only a set of measure 0 in some processes (e.g., Brownian motion), nevertheless are individually accessible to the underlying scientific theory and are known individually to satisfy Eqs. (1-1).

(1-3) *The property of consistency* For each ω in Ω such that the $z^\rho(\mathbf{t}, \omega)$ ($\rho = 1, \ldots, r$) are Lipschitzian, the model must yield a solution for the $x^i(\mathbf{t}, \omega)$ that satisfies (1-1).

The third property, though somewhat vaguely stated, is a statement of a deep-lying requirement for applications of mathematics to systems of all kinds. We first assume that we have adopted for two processes some sort of "closeness" definition that is compatible with experimental procedures, so that an extreme degree of closeness corresponds to practical impossibility of distinguishing the process by means of available experimental procedures.

(1-4) *The property of stability* The model must be such that if the noise process $z^{\rho}(\mathbf{t}, \omega)$ is replaced by another permitted process $z_0^{\rho}(\mathbf{t}, \omega)$ close to it, the corresponding solutions $x^{i}(\mathbf{t}, \omega)$, $x_0^{i}(\mathbf{t}, \omega)$ are also close to each other.

It is obvious that we need to develop some sort of calculus for stochastic processes. We shall in fact develop only an integral calculus, with no independent definition of derivative. Thus, although we shall often speak of solutions of differential equations, such as (1-1), we shall always mean solutions of the corresponding integral equation. For example, by a "solution x of Eq. (1-1)" we shall mean a process x that satisfies

$$(1\text{-}5)\quad x^i(t) = x^i(a) + \int_a^t f^i(s, x(s))\, ds + \sum_{\rho=1}^{r} \int_a^t g_\rho{}^i(s, x(s))\, dz^\rho(s).$$

For Lipschitzian z^ρ, this is a convenient and meaningful formulation; under mild continuity assumptions the last integrals in (1-5) are well-defined Riemann–Stieltjes integrals.

However, the customary definition of Stieltjes integral fails to give any meaning to the last r integrals in (1-5) even for such important processes as Brownian motion, because the customary definition applies to functions z^ρ of bounded variation, while in the Brownian motion $z(\mathbf{t}, \omega)$ is a.s. not of bounded variation. So our first task is to extend the customary definition so that such integrals are defined and are tractable enough for us to use in differential equations [as we shall persistently misname (1-5)!]. We shall therefore begin with a definition of an integral

$$\int_a^b f(s, \omega)\, dz(s, \omega)$$

in which both the integrand f and the integrator z are stochastic processes.

2 *Definition of the integral*

As a guide in defining the kind of integral that plays a major role throughout this book, we recall a definition (there are several) of the Riemann–Stieltjes integral. Let f be a real-valued function defined on a set T of real numbers, and let $[a, b]$ be an interval contained in T. To avoid confusion, from now on we reserve the letter τ for points of T and the letter t for points of $[a, b]$. By a **partition** Π of $[a, b]$ we shall mean an ordered $(2m + 1)$-tuple (m a positive integer) $\Pi = (t_1, \ldots, t_{m+1}; \tau_1, \ldots, \tau_m)$ such that all τ_j are in T and $t_1 = a < t_2 < \cdots < t_{m+1} = b$. The intervals $[t_j, t_{j+1}]$ are the intervals of the partition, the points t_j are its division points, the points τ_j are its evaluation points. Properly speaking, this Π is a partition of $[a, b]$ with evaluation points in T; but we shall ordinarily omit the last phrase. The point τ_j is **associated** with the interval $[t_j, t_{j+1}]$; we do *not* assume that τ_j is in $[t_j, t_{j+1}]$, even though this has been customary. By the **mesh** of Π we shall mean the number

$$\text{mesh } \Pi = \max_{j=1,\cdots,m} [\max\{\tau_j, t_{j+1}\} - \min\{\tau_j, t_j\}].$$

This agrees with the familiar use of the term when $t_j \leqq \tau_j \leqq t_{j+1}$ for all j.

We can now define the Riemann–Stieltjes integral of f with respect to z, as follows:

(2-1) *DEFINITION The function f has a Riemann–Stieltjes integral with respect to z over $[a, b]$ if there exists a number J such that: to each positive ϵ there corresponds a positive number δ such that for every partition $\Pi = (t_1, \ldots, t_{m+1}; \tau_1, \ldots, \tau_m)$ of $[a, b]$ with* mesh $\Pi < \delta$, *it is true that*

$$\left| \sum_{j=1}^{m} f(\tau_j)[z(t_{j+1}) - z(t_j)] - J \right| < \epsilon.$$

We then define $\int_a^b f(t)\, dz(t) = J$.

This definition differs from the more usual one by not requiring each τ_j to lie in its associated interval $[t_j, t_{j+1}]$. When z is continuous and of bounded variation, it can easily be shown that this change has no effect. For other z, the integral defined in Definition (2-1) is less general, but better behaved, than the usual; but we shall not digress to discuss this.

Obviously Definition (2-1) is lengthy. We can improve it by introducing three abbreviations.

First, for every function ϕ defined on $[a, b]$, we define

$$\Delta_j\phi = \phi(t_{j+1}) - \phi(t_j).$$

Second, for the sum written in the definition, which we call a **Riemann sum**, we introduce the symbol

$$S(\Pi; f; z) = \sum_{j=1}^{m} f(\tau_j)\, \Delta_j z.$$

Third, whenever $\Phi(\Pi)$ is any assertion concerning Π, we use this abbreviation:

(2-2) *DEFINITION The statement* **ultimately** $\Phi(\Pi)$ *means that there exists a positive number* δ *such that for every partition* Π *of* $[a, b]$ *with* mesh $\Pi < \delta$, *the sentence* $\Phi(\Pi)$ *is true.*

Now we can rephrase Definition (2-1) thus.

(2-3) *DEFINITION* (2-1) *REPHRASED The function f has a Riemann–Stieltjes integral with respect to z over* $[a, b]$ *if there exists a number J such that for each positive* ϵ, *ultimately* $|S(\Pi; f; z) - J| < \epsilon$. *Then we define*

$$\int_a^b f(t)\, dz(t) = J.$$

Let us try to extend this to stochastic processes. We can use the same definitions about partitions. If f is a real-valued process on T and z a real-valued process on $[a, b]$, as before we can define

$$\Delta_j z(\omega) = z(t_{j+1}, \omega) - z(t_j, \omega),$$

(2-4)
$$S(\Pi; f; z) = \sum_{j=1}^{m} f(\tau_j, \omega)\, \Delta_j z(\omega).$$

The most straightforward extension of Definition (2-3) would be as follows.

The integral of f with respect to z over $[a, b]$ *exists if there is a r.v.* $J(\omega)$ *such that for each positive* ϵ, *ultimately*

$$\bar{\rho}(S(\Pi; f; z), J) < \epsilon.$$

Here $\bar{\rho}$ is the pseudo-metric of convergence in probability.

However, there are well-known examples (we shall soon exhibit one) in which f and z are quite simple processes, but the integral thus defined would fail to exist. We extricate ourselves from this trouble by restricting the type of partition we shall use. Historically, the first definition of integral in the style of Definition (2-1) was devised by Cauchy, who considered only partitions in which $\tau_j = t_j$ for all j. Accordingly, we shall call such partitions **Cauchy partitions**. They have serious technical disadvantages when f is discontinuous, so that Cauchy's integral was practically totally displaced by Riemann's over a century ago. Riemann used partitions in which $t_j \leqq \tau_j \leqq t_{j+1}$ for all j, and accordingly, we shall call these **Riemann partitions**. They have been extremely useful; but in stochastic integration they cannot be used profitably. Instead, we shall use what we call **belated**

partitions, which are partitions Π such that $\tau_j \leqq t_j$ for all j. Sometimes, as a help to memory, we shall use b as a subscript to the name of a belated partition. Thus if we name a partition Π_b or $\Pi_{b,k}''$ or anything else with a subscript b, it will be understood [even if not so stated in words, as for example in Definition (2-6)], that the partition is a belated partition. However, when this aid to memory seems superfluous, we shall simply say, for example, that Π' is a belated partition, without using subscript b to emphasize the fact.

Our definition of a stochastic integral, which we call the **belated** integral, is obtained from Definition (2-1) by inserting the word *belated* before *partition*. To write this in an abbreviated form similar to Definition (2-3) we make the obvious modification of Definition (2-2).

(2-5) *DEFINITION* *The statement* **ultimately** $\Phi(\Pi_b)$ *means that there exists a positive number* δ *such that for every belated partition* Π_b *of* $[a, b]$ *with* mesh $\Pi < \delta$, *the sentence* $\Phi(\Pi_b)$ *is true.*

This permits us to phrase the definition of the belated integral thus.

(2-6) *DEFINITION* *Let* $(\Omega, \mathcal{A}, P)$ *be a probability triple,* T *a set of real numbers, and* $[a, b]$ *an interval contained in* T. *Let* $f(\boldsymbol{\tau}, \omega)$ *and* $z(\mathbf{t}, \omega)$ *be (finite) real-valued stochastic processes on* T *and* $[a, b]$ *respectively. Then* f *has a belated integral with respect to* z *over* $[a, b]$ *if there exists a r.v.* J *such that for each positive* ϵ, *ultimately*

$$\bar{p}(S(\Pi_b; f; z), J) < \epsilon.$$

If the requirement in this definition is satisfied with some one r.v. $J(\omega)$, it is also satisfied with every r.v. $J_0(\omega)$ that is a.s. equal to $J(\omega)$. Every such r.v. will be called a **version** of the belated integral of f with respect to z, and will be denoted (possibly ambiguously) by the symbol

$$\int_a^b f(t, \omega)\, dz(t, \omega),$$

from which as usual we omit the ω whenever it seems desirable to do so. We shall occasionally omit the adjective *belated* when this causes no danger of confusion.

When we wish to discuss integrals, such as ordinary Riemann–Stieltjes integrals, in which the functions are defined on some set of reals and no probability triple is involved, we call the integral **deterministic**, or sometimes **nonstochastic**. In the deterministic case, if $f(\mathbf{t})$ is bounded, we shall see in Chapter III that the belated integral $\int f(t)\,dt$ exists if and only if the Riemann integral $\int f(t)\,dt$ exists. However, in general the usual Stieltjes integral and the belated integral are not equivalent. If $z(\mathbf{t})$ is continuous on $[a, b]$, the usual Stieltjes integral $\int_a^b t\,dz(t)$ always exists; but the belated integral with the same symbols does not exist unless z is of bounded variation. On the other hand, let z be a left-continuous step function on $[a, b]$, vanishing at a and constant between points $t_1^*, \ldots, t_l^*$ $(a \leqq t_1^* < \cdots < t_l^* < b;\ l > 0)$, at which it has jumps of $+1$. If mesh $\Pi < \min(t_{j+1}^* - t_j^*\colon j = 1, \ldots, l-1)$, there will be l intervals $[t_j, t_{j+1})$ of Π that each contain a point t_h^*; and if $[t_j, t_{j-1})$ contains t_h^*, then $z(\tau_j) = h - 1$, whence

$$\sum_{j=1}^{m} z(\tau_j)\,\Delta_j z = \sum_{h=1}^{l} (h-1)1 = l(l-1)/2.$$

Hence

$$\int_a^b z(t)\,dz(t) = \tfrac{1}{2}[z(b)^2 - z(b)].$$

The usual integral $\int z\,dz$ does not exist. The Lebesgue–Stieltjes integral $\int z(t)\,dz(t)$ exists, and has the same value as the belated integral.

If $z(\mathbf{t}, \boldsymbol{\omega})$ is a Poisson process, except on a negligible subset N of Ω the sample function $z(\mathbf{t}, \omega)$ will be finite valued, and nondecreasing with unit jumps. We choose it to be left-continuous at these jumps. By the preceding paragraph, for ω in $\Omega\backslash N$, the Riemann sum $S(\Pi; f; z)$ will converge, as the mesh of the belated partition Π tends to 0, to the limit

$$\tfrac{1}{2}[z(b, \omega) - z(a, \omega)]^2 - \tfrac{1}{2}[z(b, \omega) - z(a, \omega)].$$

So the belated integral $\int z\,dz$ exists and has this as a version.

Some other examples in which the sample functions of $z(\mathbf{t}, \boldsymbol{\omega})$ are step functions will be found in Section 8 of Chapter III.

As a more interesting example, let $z(\mathbf{t}, \boldsymbol{\omega})$ be Brownian motion on $[0, \infty)$, and let $[a, b]$ be an interval with $0 \leqq a < b < \infty$. We could show directly that $f(\mathbf{t}, \boldsymbol{\omega}) = z(\mathbf{t}, \boldsymbol{\omega}) - z(a, \boldsymbol{\omega})$ has a belated integral with respect to z over $[a, b]$. But this is a consequence of a theorem [Theorem III(2-3)] that we shall prove soon, so we merely assume the existence of the integral and try to find a version of it. To do this, it is enough to use only Cauchy partitions. Let $\Pi = (t_1, \ldots, t_{m+1};\ t_1, \ldots, t_m)$ be a Cauchy partition of $[a, b]$. Then

$$S(\Pi; f; z) = \sum_{j=1}^{m} f(t_j)[z(t_j + 1) - z(t_j)], \tag{2-7}$$

and

$$\text{(2-8)} \qquad \sum_{j=1}^{m} [f(t_{j+1}) + f(t_j)][z(t_{j+1}) - z(t_j)] = f(b)^2 - f(a)^2.$$

The r.v.'s $[z(t_{j+1}) - z(t_j)]^2$ $(j = 1, \ldots, m)$ are independent, and

$$E([z(t_{j+1}) - z(t_j)]^2) = t_{j+1} - t_j,$$

$$\begin{aligned} \operatorname{var} \sum_{j=1}^{m} [z(t_{j+1}) - z(t_j)]^2 &= \sum_{1}^{m} \operatorname{var}[z(t_{j+1}) - z(t_j)]^2 \\ &\leqq \sum_{1}^{m} E[z(t_{j+1}) - z(t_j)]^4 \\ &= \sum_{1}^{m} 3(t_{j+1} - t_j)^2 \\ &\leqq 3(\text{mesh }\Pi)(b - a). \end{aligned}$$

So $\sum [z(t_{j+1}) - z(t_j)]^2$ has mean $b - a$ and variance that tends to 0 with mesh Π, whence

$$\text{(2-9)} \qquad \sum_{1}^{m} [f(t_{j+1}) - f(t_j)][z(t_{j+1}) - z(t_j)] = \sum_{1}^{m} [z(t_{j+1}) - z(t_j)]^2$$

$$\to b - a \qquad \text{in } L_2\text{-norm}.$$

Taking half the difference of (2-8) and (2-9) yields, with (2-7),

$$S(\Pi; f; z) \to [z(b) - z(a)]^2/2 - (b - a)/2,$$

the convergence being in L_2-norm and therefore in probability also. So one version of the integral is

$$\text{(2-10)} \qquad \int_a^b [z(t, \omega) - z(a, \omega)]\, dz(t, \omega)$$

$$= [z(b, \omega) - z(a, \omega)]^2/2 - (b - a)/2.$$

We here digress to remark that $\Pi = (t_1, \ldots, t_{m+1}; t_2, \ldots, t_{m+1})$ is a Riemann partition, but not a belated partition. We leave it as an exercise for the reader to show that for such partitions, as mesh $\Pi \to 0$ the Riemann sums $S(\Pi; f; z)$ converge in L_2-norm to a limit, but this limit is different from the right member of (2-10). So the Riemann integral of $z(t) - z(a)$ with respect to $z(t)$ does not exist.

Suppose now that ω_0 is a point of Ω such that $z(\mathbf{t}, \omega_0)$ is Lipschitzian. We deduce from Theorem III(2-3) that the belated integral of $f(\mathbf{t}, \omega_0)$ with respect to $z(\mathbf{t}, \omega_0)$ exists. It is easily computed; for example, (2-8) still holds, and the left member of (2-9) clearly tends to 0, so

$$\lim_{\text{mesh }\Pi\to 0} S(\Pi; f; z)(\omega_0) = [z(b, \omega_0) - z(a, \omega_0)]^2/2. \tag{2-11}$$

Thus the integral lacks the property of consistency. In order to make this example look more like the differential-equation models previously considered, we note that the solution of the differential equations

$$x^1(t, \omega) = \int_a^t 1\, dz(s, \omega) \tag{2-12}$$

$$x^2(t, \omega) = \int_a^t x^1(s, \omega)\, dz(s, \omega) \tag{2-13}$$

has $x^1(t) = z(t) - z(a)$ and

$$x^2(b, \omega) = \int_a^b [z(s, \omega) - z(a, \omega)]\, dz(s, \omega). \tag{2-14}$$

If we imagine that these differential equations furnished, for Lipschitzian $z(\mathbf{t}, \omega)$, a valid model for some physical system, consistency would require that our solution should have the property that $x^2(b, \omega)$ should be the right member of (2-11) whenever $z(\mathbf{t}, \omega)$ is Lipschitzian. (2-10) lacks this property. But it is easily recovered. If the f^i and $g_\rho{}^i$ in (1-1) satisfy the hypotheses in later chapters, which are general enough to cover applications, the $f^i(\mathbf{t}, x(\mathbf{t}))$ and $g_\rho{}^i(\mathbf{t}, x(\mathbf{t}))$ corresponding to Lipschitzian $z^\rho(\mathbf{t}, \omega)$ will be Riemann-integrable functions. (It would not be difficult to use the Lebesgue integral instead of the Riemann, but little is gained thereby.) Now, with f and z as in Definition (2-6), we let Ω_0 denote the set of all points ω_0 such that $z(\mathbf{t}, \omega_0)$ is Lipschitzian and $f(\boldsymbol{\tau}, \omega_0)$ is Riemann integrable over $[a, b]$. Then by elementary mathematics, for all ω_0 in Ω_0 the limit

$$J(\omega_0) = \lim_{\text{mesh }\Pi\to 0} S(\Pi; f(\boldsymbol{\tau}, \omega_0); z(\mathbf{t}, \omega_0)) \tag{2-15}$$

exists, Π being restricted to Riemann partitions. We shall use only Cauchy partitions in (2-15). In Definition (2-6) we defined the belated integral as the limit in probability of Riemann sums. Convergence in probability does not guarantee convergence at any one point. But it is an obvious conjecture, which we shall later prove correct, that among the versions of the integral there is at least one that coincides with the pointwise limit $J(\omega_0)$ at all points ω_0 in Ω_0. Such a version will be called a **strict** version of the integral.

(2-16) *DEFINITION* *Let* Ω_0 *be the set of all* ω_0 *in* Ω *such that* $z(\mathbf{t}, \omega_0)$ *is Lipschitzian and* $f(\boldsymbol{\tau}, \omega_0)$ *is Riemann integrable over* $[a, b]$. *A version* $J(\boldsymbol{\omega})$ *of the belated integral of* f *with respect to* z *is a* **strict** *version if for each* ω_0 *in* Ω_0,

$$J(\omega_0) = \int_a^b f(t, \omega_0)\, dz(t, \omega_0),$$

the right member being an ordinary Riemann–Stieltjes integral.

The right member of (2-10) is a version of the integral that is not strict. If we replace it by a strict version, say $F(\omega)$, this will differ from the right member of (2-10) only on a set Ω_0 of P-measure 0. But if $z(\mathbf{t}, \omega_0)$ is Lipschitzian, ω_0 is in Ω_0, and (2-12) and (2-13) are satisfied and the solution of (2-14) has the consistency property (1-3).

If we use strict versions for integrals, our solutions of differential equations will have the property of consistency. But as our example shows, stability may still be lacking. Suppose that we subdivide $[a, b]$ by points $t_1', t_2', \ldots, t_l'$ with $a = t_1' < \cdots < t_l' = b$, and with all $t_{i+1}' - t_i'$ less than the thousandth part of the shortest time interval that can be detected by any known experimental procedure. For each ω, in the Brownian-motion path $z(\mathbf{t}, \omega)$ we inscribe the polygon $y(\mathbf{t}, \omega)$ with vertices $(t_i', z(t_i', \omega))$, $i = 1, \ldots, l$. Then the processes $z(\mathbf{t}, \boldsymbol{\omega})$ and $y(\mathbf{t}, \boldsymbol{\omega})$ will be experimentally indistinguishable, and yet a.s.

$$\int_a^b [y(t, \boldsymbol{\omega}) - y(a, \boldsymbol{\omega})]\, dy(t, \boldsymbol{\omega}) - \int_a^b [z(t, \boldsymbol{\omega}) - z(a, \boldsymbol{\omega})]\, dz(t, \boldsymbol{\omega})$$
$$= (b - a)/2.$$

It is easy to locate the source of the instability. The sums

$$\sum [z(t_{j+1}, \boldsymbol{\omega}) - z(t_j, \boldsymbol{\omega})]^2$$

converged in L_2-norm to 0 for Lipschitzian z and to $b - a$ for Brownian-motion z. Our procedure for restoring stability involves converting such liabilities into assets. Along with the sums $S(\Pi; f; z)$ we shall consider sums in which each term has two factors of the type $z(t_{j+1}) - z(t_j)$, corresponding to possibly different noises. In fact, there is no difficulty in defining such sums with more than two noise factors, thus:

(2-17) *DEFINITION* *If* $z^1(\mathbf{t}, \boldsymbol{\omega}), \ldots, z^q(\mathbf{t}, \boldsymbol{\omega})$ *are stochastic processes all defined for* t *in an interval* $[a, b]$ *and* ω *in a set* Ω, *and* $f(\boldsymbol{\tau}, \boldsymbol{\omega})$ *is a finite-real-valued process defined on a set* T *that contains* $[a, b]$, *and* Π *is a partition* $(t_1, \ldots, t_{m+1}; \tau_1, \ldots, \tau_m)$ *of* $[a, b]$, *with evaluation points in* T, *the corresponding Riemann sum is*

$$S(\Pi; f; z^1, \ldots, z^q)(\omega)$$

$$= \sum_{j=1}^{m} f(\tau_j, \omega)[z^1(t_{j+1}, \omega) - z^1(t_j, \omega)] \cdots [z^q(t_{j+1}, \omega) - z^q(t_j, \omega)].$$

This leads to a definition of a new type of integral.

(2-18) *DEFINITION With the notation of Definition* (2-17), *the statement that* f *is (belatedly) integrable over* $[a, b]$ *with respect to* $(z^1, \ldots, z^q)$ *means that, with* Π_b *restricted to the class of belated partitions of* $[a, b]$, $S(\Pi_b; f; z^1, \ldots, z^q)$ *converges in probability as* mesh Π_b *tends to zero. Any r.v.* J *such that* $S(\Pi_b; f; z^1, \ldots, z^q)$ *converges in probability to* J *as* mesh $\Pi_b \to 0$ *is a* **version** *of the integral, and is denoted (possibly ambiguously) by*

$$\int_a^b f(t, \omega)\, dz^1(t, \omega) \cdots dz^q(t, \omega),$$

from which we omit the ω *when convenient. The r.v.* J *is a* **strict** *version of the integral if for each* ω_0 *such that the* $z^k(\mathbf{t}, \omega_0)$ *are Lipschitzian and* $f(\mathbf{t}, \omega_0)$ *is Riemann integrable over* $[a, b]$, $J(\omega_0)$ *is equal to the integral*

$$\int_a^b f(t, \omega_0)\, dz^1(t, \omega_0) \cdots dz^q(t, \omega_0),$$

defined as the limit of Riemann sums.

For $q = 1$, this is the integral already defined in Definitions (2-6) and (2-16). For $q \geqq 2$, it is a new type, except that when $q = 2$ and z^1 and z^2 are the same Wiener process, analogous sums occur in proving Itô's differentiation lemma. In this case the limits of the sums, for properly chosen Π and f, have been called "second-order integrals." We shall adopt and extend this terminology, and give the name "qth order integrals" to the integrals defined in Definition (2-18), for $q = 1, 2, 3, \ldots$.

If $q > 1$, it is quite easy to see that the integral in the last sentence of Definition (2-18) has the value 0.

The integral just defined is ambiguous in that it can be replaced by any equivalent r.v. But this is all the ambiguity there is. If $J_1(\omega)$ and $J_2(\omega)$ are both versions of the integral in Definition (2-18), and $\epsilon > 0$, there are positive numbers δ_1, δ_2 such that if Π_b is any belated partition of $[a, b]$

with mesh less than δ_i, then

$$\bar{\rho}(S(\Pi_b; f; z^1, \ldots, z^q), J_i) < \epsilon/2 \qquad (i = 1, 2).$$

We choose a Π_b with mesh $\Pi_b < \min\{\delta_1, \delta_2\}$. Then this inequality holds both for J_1 and for J_2, so $\bar{\rho}(J_1, J_2) < \epsilon$. Now $\bar{\rho}(J_1, J_2)$ is a nonnegative number less than any positive ϵ, so it is 0, and J_1 and J_2 are equivalent.

As an easy but for us rather unimportant example, consider a process $z(\mathbf{t}, \omega)$ (for example, a Poisson process) with the property that almost all sample curves $z(\mathbf{t}, \omega)$ are constant between jumps, the jumps being finite in number and all of the same height λ. If $f(\mathbf{t}, \omega)$ is a.s. continuous in t, it is easy to see that for almost all ω, f has a Stieltjes integral with respect to z, and

$$\int_a^b f(t, \omega)\, dz(t, \omega) = \sum_{t \in [a,b]} f(t, \omega)[z(t+, \omega) - z(t-, \omega)].$$

But if $z(\mathbf{t}, \omega)$ has only finitely many jumps and those are all of height λ, for all Π of sufficiently fine mesh we have

$$[z(t_{j+1}, \omega) - z(t_j, \omega)]^q = \lambda^{q-1}[z(t_{j+1}, \omega) - z(t_j, \omega)].$$

It follows readily that

$$\text{(2-19)} \quad \int_a^b f(t, \omega)\, [dz(t, \omega)]^q = \lambda^{q-1} \sum_t f(t, \omega)[z(t+, \omega) - z(t-, \omega)].$$

Since the property of strictness is a desirable one, it is useful for us to know that whenever f is integrable with respect to $(z^1, \ldots, z^q)$ over $[a, b]$, a strict version of the integral exists. However, there are other useful functions of partitions besides Riemann sums; for example, the approximations to solutions of differential equations by the Cauchy or Runge–Kutta methods are functions of (Cauchy) partitions. So we shall prove a more general statement.

(2-20) *LEMMA For each Π in a class Q of partitions of an interval $[a, b]$, let $F(\Pi, \omega)$ be a r.v. such that as* mesh $\Pi \to 0$ *(Π in Q), $F(\Pi, \omega)$ converges in probability to a r.v. $F(\omega)$. Assume that there is a sequence $\Pi(1), \Pi(2), \ldots$ of partitions in the class Q, with* mesh $\Pi(n) \to 0$, *and a subset Ω_0 of Ω, such that for every ω_0 in Ω_0, $F(\Pi(n), \omega_0)$ converges to a number $J(\omega_0)$ in $\bar{R}$ as $n \to \infty$. Then there is a negligible set N such that*

$$F(\omega_0) = J(\omega_0)$$

for all ω_0 in $\Omega_0 \backslash N$.

By hypothesis, mesh $\Pi(n) \to 0$, and so $F(\Pi(n), \boldsymbol{\omega})$ converges in probability to $F(\boldsymbol{\omega})$. As remarked after I(1-5), we can extract a subsequence that converges a.s. to $F(\boldsymbol{\omega})$. To simplify notation we assume that $F(\Pi(n), \boldsymbol{\omega})$ $(n = 1, 2, 3, \ldots)$ is already such a sequence. Define, for all ω in Ω,

$$F^*(\omega) = \limsup_{n\to\infty} F(\Pi(n), \omega).$$

Except on a negligible set N, the right member is equal to $F(\omega)$, so $F(\omega) = F^*(\omega)$ on $\Omega\backslash N$; and for all ω in Ω_0, the right member is equal to $J(\omega)$ by hypothesis.

(2-21) *COROLLARY Let $f(\boldsymbol{\tau}, \boldsymbol{\omega})$ be a real-valued process on a set $T \subset R$, and let $z^1(\mathbf{t}, \boldsymbol{\omega}), \ldots, z^q(\mathbf{t}, \boldsymbol{\omega})$ be real-valued processes on $[a, b] \subset T$. If the integral*

$$(*) \qquad \int_a^b f(t, \omega)\, dz^1(t, \omega) \cdots dz^q(t, \omega)$$

exists, there is a r.v. F^* that is a strict version of the integral (*).

Let $F(\boldsymbol{\omega})$ be any version of the integral (*). Let Q be the class of all belated partitions of $[a, b]$, and for each Π in Q define $F(\Pi, \boldsymbol{\omega}) = S(\Pi; f; z^1, \ldots, z^q)$. Let Ω_0 be the set of all ω_0 in Ω for which $z^1(\mathbf{t}, \omega_0), \ldots, z^q(\mathbf{t}, \omega_0)$ are Lipschitzian and $f(\mathbf{t}, \omega_0)$ is Riemann integrable, over $[a, b]$. Then for all ω_0 in Ω_0, the integral

$$J(\omega_0) = \int_a^b f(t, \omega_0)\, dz^1(t, \omega_0) \cdots dz^q(t, \omega_0)$$

exists. Let $\Pi(1), \Pi(2), \ldots$ be Cauchy partitions of $[a, b]$ with mesh $\Pi(n) < 1/n$. Then for each ω_0 in Ω_0, $S(\Pi(n); f; z^1, \ldots, z^q)(\omega_0)$ converges to $J(\omega_0)$ as $n \to \infty$. By Lemma (2-20) there is a negligible set N such that $F(\omega) = J(\omega)$ for all ω in $\Omega_0\backslash N$. We define

$$\begin{aligned} F^*(\omega) &= J(\omega) \qquad (\omega \in \Omega_0) \\ &= F(\omega) \qquad (\omega \in \Omega\backslash\Omega_0). \end{aligned}$$

Then $F^*(\omega) = F(\omega)$ except on N, and $F^*(\boldsymbol{\omega})$ is a strict version of the integral (*).

The next corollary is merely a trivial observation; we record it only to be able to refer to it when convenient.

(2-22) *COROLLARY Let $f_1(\boldsymbol{\tau}, \boldsymbol{\omega}), f_2(\boldsymbol{\tau}, \boldsymbol{\omega})$ be processes on $T \subset R$ and let $z^1(\mathbf{t}, \boldsymbol{\omega}), \ldots, z^q(\mathbf{t}, \boldsymbol{\omega})$ be processes on $[a, b] \subset T$. Let Ω_1 be a* *subset of* Ω *such that*

$$f_1(\tau, \omega) = f_2(\tau, \omega)$$

for all ω *in* Ω_1. *If* f_1 *and* f_2 *are both integrable with respect to* $(z^1, \ldots, z^q)$ *over* $[a, b]$, *and* $F_1(\omega)$ *and* $F_2(\omega)$ *are their respective integrals, then* $F_1(\omega) = F_2(\omega)$ *for all* ω *in* $\Omega_1 \backslash N$, *where* N *is a negligible set.*

The Riemann sums $S(\Pi_b; f_1 - f_2; z^1, \ldots, z^q)$ converge in probability to the integral

$$\int_a^b [f_1(t) - f_2(t)]\, dz^1(t) \cdots dz^q(t)$$

as mesh $\Pi_b \to 0$; let $F_3(\omega)$ be a version of that integral. The Riemann sums have value 0 for all ω in Ω_1. So by Lemma (2-20) there is a negligible set N_1 such that

$$F_3(\omega) = 0 \qquad (\omega \in \Omega_1 \backslash N_1).$$

Since

$$S(\Pi_b; f_1; z^1, \ldots, z^q) = S(\Pi_b; f_2; z^1, \ldots, z^q) + S(\Pi_b; f_1 - f_2; z^1, \ldots, z^q)$$

for all belated partitions Π_b, and the three sums converge in probability to $F_1(\omega)$, $F_2(\omega)$, $F_3(\omega)$ as mesh $\Pi_b \to 0$, it follows that there is a negligible set N_2 such that

$$F_1(\omega) = F_2(\omega) + F_3(\omega) \qquad (\omega \in \Omega \backslash N_2).$$

Define $N = N_1 \cup N_2$. Then for all ω in $\Omega \backslash N$ we have

$$F_1(\omega) = F_2(\omega) + 0.$$

In some applications of stochastic processes it is useful to consider different probability measures P, P' on the same σ-algebra $\mathfrak{A}$. Since convergence in probability depends on the measure, the existence and value of the stochastic integral are not invariant under general changes of measure. It can be shown [it will follow from Theorem (2-23)] that if P and P' are equivalent in the sense that for every A in $\mathfrak{A}$, $P(A) = 0$ if and only if $P'(A) = 0$, then every version of the integral when the measure is P remains a version when the measure is P'. However, since we are especially interested in strict versions, it is advantageous to prove the following stronger statement.

(2-23) *THEOREM Let* $(\Omega, \mathfrak{A}, P)$ *and* $(\Omega, \mathfrak{A}, P')$ *be probability triples with the same* Ω *and* $\mathfrak{A}$. *Let* $z^1, \ldots, z^q, f$ *be as in Definition* (2-17). *Let* Ω_0 *be a subset of* Ω *such that for all* ω_0 *in* Ω_0, *the value of*

$S(\Pi_b; f; z^1, \ldots, z^q)$ at ω_0 converges to a limit $J(\omega_0)$ as the mesh of the (belated) partition Π_b tends to 0. Assume that for every set A in $\mathfrak{A}$ contained in $\Omega \backslash \Omega_0$ and having $P'(A) = 0$, it is also true that $P(A) = 0$. If f has an integral over $[a, b]$ with respect to $(z^1, \ldots, z^q)$ when P' is used as probability measure, and $F(\boldsymbol{\omega})$ is a version of the integral such that $F(\omega_0) = J(\omega_0)$ for all ω_0 in Ω_0, then $F(\boldsymbol{\omega})$ is also a version of the integral when P is used as the probability measure.

Let $\bar{\rho}$, $\bar{\rho}'$ denote the metrics of convergence in probability corresponding to the respective measures P, P' [cf. I(1-6)]. Let $\Pi_b(1), \Pi_b(2), \ldots$ be any sequence of belated partitions of $[a, b]$ with mesh $\Pi_b(n) \to 0$. By hypothesis,

$$\lim_{n\to\infty} \bar{\rho}'(S(\Pi_b(n); f; z^1, \ldots, z^q), F(\boldsymbol{\omega})) = 0.$$

So there exists a subsequence $\Pi_b(n_1), \Pi_b(n_2), \ldots$ such that

$$\lim_{j\to\infty} S(\Pi_b(n_j); f; z^1, \ldots, z^q)(\omega) = F(\omega) \tag{2-24}$$

for all ω except those in a set N with $P'(N) = 0$. By hypothesis (2-24) holds for all ω in Ω_0, so $N \subset \Omega - \Omega_0$. Then by hypothesis $P(N) = 0$. That is, (2-24) holds except on a set N with P-measure 0, whence

$$\lim_{j\to\infty} \bar{\rho}(S(\Pi_b(n_j); f; z^1, \ldots, z^q), F(\boldsymbol{\omega})) = 0. \tag{2-25}$$

This implies

$$\lim_{\text{mesh } \Pi_b\to 0} \bar{\rho}(S(\Pi_b; f; z^1, \ldots, z^q), F(\boldsymbol{\omega})) = 0;$$

for otherwise we could find a positive ϵ and a sequence $\Pi_b(1), \Pi_b(2), \ldots$ of belated partitions such that mesh $\Pi_b(n) \to 0$ and

$$\bar{\rho}(S(\Pi_b(n); f; z^1, \ldots, z^q), F(\boldsymbol{\omega})) > \epsilon$$

for all n, which is incompatible with (2-25). This completes the proof.

From Theorem (2-23) we deduce that by restricting ourselves to strict versions of integrals we gain an important consistency property. If a physical quantity is affected by random noises $z^1, \ldots, z^q$, and its value corresponding to each Lipschitzian $z^1, \ldots, z^q$ is given by an integral or sum of integrals, we know that the distribution of the physical quantity is closely represented by the value of the integrals corresponding to some (usually complicated) probability measure P on the space Ω for the z^k, a set Ω_0 with P-measure 1 furnishing Lipschitzian $z^k(\mathbf{t}, \omega_0)$. If we assign another measure P' on $\mathcal{A}$, for example, so that the z^k become Wiener processes, we may thereby depart from physical plausibility for the z^k. Nevertheless, by Theorem (2-23) every strict version of the integrals corresponding to P' will still be a version when we revert to the measure P.

In later theorems establishing the existence of stochastic integrals [e.g., Theorem III(2-3)] we shall use hypotheses that are not invariant under even rather simple changes of probability measure P. By Theorem (2-23), each such existence theorem has an obvious generalization; instead of assuming its hypotheses as stated, we need only assume that the hypotheses hold with some measure P' on $\mathcal{A}$ such that if $A \in \mathcal{A}$ and $P'(A) = 0$ and the $z^k(\mathbf{t}, \omega)$ are not all Lipschitzian for any ω in A, then $P(A) = 0$. Since this generalization is not essential for our present purposes, we pursue it no further.

3 The canonical form

We are now able to describe the method by which we shall construct stochastic models of physical systems which in the physically realizable case of Lipschitzian noises are known to satisfy differential equations (1-1) or (1-5). Suppose that $g^i_{\rho\sigma}(\mathbf{t}, \mathbf{x})$ are functions defined for t in $[a, b]$ and x in R^n and bounded on bounded sets of (t, x). Then for all Lipschitzian processes the solution $x^i(t, \omega)$ of (1-5) is also a solution of

$$(3\text{-}1) \qquad x^i(t, \omega) = x^i(a, \omega) + \int_a^t f^i(s, x(s, \omega))\, ds + \sum_{\rho=1}^{r} \int_a^t g_\rho{}^i(s, x(s, \omega))\, dz^\rho(s, \omega) + \sum_{\rho,\sigma=1}^{r} \int_a^t g^i_{\rho\sigma}(s, x(s, \omega))\, dz^\rho(s, \omega)\, dz^\sigma(s, \omega)/2,$$

since the last integral vanishes for Lipschitzian noises. There are infinitely many such systems (3-1); (1-5) is one of them, distinguished only by being typographically simplest. Apart from this triviality, the infinitely many possible forms of (3-1), with various $g^i_{\rho\sigma}$, all stand on equal footing as far as the experimentally crucial Lipschitzian case is concerned. But when we extend to the larger class of processes required by the inclusiveness property (1-2), the systems (3-1) are no longer indistinguishable; the second-order integrals need not vanish. So to construct a model in which the noises belong to this larger class we must make a choice among the possible $g^i_{\rho\sigma}$. We propose the following method of selecting the $g^i_{\rho\sigma}$.

(3-2) *DEFINITION The* **canonical extension** *of equations* (1-5) *is the set of equations* (3-1) *in which*

$$(3\text{-}3) \qquad g^i_{\rho\sigma}(t, x) = \sum_{k=1}^{n} [\partial g_\rho{}^i(t, x)/\partial x^k]\, g_\sigma{}^k(t, x)$$

$$(i = 1, \ldots, n; \quad \rho, \sigma = 1, \ldots, q; \quad t \in T; \quad x \in R^n).$$

Any set of equations (3-1) *in which the* $g^i_{\rho\sigma}$ *satisfy* (3-3) *will be said to be in* **canonical form**, *or to be a* **canonical system** *of stochastic differential equations.*

We shall also use this definition of the canonical form in the more general situation in which the coefficient functions $g_\sigma{}^i$ are random for each (t, x); that is, they depend directly on ω, as well as on t and x. We do not here justify this choice. A discussion of its justification will be found in Chapter VI. In this section we shall merely exhibit one example in which there is an obvious advantage in using it.

Let us emphasize that we are not adding a "correction term" to the "right" set of equations (1-5). Equations (3-1) with any $g^i_{\rho\sigma}$ serve for the Lipschitzian case. When the second-order integral is defined, the old choice $g_{\rho\sigma} = 0$ has no longer any virtue other than brevity. (Of course, if we have not defined second-order integrals, the choice $g^i_{\rho\sigma} = 0$ is forced on us.) We do not consider (1-5) the fundamental case and (3-1) a patched-up modification. Instead, (3-1) with (3-3) is the distinguished form, and (1-5) merely a typographically simplified version available for use when the noises are Lipschitzian, but not otherwise.

Before we begin the somewhat lengthy proofs needed for establishing existence and properties of solutions of stochastic differential equations, we shall look at the canonical extension of the simple example (2-12) and (2-13). However, to avoid confusion with exponents we denote the left members of Eqs. (2-12) and (2-13) by $x^{(1)}(t, \omega)$, $x^{(2)}(t, \omega)$. There is only one noise $z^1 = z$; the example in (2-12) and (2-13) has the form (1-5) with $f^1 = f^2 = 0$, $g_1{}^1(t, x) = 1$, $g_1{}^2(t, x) = x^{(1)}$. Then by (3-3) we compute

$$g^1_{1,1}(t, x) = 0, \qquad g^2_{1,1}(t, x) = 1.$$

The canonical extension of Eqs. (2-12) and (2-13) is then

$$x^{(1)}(t, \omega) = \int_a^t 1 \, dz(s, \omega),$$

$$x^{(2)}(t, \omega) = \int_a^t x^{(1)}(s, \omega) \, dz(s, \omega) + \int_a^t \tfrac{1}{2} \, dz(s, \omega) \, dz(s, \omega). \tag{3-4}$$

For the Brownian motion, we have already evaluated the first-order integrals and could similarly evaluate the second-order integral without anticipating the proof in Theorem III(3-2) that those integrals exist. However, if we assume the existence of the integrals, the calculations made for the Brownian-motion case become entirely unnecessary, and the solution of (3-4) is obtained without effort. Let z be any process for which the integrals in (3-4) exist. To evaluate the right members of (3-4) it is enough to use Cauchy partitions. So let $\Pi = (t_1, \ldots, t_{m+1}; t_1, \ldots, t_m)$ be a Cauchy partition of $[a, t]$. Then, omitting the variable ω, we have first

$$\begin{aligned} S(\Pi; 1; z) &= \sum_1^m 1[z(t_{j+1}) - z(t_j)] \\ &= z(t) - z(a), \end{aligned}$$

so by (3-4) we have $x^{(1)}(t) = z(t) - z(a)$. Next,

$$\begin{aligned} S(\Pi; x^{(1)}; z) &= \sum_{j=1}^m x^{(1)}(t_j)[z(t_{j+1}) - z(t_j)] \\ &= \sum_{j=1}^m z(t_j)[z(t_{j+1}) - z(t_j)] - z(a)[z(t) - z(a)], \end{aligned}$$

$$S(\Pi; \tfrac{1}{2}; z, z) = \sum_{j=1}^m [z(t_{j+1}) - z(t_j)]^2/2.$$

By adding the last two we obtain

$$\begin{aligned} &S(\Pi; x^{(1)}; z) + S(\Pi; \tfrac{1}{2}; z, z) \\ &\quad = \sum_j [z(t_{j+1}) + z(t_j)][z(t_{j+1}) - z(t_j)]/2 - z(a)[z(t) - z(a)] \\ &\quad = [z(t) - z(a)]^2/2. \end{aligned}$$

Now, letting mesh $\Pi \to 0$, we find that (3-4) has the solution

$$x^{(2)}(t) = [z(t) - z(a)]^2/2. \tag{3-5}$$

This solution has the property of inclusiveness, since it applies to all processes z for which the first- and second-order integrals exist. It has the property of consistency, since its right member agrees with the familiar solution of (2-12) and (2-13) when z is Lipschitzian; moreover, this consistency is the property of that version of the solution that presents itself most naturally, and does not require special treatment for those ω_0 for which $z(\mathbf{t}, \omega_0)$ is Lipschitzian. Finally, it has stability. Given two noise processes z and z_0 such that $(z(a), z(t))$ and $(z_0(a), z_0(t))$ have two-dimensional distributions that are in any reasonable sense almost the same, the two corresponding solutions for $x^2(t)$ will also have nearly the same distribution.

Our task is now to establish properties of stochastic integrals and stochastic differential equations that will allow us to make somewhat similar assertions about solutions of (3-1), in particular when (3-3) is satisfied. These theorems are essential for mathematical soundness. Most of them have statements that are rather easy to understand and to believe, accompanied by annoyingly tedious proofs. They will be the subject matter of most of the remainder of this book.

4 *Elementary properties of the integral*

It is evident that if in computing the integral of f with respect to $(z^1, \ldots, z^q)$ we replace $f, z^1, \ldots, z^q$ by equivalent processes $g, y^1, \ldots, y^q$, neither the existence nor the value of the integral is changed. For if Π is any belated partition, a.s.

$$S(\Pi; f; z^1, \ldots, z^q) = S(\Pi; g; y^1, \ldots, y^q),$$

so if either member of this equation converges in probability to a limit J, so does the other. In later theorems on the existence and properties of integrals we assume properties of $f, z^1, \ldots, z^q$. It would be enough to assume that $f, z^1, \ldots, z^q$ are equivalent to processes satisfying the conditions of the theorem.

The first theorem on the integrals is an extension of the completeness property defined in Chapter I, following the definitions of $\|f\|_p$ and $\bar{\rho}(f_1, f_2)$.

(4-1) *THEOREM* *If $f(\boldsymbol{\tau}, \omega)$ is a process defined on a set T that contains an interval $[a, b]$, and $z^1(\mathbf{t}, \omega), \ldots, z^q(\mathbf{t}, \omega)$ are processes on $[a, b]$,* *the Riemann sums* $S(\Pi_b; f; z^1, \ldots, z^q)$ *for belated partitions* Π_b *of* $[a, b]$ *converge in probability to a r.v.* $J(\omega)$ *(necessarily a version of the integral of* f *with respect to* $z^1, \ldots, z^q$*) if and only if to each positive* ϵ *there corresponds a positive* δ *such that [cf.* I(1-6)]

$$\bar{\rho}(S(\Pi_b'; f; z^1, \ldots, z^q),\ S(\Pi_b''; f; z^1, \ldots, z^q)) < \epsilon$$

whenever Π_b' *and* Π_b'' *are belated partitions of* $[a, b]$ *with mesh less than* δ. *Moreover, a corresponding statement holds if the pseudo-metric* $\bar{\rho}(x, y)$ *is replaced by the distance* $\| x - y \|_p$, *where* $p \geqq 1$.

The "only if" part is trivial. To simplify notation we shorten $S(\Pi; f; z^1, \ldots, z^q)$ to $S(\Pi)$. To prove the "if" part, for $n = 1, 2, 3, \ldots$ let δ_n be a positive number such that if Π' and Π'' are belated partitions with mesh less than δ_n,

$$\bar{\rho}(S(\Pi'), S(\Pi'')) < 2^{-n}.$$

There is no loss of generality in supposing $\delta_1 \geqq \delta_2 \geqq \delta_3 \geqq \cdots$. For each n, let $\Pi(n)$ be a belated partition of $[a, b]$ with mesh $\Pi(n) < \delta_n$. If $\epsilon > 0$, and $n(\epsilon)$ is a number such that $2^{-n(\epsilon)} < \epsilon$, we have

$$\bar{\rho}(S(\Pi(n)), S(\Pi)(m))) < \epsilon$$

whenever $m, n > n(\epsilon)$. So by completeness, the $S(\Pi(n))$ converge in probability to a limit r.v. $J(\omega)$.

Now let ϵ be positive; let k be such that $2^{-k} < \epsilon/2$; and let n be a number greater than k such that $\bar{\rho}(S(\Pi(n)), J) < \epsilon/2$. For all Π with mesh $\Pi < \delta_n$, we have

$$\bar{\rho}(S(\Pi), J) \leqq \bar{\rho}(S(\Pi), S(\Pi(n))) + \bar{\rho}(S(\Pi(n)), J) < \epsilon,$$

so $S(\Pi)$ converges to J in probability as mesh $\Pi \to 0$.

The proof for the L_p-distance is the same except for obvious notational changes.

Belated integrals have obvious linearity properties, as we now show.

(4-2) *THEOREM*

(i) *Let* $f_1(\tau)$ *and* $f_2(\tau)$ *be integrable with respect to* $(z^1, \ldots, z^q)$ *over* $[a, b]$, *and let* c *be a (finite real-valued) random variable. Then* cf_1 *and* $f_1 + f_2$ *are integrable with respect to* $(z^1, \ldots, z^q)$

over $[a, b]$. *Also, if* J_1 *and* J_2 *are (strict) versions of the respective integrals*

$$\int_a^b f_i(t)\, dz^1(t) \cdots dz^q(t) \qquad (i = 1, 2)$$

then cJ_1 *and* $J_1 + J_2$ *are (strict) versions of the respective integrals*

$$\int_a^b cf_1(t)\, dz^1(t) \cdots dz^q(t), \quad \int_a^b [f_1(t) + f_2(t)]\, dz^1(t) \cdots dz^q(t).$$

(ii) *Let* f *be integrable over* $[a, b]$ *with respect to* z^1 *and to* z^2, *and let* c *be a (finite real-valued) r.v. Then* f *is integrable with respect to* cz^1 *and to* $z^1 + z^2$. *Also if* J_1 *and* J_2 *are versions of the respective integrals*

$$\int_a^b f(t)\, dz^1(t), \qquad \int_a^b f(t)\, dz^2(t),$$

then cJ_1 *and* $J_1 + J_2$ *are versions of the respective integrals*

$$\int_a^b f(t)\, d(cz^1(t)), \qquad \int_a^b f(t)\, d[z^1(t) + z^2(t)].$$

Let ϵ be positive. There exists an $M > 1$ such that the set

$$\Omega_M' = \{\omega : | c(\omega) | > M\}$$

has $P(\Omega_M') < \epsilon/2$. Since f_1 is integrable, there exists a positive δ such that if Π is any belated partition of $[a, b]$ with mesh $\Pi < \delta$,

$$| S(\Pi; f_1; z^1, \ldots, z^q)(\omega) - J_1(\omega) | < \epsilon/2M$$

except on a set Ω_0 with $P(\Omega_0) < \epsilon/2M$. Then for all such Π we have

$$| S(\Pi; cf_1; z^1, \ldots, z^q)(\omega) - c(\omega)J_1(\omega) |$$
$$= | c(\omega) | \cdot | S(\Pi; f_1; z^1, \ldots, z^q) - J_1(\omega) | < \epsilon/2$$

except on the set $\Omega_M' \cup \Omega_0$, whose P-measure is less than ϵ. The other conclusions of the theorem are even easier to prove.

In the next theorem $\bar{\rho}$ denotes the pseudo-metric of convergence in probability [cf. I(1-6)].

(4-3) *THEOREM* *Let f be integrable with respect to $(z^1, \ldots, z^q)$ over $[a, b]$. If $a < c < b$, f is integrable with respect to $(z^1, \ldots, z^q)$ over $[a, c]$ and over $[c, b]$ and a.s.*

$$(*) \qquad \int_a^b f(t)\, dz^1(t) \cdots dz^q(t) = \int_a^c f(t)\, dz^1(t) \cdots dz^q(t) + \int_c^b f(t)\, dz^1(t) \cdots dz^q(t).$$

Moreover, for each positive ϵ, if δ is a number such that for every belated partition Π_b of $[a, b]$ with mesh $\Pi_b < \delta$ *it is true that*

$$(**) \qquad \bar{\rho}\left(S(\Pi_b; f; z^1, \ldots, z^q), \int_a^b f(t)\, dz^1(t) \cdots dz^q(t)\right) \leqq \epsilon,$$

then for every belated partition Π_b of $[a, c]$ with mesh $\Pi_b < \delta$ *the inequality analogous to* (**) *holds, and likewise for $[c, b]$.*

Let ϵ be positive, and let δ correspond to ϵ as in the statement of the theorem. Let

$$\Pi' = (t_1', \ldots, t_{k+1}'; \tau_1', \ldots, \tau_k'),$$
$$\Pi'' = (t_1'', \ldots, t_{l+1}''; \tau_1'', \ldots, \tau_l'')$$

be two belated partitions of $[a, c]$, both with mesh less than δ. Let

$$\Pi^* = (t_1^*, \ldots, t_{m+1}^*; \tau_1^*, \ldots, \tau_m^*)$$

be a belated partition of $[c, b]$ with mesh less than δ. Then the partitions

$$\Pi_1 = (t_1', \ldots, t_k', t_1^*, \ldots, t_{m+1}^*; \tau_1', \ldots, \tau_k', \tau_1^*, \ldots, \tau_m^*),$$
$$\Pi_2 = (t_1'', \ldots, t_l'', t_1^*, \ldots, t_{m+1}^*; \tau_1'', \ldots, \tau_l'', \tau_1^*, \ldots, \tau_m^*),$$

are belated partitions of $[a, b]$ with mesh less than δ. So the Riemann sums corresponding to Π_1 and Π_2 both satisfy (**), and therefore

$$(4\text{-}4) \qquad \bar{\rho}(S(\Pi_1; f; z^1, \ldots, z^q),\ S(\Pi_2; f; z^1, \ldots, z^q)) \leqq 2\epsilon.$$

If from both Riemann sums we discard the last m terms, which are the same in both sums, the $\bar{\rho}$-distance is unchanged, and from (4-4) we obtain

$$\bar{\rho}(S(\Pi'; f; z^1, \ldots, z^q),\ S(\Pi''; f; z^1, \ldots, z^q)) \leqq 2\epsilon.$$

Therefore the Riemann sums corresponding to belated partitions of $[a, c]$ satisfy the Cauchy condition with respect to the pseudo-metric $\bar{\rho}$. By Theorem (4-1), the Riemann sums converge in $\bar{\rho}$-pseudo-metric to some r.v. By definition, the first integral in the right member of (*) exists. In the same way the second integral exists.

With the same notation as before we have

$$(4\text{-}5)\quad S(\Pi_1; f; z^1, \ldots, z^q) = S(\Pi'; f; z^1, \ldots, z^q) + S(\Pi^*; f; z^1, \ldots, z^q).$$

By choosing δ small enough we can make the three terms in this equation arbitrarily close to the corresponding terms in (*), so (*) is established.

If we subtract (*) member by member from (4-5), by the triangle inequality we obtain

$$(4\text{-}6)\quad \bar{\rho}\left(S(\Pi'; f; z^1, \ldots, z^q), \int_a^c f(t)\, dz^1(t) \cdots dz^q(t)\right)$$
$$\leqq \bar{\rho}\left(S(\Pi_1; f; z^1, \ldots, z^q), \int_a^b f(t)\, dz^1(t) \cdots dz^q(t)\right)$$
$$+ \bar{\rho}\left(\int_c^b f(t)\, dz^1(t) \cdots dz^q(t),\ S(\Pi^*; f; z^1, \ldots, z^q)\right).$$

The first term in the right member is at most ϵ, and we can choose Π^* to make the second term arbitrarily small, so the left member of (4-6) is at most ϵ. With the analogous estimate for $[c, b]$, this completes the proof.

REMARK If the Riemann sums $S(\Pi; f; z^1, \ldots, z^q)$, for belated partitions Π of $[a, b]$, converge in L_p-norm to the left member of (*) for some $p \geqq 1$, trivial modification of the preceding proof shows that the same holds for partitions of $[a, c]$ and of $[c, b]$, and that we can replace $\bar{\rho}$ by the L_p-norm in the last sentence of the theorem.

We leave the following easy proof as an exercise.

(4-7) Let $f(\mathbf{t}, \omega)$ be a simple process with values $v_1(\omega), \ldots, v_h(\omega)$ on the respective intervals $[t_1, t_2), \ldots, [t_h, t_{h+1}]$, where $t_1 = a$ and $t_{h+1} = b$. If the integral

$$\int_a^b f(t, \omega)\, dz(t, \omega)$$

exists, the sum

$$\sum_{i=1}^{h} v_i(\omega)[z(t_{i+1}, \omega) - z(t_i, \omega)]$$

is a version of the integral.

[If $h = 1$, all Riemann sums are equal. For general h, use Theorem (4-3).]

Finally we state two theorems that have some intrinsic interest, although it happens that we do not need them in any later proofs. We leave their proofs as exercises.

(4-8) *THEOREM* *Let the integral*

$$\int_a^t f(s)\, dz^1(s) \cdots dz^q(s)$$

exist for $t = b$, *and therefore by Theorem* (4-3) *for all* t *in* $[a, b]$; *let* $F(t, \omega)$ *be a version of the integral. Then at each* t_0 *in* $[a, b]$ *such that*

$$(**) \qquad \lim_{t \to t_0} \bar{\rho}\left(\prod_{k=1}^{q} [z^k(t) - z^k(t_0)], 0\right) = 0$$

the process $F(\mathbf{t}, \omega)$ *is continuous in probability.*

[Suggestion: Because of the uniform convergence in the conclusion of Theorem (4-3), it is enough to prove this for f constant between jumps, hence for f constant on $[a, b]$. For $\epsilon > 0$, choose $\delta > 0$ such that for all Π_b with mesh $< \delta$, the Riemann sum approximates the integral to within ϵ in $\bar{\rho}$-distance. Insert t, t_0 as division points. If t is near enough to t_0,

$$\bar{\rho}\left(f \prod [z^k(t) - z^k(t_0)], 0\right) < \epsilon,$$

so by Theorem (4-3)

$$\bar{\rho}\left(\int_{t_0}^t f\, dz^1 \cdots dz^q, 0\right) < 2\epsilon.]$$

(4-9) *THEOREM* *Let* $z^1, \ldots, z^q$ *be a set of processes on* $[a, b]$ *such that for each* t_0 *in* $[a, b]$,

$$\prod_{k=1}^{q} [z^k(t) - z^k(t_0)]$$

tends to 0 *in probability as the point* t *of* $[a, b]$ *tends to* t_0. *Let* $f(\boldsymbol{\tau})$ *be a process such that the integral*

$$(*) \qquad \int_a^t f(s)\, dz^1(s) \cdots dz^q(s)$$

exists for $t = b$, *and hence for all* t *in* $[a, b]$. *Then there is a separable measurable process* $F(\mathbf{t}, \omega)$ *such that for each* t *in* $[a, b]$, $F(t, \omega)$ *is a strict version of the integral* (*).

5 The Itô-belated integral

Itô has defined a (first-order) stochastic integral that has justly become widely known and frequently used. It differs from the belated integral defined in Section 2 by having stronger hypotheses on the probabilistic properties of the $z(\mathbf{t}, \omega)$, but much weaker hypotheses concerning continuity of the integrand $f(\tau, \omega)$. We shall now define an integral that generalizes Itô's integral when $E(f^2) < \infty$ and also generalizes the belated integral insofar as the belated integral has proved useful. For this reason the new integral will be called the "Itô-belated" integral. From a technical point of view, this new integral is only a little more complicated than the belated integral. Nevertheless, some experience indicates that the familiar almost-Riemann form of the definition of the belated integral is an aid in comprehending its meaning and uses. So, in spite of the added power of the Itô-belated integral, we shall develop our calculus primarily for the belated integral; extensions of the calculus to the Itô-belated integral will be indicated in sections written for that purpose.

Let us change the notation for partitions slightly. If

$$\Pi = (t_1, \ldots, t_{m+1}; \tau_1, \ldots, \tau_m),$$

we can denote $[t_j, t_{j+1})$ by A_j and pair each A_j with the corresponding τ_j; then Π can be specified by the set of pairs $\{(A_1, \tau_1), \ldots, (A_m, \tau_m)\}$. Such a set of pairs, with all τ_j in $[a, b]$, and with the A_j pairwise disjoint right-open intervals whose union is $[a, b)$, will be called a partition **of** $[a, b)$. If the τ_j are in $[a, b]$, and the A_j are pairwise disjoint right-open subintervals of $[a, b)$, the set of pairs will be called a partition **in** $[a, b)$, even though $\cup A_i$ is not all of $[a, b)$.

As before, if ϕ is any function on $[a, b)$, and the interval A_j of a partition $\Pi = \{(A_1, \tau_1), \ldots, (A_k, \tau_k)\}$ has ends u_j, v_j, we define $\Delta_j\phi$ to be $\phi(v_j) - \phi(u_j)$. Also, as before,

$$S(\Pi; f; z^1, \ldots, z^q) = \sum_{j=1}^{k} f(\tau_j)\, \Delta_j z^1 \cdots \Delta_j z^q.$$

For a belated partition Π of $[a, b]$, the statement, "mesh $\Pi < \delta$" is equivalent to "$A_j \subset [\tau_j, \tau_j + \delta)$ for $j = 1, \ldots, k$." We now make the decisive step of replacing the positive constant δ by an a.e. positive function $\delta(\cdot)$ on $[a, b)$; we shall be interested in partitions such that $A_j \subset [\tau_j, \tau_j + \delta(\tau_j))$ for each j. However, a difficulty arises. Given a positive-valued δ on $[a, b)$, there may not exist any partition Π *of* $[a, b)$ that satisfies this condition; for example, this happens if $\delta(\tau) = (b - \tau)/2$. On the other hand, we can and soon shall show that there exist partitions

Π *in* $[a, b)$ that satisfy this condition and have $\sum mA_j$ as close to $b - a$ as desired. Accordingly, for each function $\delta > 0$ a.e. on $[a, b)$ and each positive number δ^*, we define **a** (δ, δ^*) **partition** in $[a, b)$ to be a partition $\Pi = \{(A_1, \tau_1), \ldots, (A_k, \tau_k)\}$ in $[a, b)$ such that

(5-1) (i) $A_j \subset (\tau_j - \delta(\tau_j), \tau_j + \delta(\tau_j)) \qquad (j = 1, \ldots, k)$,

(ii) $m\left[[a, b) \setminus \bigcup_{j=1}^{k} A_j\right] < \delta^*$;

and we define **a** (δ, δ^*) **belated partition** in $[a, b)$ to be a partition $\Pi = \{(A_1, \tau_1), \ldots, (A_k, \tau_k)\}$ in $[a, b)$ such that

(5-2) (i) $A_j \subset [\tau_j, \tau_j + \delta(\tau_j)) \qquad (j = 1, \ldots, k)$,

(ii) $m\left[[a, b) \setminus \bigcup_{j=1}^{k} A_j\right] < \delta^*$.

In Section 2 we found it convenient to define the word *ultimately*, for use in defining the belated integral. For the Itô-belated integral, we give a new meaning to the word, as follows.

(5-3) *DEFINITION If* $\Phi(\Pi)$ *is any assertion concerning partitions* Π *in* $[a, b)$, *the statement* **ultimately** $\Phi(\Pi_b)$ *means that there exists a function* $\delta(\mathbf{t})$ *a.e. positive on* $[a, b)$ *and a positive number* δ^* *such that whenever* Π_b *is a* (δ, δ^*) *belated partition in* $[a, b)$, $\Phi(\Pi_b)$ *is a valid statement.*

Now to define the Itô-belated integral we can copy Definition (2-6), with no visible change except replacing *belated* by *Itô-belated.* Of course there is the invisible but important change that the word *ultimately* is given the meaning in Definition (5-3) instead of the meaning in Definition (2-5).

In order that this definition be useful, it is essential that we prove that for every δ positive a.e. on $[a, b)$ and every positive number δ^*, there is a (δ, δ^*) belated partition in $[a, b)$. We shall prove more than this. We shall show that for every function $\delta(\mathbf{t}) > 0$ a.e. on $[a, b)$ and every positive δ^* there is a (δ, δ^*) belated partition $\Pi = \{(A_1, \tau_1), \ldots, (A_k, \tau_k)\}$ in $[a, b)$ such that for $j = 1, \ldots, k$, the point τ_k is the left end-point of A_j. Such partitions we shall call (δ, δ^*) Cauchy partitions in $[a, b)$. Let $\delta(\mathbf{t})$ be positive a.e. on a set $E \subset [a, b)$ with $mE = b - a$, and let δ^* be a positive number. Denote by $\mathcal{K}$ the set of all closed subintervals $[t, \beta)$ of $[a, b)$ with t in E and $\beta - t < \delta(t)$. These form a Vitali covering of E, so by Vitali's theorem there exists a sequence $A_1, A_2, \ldots$ of pairwise disjoint members of

$\mathcal{K}$ whose union contains almost all points of E, and therefore has measure $b - a$. If k is large enough,

$$m(A_1 \cup \cdots \cup A_k) > b - a - \delta^*.$$

Let t_j be the left end point of A_j; then

$$\Pi = \{(A_1, t_1), \ldots, (A_k, t_k)\}$$

is a (δ, δ^*) belated Cauchy partition in $[a, b)$.

We can now repeat the proof given just after Definition (2-18) to conclude that the Itô-belated integral is unique up to equivalence.

We shall use a definition of *strict-version* a trifle weaker than that in Definition (2-16). A version $J(\boldsymbol{\omega})$ of the integral is strict if for each ω_0 such that the $z^k(\mathbf{t}, \omega_0)$ are Lipschitzian and $f(\boldsymbol{\tau}, \omega_0)$ is bounded and has a Cauchy (or, more particularly, a Riemann) integral with respect to $(z^1, \ldots, z^q)$, $J(\omega_0)$ coincides with that Cauchy integral. We now show that if the integral exists it has a strict version. Let $J(\boldsymbol{\omega})$ be a version of

$$\int_a^b f(t)\, dz^1(t) \cdots dz^q(t)$$

and let Ω_0 be the set of ω such that the $z^k(\mathbf{t}, \omega)$ are Lipschitzian and $f(\mathbf{t}, \omega)$ is bounded and has a Cauchy integral $F(\omega)$ with respect to $(z^1, \ldots, z^q)$. For each positive integer n, there exist an a.e. positive $\delta_n(\mathbf{t})$ and a positive number δ_n^* such that for every (δ_n, δ_n^*) belated partition Π in $[a, b)$,

$$\bar{\rho}(S(\Pi; f; z^1, \ldots, z^q), J) < 1/n.$$

We may suppose $\delta_n(t) < 1/n$ $(a \leqq t < b)$ and $\delta_n^* < 1/n$.

As we have just shown, for each n we can and do choose a (δ_n, δ_n^*) Cauchy partition Π_n in $[a, b)$; then

$$\bar{\rho}(S(\Pi_n; f; z^1, \ldots, z^q), J) < 1/n.$$

From the Π_n we can choose a subsequence, which without loss of generality we may consider to be the whole sequence, such that

$$\lim_{n\to\infty} S(\Pi_n; f; z^1, \ldots, z^q)(\omega) = J(\omega)$$

for all ω in Ω except those in a negligible set N.

Let ω_0 be in $\Omega_0 \backslash N$; denote Π_n by $\{(A_1, \tau_1), \ldots, (A_k, \tau_k)\}$. The set $[a, b) \backslash (A_1 \cup \cdots \cup A_k)$ consists of intervals of total length less than δ_n^*, which is less than $1/n$. We name them $A_{k+1}, \ldots, A_h$, and define τ_j to be the left end-point of A_j $(j = k + 1, \ldots, h)$. Then $\Pi_n^* = \{(A_1, \tau_1), \ldots, (A_h, \tau_h)\}$ is a Cauchy partition of $[a, b)$ with mesh less than $1/n$. If M

is an upper bound for $|f(\boldsymbol{\tau}, \omega_0)|, |z^1(\mathbf{t}, \omega_0)|, \ldots, |z^{q-1}(\mathbf{t}, \omega_0)|$, and L is a Lipschitz constant for $z^q(\mathbf{t}, \omega_0)$,

$$|S(\Pi_n^*; f; z^1, \ldots, z^q)(\omega_0) - S(\Pi_n; f; z^1, \ldots, z^q)(\omega_0)|$$

$$= \left| \sum_{j=k+1}^{h} f(\tau_j, \omega_0) \, \Delta_j z^1 \cdots \Delta_j z^q \right|$$

$$\leqq \sum_{j=k+1}^{h} M(2M)^{q-1} L \, \Delta_j t$$

$$< 2^{q-1} M^q L \delta_n^*.$$

By definition of Ω_0,

$$\lim_{n\to\infty} S(\Pi_n; f; z^1, \ldots, z^q)(\omega_0) = \lim_{n\to\infty} S(\Pi_n^*; f; z^1, \ldots, z^q)(\omega_0)$$

$$= F(\omega_0).$$

From this and the preceding equation we have $J(\omega_0) = F(\omega_0)$ for all ω_0 in $\Omega_0 \backslash N$. If we now replace $J(\omega)$ by $F(\omega)$ for all ω in N, we obtain a strict version of the integral.

Given a r.v.-valued function $F(\Pi)$ of partitions in an interval $[a, b)$ and a r.v. $J(\boldsymbol{\omega})$, we shall say that $F(\Pi_b)$ converges to J in probability **as** (δ, δ^*) **shrinks** if for each positive ϵ there is a $\delta(\mathbf{t})$ a.e. positive on $[a, b)$ and a positive δ^* such that $\bar{\rho}(F(\Pi), J) < \epsilon$ for every (δ, δ^*) belated partition Π of $[a, b)$. Similar definitions are used for the convergence of $F(\Pi)$ to J in L_p-norm "as (δ, δ^*) shrinks." In adapting to Itô-belated integrals statements and proofs already made for belated integrals, usually the phrase "as mesh Π tends to 0" will be replaced by "as (δ, δ^*) shrinks."

Theorem (4-1) extends to the Itô-belated integral with only trivial change in the proof. For $n = 1, 2, 3, \ldots$, let δ_n be a positive a.e. on $[a, b)$ and δ_n^* a positive number such that if Π', Π'' are both (δ_n, δ_n^*) belated partitions of $[a, b)$,

$$\bar{\rho}(S(\Pi'), S(\Pi'')) < 2^{-n}.$$

There is no loss of generality in assuming $\delta_1 \geqq \delta_2 \geqq \delta_3 \geqq \cdots$ and $\delta_1^* \geqq \delta_2^* \geqq \cdots$; for otherwise we replace $\delta_n(t)$ by $\min\{\delta_1(t), \ldots, \delta_n(t))\}$ and δ_n^* by $\min\{\delta_1^*, \ldots, \delta_n^*\}$. The rest of the proof requires only obvious and trivial changes.

Theorem (4-2) extends with only obvious notational changes in proof. Theorem (4-3) also extends, if we rephrase the second part to read, "Moreover, for each positive ϵ, if δ is positive a.e. on $[a, b)$ and δ^* a positive number such that for every (δ, δ^*) belated partition Π in $[a, b)$

$$(**) \qquad \bar{\rho}\left(S(\Pi; f; z^1, \ldots, z^q), \int_a^b f(t)\, dz^1(t) \cdots dz^q(t)\right) \leqq \epsilon,$$

then for every (δ, δ^*) belated partition in $[a, c)$ the inequality analogous to $(**)$ holds, and likewise for $[c, b)$." The proof of the first part of the theorem is practically the same as in Section 4, but the proof of the second part needs the following revision:

Let $\Pi' = \{(A_1, \tau_1), \ldots, (A_k, \tau_k)\}$ be a (δ, δ^*) belated partition in $[a, c)$. Then

$$0 \leqq \gamma = (c - a) - \sum mA_j < \delta^*.$$

Let ϵ' be positive. Since the integral over $[c, b)$ exists it is possible to choose a $(\delta, \delta^* - \gamma)$ belated partition Π'' in $[c, b)$ for which

$$\bar{\rho}\left(S(\Pi''); f; z^1, \ldots, z^q), \int_c^b f\, dz^1 \cdots dz^q\right) < \epsilon'.$$

Then $\Pi' \cup \Pi''$ is a (δ, δ^*) belated partition in $[a, b)$, so

$$\bar{\rho}\left(S(\Pi' \cup \Pi''; f; z^1, \ldots, z^q), \int_a^b f\, dz^1 \cdots dz^q\right) < \epsilon.$$

By the part of the theorem already proved,

$$\bar{\rho}\left(S(\Pi'; f; z^1, \ldots, z^q), \int_a^c f\, dz^1 \cdots dz^q\right)$$

$$\leqq \bar{\rho}\left(S(\Pi' \cup \Pi''; f; z^1, \ldots, z^q), \int_a^b f\, dz^1 \cdots dz^q\right)$$

$$+ \bar{\rho}\left(S(\Pi''; f; z^1, \ldots, z^q), \int_c^b f\, dz^1 \cdots dz^q\right) < \epsilon + \epsilon'.$$

The first member of this sequence of inequalities does not depend on the arbitrary positive ϵ', so it is at most equal to ϵ.

The following theorem has an analog for the belated integral, but only with some hypotheses on the behavior of z and f near the point c.

(5-4) *THEOREM If $a < c < b$, and f has an Itô-belated integral with respect to $(z^1, \ldots, z^q)$ over $[a, c)$ and over $[c, b)$, it has an Itô-belated integral with respect to $(z^1, \ldots, z^q)$ over $[a, b)$.*

Let ϵ be positive. There exist a $\delta_1(\mathbf{t})$ a.e. positive on $[a, c)$ and a positive δ_1^* such that for each (δ_1, δ_1^*) belated partition Π_1 in $[a, c)$,

$$\bar{\rho}\left(S(\Pi_1; f; z^1, \ldots, z^q), \int_a^c f(t)\, dz^1(t) \cdots dz^q(t)\right) < \epsilon/2.$$

There exists a $\delta_2(\mathbf{t})$ a.e. positive on $[c, b)$ and a positive δ_2^* that are similarly related to the sums and integral over $[c, b)$. Now define

$$\begin{aligned} \delta(t) &= \min\{\delta_1(t), c - t\} && \text{if } a \leqq t < c, \\ &= \delta_2(t) && \text{if } c \leqq t < b, \\ \delta^* &= \min\{\delta_1^*, \delta_2^*\}. \end{aligned}$$

If Π is any (δ, δ^*) belated partition in $[a, b)$, it is the union of a (δ_1, δ_1^*) belated partition Π_1 in $[a, c)$ and a (δ_2, δ_2^*) belated partition Π_2 in $[c, b)$. Therefore

$$\begin{aligned} &\bar{\rho}\left(S(\Pi; f; z^1, \ldots, z^q), \int_a^c + \int_c^b f(t)\, dz^1(t) \cdots dz^q(t)\right) \\ &\leqq \bar{\rho}\left(S(\Pi_1; f; z^1, \ldots, z^q), \int_a^c f(t)\, dz^1(t) \cdots dz^q(t)\right) \\ &\quad + \bar{\rho}\left(S(\Pi_2; f; z^1, \ldots, z^q), \int_c^b f(t)\, dz^1(t) \cdots dz^q(t)\right) < \epsilon. \end{aligned}$$

Theorem (4-7) extends at once to Itô-belated integrals. Theorem (4-8) has what seems to be a strengthened form for such integrals as follows:

(5-5) *THEOREM* *Let the Itô-belated integral*

$$F(t) = \int_a^t f(s)\, dz^1(s) \cdots dz^q(s)$$

exist for $t = b$, *and therefore for* $a \leqq t \leqq b$. *Then* F *is continuous in probability on* $[a, b]$.

Let ϵ be positive; let δ be positive a.e. on $[a, b)$ and δ^* a positive number such that for every (δ, δ^*) belated partition Π in $[a, b)$,

$$\bar{\rho}\left(S(\Pi; f; z^1, \ldots, z^q), \int_a^b f(s)\, dz^1(s) \cdots dz^q(s)\right) < \epsilon/2.$$

Let t_1 be any point in $[a, b)$; define $t_0 = \max\{t_1 - \delta^*/3, a\}$, $t_2 = \min\{t_1 + \delta^*/3, b\}$. Let Π' be a $(\delta, \delta^*/3)$ belated partition in $[a, t_0)$. Then Π' is also a (δ, δ^*) belated partition in $[a, t')$ whenever $t_0 \leqq t' \leqq t_2$, so by the extension of Theorem (4-3)

$$\bar{\rho}(S(\Pi'; f; z^1, \ldots, z^q), F(t')) \leqq \epsilon/2$$

for all such t'. The Riemann sum is independent of t', so for all t' in $[t_0, t_2]$ we have $\bar{\rho}(F(t'), F(t_1)) \leqq \epsilon$, and F is continuous in probability at t'.

This theorem seems more general than (4-8), since hypothesis (**) of Theorem (4-8) is not required. But this is largely spurious. If 1 is integrable with respect to $(z^1, \ldots, z^q)$ over $[a, b)$, hypothesis (**) has to hold.

(5-6) *COROLLARY Let f be real-valued (nonstochastic) on a set T in R, and let* $z^1, \ldots, z^q$ *be real-valued (nonstochastic) on* $[a, b] \subset T$. *If f has an Itô-belated integral with respect to* $(z^1, \ldots, z^q)$ *over* $[a, b)$, *the indefinite integral F, defined as in Theorem* (5-5), *is continuous on* $[a, b)$.

The process $F(\mathbf{t}, \omega)$ defined in Theorem (5-5) is unique except that for each t, we are at liberty to change $F(t, \omega)$ on any negligible set. We have already seen that we can use this freedom to select a strict version. We shall now show that we can go so far as to select a version that has all three desirable properties of strictness, separability, and measurability.

(5-7) *THEOREM Let* $f(\boldsymbol{\tau}, \omega)$ *be a real-valued process on a set T in R, and let* $z^1(\mathbf{t}, \omega), \ldots, z^q(\mathbf{t}, \omega)$ *be real-valued on* $[a, b] \subset T$. *Let f have an Itô-belated integral with respect to* $(z^1, \ldots, z^q)$ *over* $[a, b]$. *Then there is a separable measurable process* $F(\mathbf{t}, \omega)$ *on* $[a, b]$ *such that for each t in* $[a, b]$, $F(t, \omega)$ *is a strict version of the integral*

$$(*) \qquad \int_a^t f(s, \omega)\, dz^1(s, \omega) \cdots dz^q(s, \omega).$$

By Corollary (2-21), for each t in $[a, b]$ there is a r.v. $X(t, \omega)$ which is a strict version of the integral (*). By Theorem (5-5), $X(\mathbf{t}, \omega)$ is continuous in probability on $[a, b]$. Let $\tilde{X}(\mathbf{t}, \omega)$ be the separable measurable process equivalent to $X(\mathbf{t}, \omega)$ with the properties specified in Theorem I(3-14). We define $F(\mathbf{t}, \omega) = \tilde{X}(\mathbf{t}, \omega)$. Since $F(t, \omega)$ is equivalent to $X(t, \omega)$ for all t in $[a, b]$, it is a version of the integral (*). It remains only to prove that for each t in $[a, b]$, $F(t, \omega)$ is a strict version of (*).

For each c in $[a, b]$, let Ω_c be the set of all ω_c in Ω such that the $z^k(\mathbf{t}, \omega_c)$ $(k = 1, \ldots, q)$ are Lipschitzian and $f(\mathbf{t}, \omega_c)$ is bounded on $[a, c]$, and the (deterministic Cauchy) integral

$$J(t, \omega_c) = \int_a^t f(s, \omega_c)\, dz^1(s, \omega_c) \cdots dz^q(s, \omega_c)$$

exists for $t = c$, and therefore for all t in $[a, c]$. It is obvious that for each ω_c in Ω_c the function $J(\mathbf{t}, \omega_c)$ is continuous on $[a, c]$. Because $X(t, \boldsymbol{\omega})$ is a strict version of the integral (*) for each t in $[a, b]$,

$$X(t, \omega_c) = J(t, \omega_c) \qquad (\omega_c \in \Omega_c; \quad a \leqq t < c).$$

By Theorem I(3-14), there is a sequence of points $t_1, t_2, \ldots$ of $[a, c]$ tending to c such that

$$\begin{aligned} F(c, \omega_c) &= \limsup_{n\to\infty} X(t_n, \omega_c) \\ &= \limsup_{n\to\infty} J(t_n, \omega_c) \\ &= J(c, \omega_c). \end{aligned}$$

This holds for all ω_c in Ω_c, so $F(c, \boldsymbol{\omega})$ is a strict version of the integral (*) with $t = c$.

III Existence of Stochastic Integrals

1 Fundamental lemma

In the preceding chapter we proved some theorems on stochastic integrals; but we have not yet shown that there are any such integrals except trivial ones. In this chapter we shall show that belated integrals do exist under hypotheses reminiscent of those for the Riemann integral, and shall provide some useful estimates for the norms of such integrals.

Our existence and continuity theorems will make repeated use of the following basic lemma. In its statement we slightly extend the use of the symbol $\| \cdot \|$; if v is a r.v. but v^2 does not have a finite expectation, we define $\| v \| = \infty$. We also define $0 \cdot \infty = 0$.

(1-1) *LEMMA* *Let $\mathfrak{F}_1, \mathfrak{F}_2, \ldots, \mathfrak{F}_m$ be σ-subalgebras of $\mathfrak{A}$. Let $u_1, \ldots, u_m, \Delta_1, \ldots, \Delta_m$ be r.v.'s such that for each k in $\{1, \ldots, m\}$, all u_j with $j \leqq k$ and all Δ_j with $j < k$ are $\mathfrak{F}_k$-measurable. Let $C_1, \ldots, C_m, D_1, \ldots, D_m$ be numbers such that a.s.*

$$| E(\Delta_j \mid \mathfrak{F}_j) | \leqq C_j, \qquad E(\Delta_j^2 \mid \mathfrak{F}_j) \leqq D_j \qquad (j = 1, \ldots, m).$$

Then

$$\left\| \sum_{j=1}^{m} u_j \Delta_j \right\| \leqq 2 \sum_{j=1}^{m} \| u_j \| C_j + \left\{ \sum_{j=1}^{m} \| u_j \|^2 D_j \right\}^{1/2}.$$

We assume the right member finite; otherwise there is nothing to prove. We define

$$S_0 = 0, \qquad T_0 = 0,$$

$$S_k = \sum_{j=1}^{k} u_j \Delta_j, \qquad T_k = \sum_{j=1}^{k} \| u_j \| C_j \qquad (k = 1, \ldots, m),$$

$$v_k = \sup\{ \| S_0 \|, \ldots, \| S_k \| \} \qquad (k = 0, 1, \ldots, m).$$

Clearly $v_0 \leqq v_1 \leqq \cdots \leqq v_m$. We shall first prove

(1-2) $$v_k^2 \leqq v_{k-1}^2 + 2 v_{k-1} \| u_k \| C_k + \| u_k \|^2 D_k.$$

We assume $v_{k-1} < v_k$ (hence $v_{k-1} < \infty$), since otherwise the inequality is trivial. Then

$$\begin{aligned} v_k^2 &= \| S_k \|^2 \\ &= E([S_{k-1} + u_k \Delta_k]^2) \\ &\leqq v_{k-1}^2 + 2 \,| E(S_{k-1} u_k \Delta_k) | + E(u_k^2 \Delta_k^2). \end{aligned}$$

Since S_{k-1} and u_k are $\mathfrak{F}_k$-measurable,

$$\begin{aligned} | E(S_{k-1} u_k \Delta_k) | &= \left| E(E(S_{k-1} u_k \Delta_k \mid \mathfrak{F}_k)) \right| \\ &= \left| E(S_{k-1} u_k E(\Delta_k \mid \mathfrak{F}_k)) \right| \\ &\leqq C_k E(| S_{k-1} u_k |) \\ &\leqq C_k \| S_{k-1} \| \cdot \| u_k \| \\ &\leqq C_k v_{k-1} \| u_k \|. \end{aligned}$$

Likewise,

$$\begin{aligned} E(u_k^2 \Delta_k^2) &= E(E(u_k^2 \Delta_k^2 \mid \mathfrak{F}_k)) \\ &= E(u_k^2 E(\Delta_k^2 \mid \mathfrak{F}_k)) \\ &\leqq D_k E(u_k^2) \\ &= D_k \| u_k \|^2. \end{aligned}$$

These inequalities imply (1-2). This, in turn, implies by induction that all v_k are finite, since $v_0 = 0$. Since the v_j are nondecreasing and the T_j are finite and nonnegative, (1-2) implies

$$(v_k - T_k)^2 \leqq (v_{k-1} - T_{k-1})^2 + \| u_k \|^2 D_k + T_k^2 - T_{k-1}^2.$$

Adding these inequalities for $k = 1, \ldots, m$ yields

$$(v_m - T_m)^2 \leqq \sum_{k=1}^{m} \| u_k \|^2 D_k + T_m^2,$$

whence

$$v_m \leqq T_m + \left\{ \sum_{k=1}^{m} \| u_k \|^2 D_k + T_m^2 \right\}^{1/2}$$

$$\leqq 2T_m + \left\{ \sum_{k=1}^{m} \| u_k \|^2 D_k \right\}^{1/2},$$

which establishes the lemma.

2 Existence of the stochastic integral: first theorem

Since we are not striving for maximum generality, but for as simple a theory as will be general enough, we shall adopt a set of standing hypotheses that are stronger than are really needed to establish the existence of stochastic integrals. They are still general enough to cover all applications known to me.

Most of our later theorems will assume at least a certain set of properties of the integrands f, etc. These we now list as our "standing hypotheses," and we then will introduce some notation that we also standardize.

(2-1) *STANDING HYPOTHESES AND NOTATION*

(i) $(\Omega, \mathcal{A}, P)$ *is a probability triple; T is a set of real numbers; and $[a, b]$ is a closed interval contained in T.*

(ii) $\{\mathcal{F}_\tau : \tau \in T\}$ *is a family of σ-subalgebras of $\mathcal{A}$ such that if τ and τ' are in T and $\tau \leqq \tau'$, then $\mathcal{F}_\tau \subset \mathcal{F}_{\tau'}$. When convenient, we shall use $\mathcal{F}(\tau)$ to denote $\mathcal{F}_\tau$.*

(iii) *Every process denoted by z, with or without affix, will be a real-valued process on $[a, b]$ adapted to $\mathcal{F}_\tau$, and satisfying with some* *positive K the inequalities*

$$| E([z(t, \omega) - z(s, \omega)] \mid \mathfrak{F}_s) | \leqq K(t - s),$$

$$E([z(t, \omega) - z(s, \omega)]^2 \mid \mathfrak{F}_s) \leqq K(t - s),$$

a.s. whenever $a \leqq s \leqq t \leqq b$.

(iv) $f(\boldsymbol{\tau}, \omega)$ *is a real-valued process on T adapted to* $\mathfrak{F}_{\boldsymbol{\tau}}$.

By easy calculations, if z is adapted to $\mathfrak{F}_{\boldsymbol{\tau}}$ and $z(\mathbf{t}, \omega)$ a.s. satisfies a Lipschitz condition of constant L, hypothesis (2-1,iii) holds with $K = L \max(1, b - a)$. Moreover, if z is a standard Brownian motion adapted to $\mathfrak{F}_{\boldsymbol{\tau}}$, it satisfies (2-1,iii).

Whenever we assume that the standing hypotheses (2-1) are satisfied, every process named z, with or without affixes, is assumed to satisfy (2-1,iii) with some K. If the statement involves several z^k, each satisfies (2-1,iii) with its own K. We shall without further mention fix a K large enough so that (2-1,iii) holds for all the z involved.

As usual, we say that a set B of real numbers has Lebesgue measure 0 if for every positive ϵ, B is contained in a countable set of open intervals $(\alpha_1, \beta_1)(\alpha_2, \beta_2), \ldots$ with total length $\sum (\beta_i - \alpha_i) < \epsilon$. If a certain statement involving t is true for all points t in a set A except those t in a subset of measure 0, we say that the statement holds at almost all points of A, or that it holds almost everywhere (abbreviated a.e.) in A.

If $f(\boldsymbol{\tau}, \omega)$ is a process on a set T, we say that f is L_p-bounded if there is a number M such that $\| f(\tau, \omega) \|_p \leqq M$ for all τ in T. The self-suggesting meaning of the expression "f is a.s. sample-function bounded" would be that for all ω except those in a negligible set N, $| f(\boldsymbol{\tau}, \omega) |$ is bounded on T. But we give the expression a meaning that is slightly stronger when f is not separable (it is equivalent when f is separable). The process $f(\boldsymbol{\tau}, \omega)$ is **a.s. sample-function bounded** if there exists a real-valued r.v. $B(\omega)$ such that for all ω except those in a negligible set N, $| f(\tau, \omega) | \leqq B(\omega)$ for all τ in T. [$B(\omega)$ is finite for each ω, but is not necessarily bounded on Ω.] The process $f(\boldsymbol{\tau}, \omega)$ is L_p-continuous at a point τ_0 of T if as the point τ of T tends to τ_0, $\| f(\tau, \omega) - f(\tau_0, \omega) \|_p$ tends to 0; and f is continuous in probability at τ_0 if as τ (in T) tends to τ_0, $f(\tau, \omega)$ converges in probability to $f(\tau_0, \omega)$.

The next lemma is merely a computational convenience.

(2-2) *LEMMA* *Let the standing hypotheses* (2-1) *be satisfied; define*

$$C = 2K(b - a)^{1/2} + K^{1/2}.$$

Let $[s_1, t_1), \ldots, [s_m, t_m)$ *be disjoint subintervals of* $[a, b)$, *and for* $j = 1$,

. . ., m let u_j be an $\mathfrak{F}(s_j)$-measurable r.v. with $\| u_j \| < \infty$. Then

$$\left\| \sum_{j=1}^{m} u_j[z(t_j) - z(s_j)] \right\| \leqq C\left\{ \sum_{j=1}^{m} \| u_j \|^2 (t_j - s_j) \right\}^{1/2}.$$

The hypotheses of Lemma (1-1) are satisfied if we define $\Delta_j = z(t_j) - z(s_j)$, $C_j = D_j = K(t_j - s_j)$. By Lemma (1-1) and the Cauchy inequality,

$$\begin{aligned}\left\| \sum_{j=1}^{m} u_j[z(t_j) - z(s_j))] \right\| &\leq 2K \sum_{j=1}^{m} \{\| u_j \| [t_j - s_j]^{1/2}\}\{t_j - s_j\}^{1/2} \\ &\quad + \left\{ \sum_{j=1}^{m} \| u_j \|^2 K(t_j - s_j) \right\}^{1/2} \\ &\leqq 2K\left\{ \sum_{j=1}^{m} \| u_j \|^2 (t_j - s_j) \right\}^{1/2} \left\{ \sum_{j=1}^{m} (t_j - s_j) \right\}^{1/2} \\ &\quad + K^{1/2}\left\{ \sum_{j=1}^{m} \| u_j \|^2 (t_j - s_j) \right\}^{1/2} \\ &\leqq C\left\{ \sum_{j=1}^{m} \| u_j \|^2 (t_j - s_j) \right\}^{1/2}.\end{aligned}$$

We can now prove a fundamental existence theorem.

(2-3) *THEOREM Let the standing hypotheses (2-1) be satisfied. Let f be L_2-bounded and be L_2-continuous a.e. in $[a, b]$. Then the belated integral*

$$\int_a^b f(t)\, dz(t)$$

exists, and $S(\Pi_b; f; z)$ converges to the integral in L_2-norm as the mesh of the belated partition Π_b of $[a, b]$ tends to 0.

Let ϵ be positive, and let M be an upper bound for $\| f(\tau, \omega) \|$ on T. With the C of Lemma (2-2), we define

$$\gamma = \epsilon / C(4M^2 + b - a)^{1/2}. \tag{2-4}$$

By hypothesis the set N at which f fails to be continuous in L_2-norm can be covered by a countable set of open intervals (α_1, β_1), (α_2, β_2), . . . such that

$$\sum (\beta_i - \alpha_i) < \gamma^2. \tag{2-5}$$

We denote the union of the (α_i, β_i) by G; this is open. There is a positive number δ such that

(2-6) if $t_0 \in [a, b]\backslash G$ and $\tau \in T$ and $|\tau - t_0| < \delta$, then $\| f(\tau) - f(t_0) \| < \gamma/2$.

For if this were not so, there would be points $t_1, t_2, t_3, \ldots$ in $[a, b]\backslash G$ and $\tau_1, \tau_2, \tau_3, \ldots$ in T with $|t_n - \tau_n|$ tending to 0 and $\| f(\tau_n) - f(t_n) \| \geqq \gamma/2$. From the t_n we could select a sequence converging to a limit t_0; to simplify the notation we suppose that $t_1, t_2, \ldots$ is already such a sequence. Since $[a, b]\backslash G$ is a closed set, t_0 is in $[a, b]\backslash G$, so it is not in N, and f is continuous in L_2-norm at t_0. But τ_n also tends to t_0, and $\| f(\tau_n) - f(t_n) \|$ does not tend to 0, which is a contradiction. So (2-6) is established.

If $\Pi = (t_1, \ldots, t_{m+1}; \tau_1, \ldots, \tau_m)$ is any belated partition of $[a, b]$ and t^* is any point of $[a, b]$ which is not a division point of Π (say $t_j < t^* < t_{j+1}$), we can form a new partition Π^* by inserting t^* as a new division point, τ_j being the evaluation point for both intervals $[t_j, t^*]$ and $[t^*, t_{j+1}]$; thus

$$\Pi^* = (t_1, \ldots, t_j, t^*, t_{j+1}, \ldots, t_{m+1}; \tau_1, \ldots, \tau_j, \tau_j, \tau_{j+1}, \ldots, \tau_m).$$

This is a belated partition of $[a, b]$, and mesh $\Pi^* \leqq$ mesh Π. All terms in their Riemann sums are the same except that where $S(\Pi; f; z)$ has one term $f(\tau_j)[z(t_{j+1}) - z(t_j)]$, in $S(\Pi^*; f; z)$ this one term is replaced by two terms $f(\tau_j)[z(t^*) - z(t_j)]$, $f(\tau_j)[z(t_{j+1}) - z(t^*)]$. So $S(\Pi^*; f; z) = S(\Pi; f; z)$.

Now let Π' and Π'' be two belated partitions of $[a, b]$, both with mesh less than the δ of (2-6). By the method described in the preceding paragraph, we can insert the division points of Π'' one by one among the division points of Π', obtaining a new belated partition Π_1' with mesh less than δ and

$$S(\Pi_1'; f; z) = S(\Pi'; f; z). \tag{2-7}$$

In the same way we can insert the division points of Π' one by one among the division points of Π'', obtaining a new belated partition Π_1'' with mesh less than δ and

$$S(\Pi_1''; f; z) = S(\Pi''; f; z). \tag{2-8}$$

Since Π_1' and Π_2'' have the same division points, they can be written as

$$\Pi_1' = (t_1, \ldots, t_{m+1}; \tau_1', \ldots, \tau_m'),$$

$$\Pi_1'' = (t_1, \ldots, t_{m+1}; \tau_1'', \ldots, \tau_m'').$$

Then by Lemma (2-2)

$$\text{(2-9)}\qquad \| S(\Pi_1') - S(\Pi_1'') \| = \left\| \sum_{j=1}^{m} [f(\tau_j') - f(\tau_j'')] \Delta_j z \right\|$$

$$\leqq C\left\{ \sum_{j=1}^{m} \| f(\tau_j') - f(\tau_j'') \|^2 \Delta_j t \right\}^{1/2}.$$

We split the sum from $j = 1$ to $j = m$ into two parts. The first, which we denote by $\sum_G$, is the sum over all j such that $[t_j, t_{j+1}] \subset G$. The second, which we denote by $\sum_{\text{not}}$, is the sum over all j such that $[t_j, t_{j+1}]$ is not contained in G. By (2-5)

$$\text{(2-10)}\qquad \sum_G \Delta_j t < \gamma^2.$$

If $[t_j, t_{j+1}]$ is not contained in G, it contains a point t_0 in $[a, b]\backslash G$. Because Π_1' and Π_1'' have mesh less than δ, both $| t_0 - \tau_j' |$ and $| t_0 - \tau_j'' |$ are less than δ, so by (2-6)

$$\| f(\tau_j') - f(\tau_j'') \| \leqq \| f(\tau_j') - f(t_0) \| + \| f(\tau_j'') - f(t_0)) \| < \gamma.$$

By this and (2-10)

$$\left\{ \sum_{j=1}^{m} \| f(\tau_j') - f(\tau_j'') \|^2 \Delta_j t \right\}^{1/2} \leqq \left\{ \sum_G (2M)^2 \Delta_j t + \sum_{\text{not}} \gamma^2 \Delta_j t \right\}^{1/2}$$

$$< \{4M^2\gamma^2 + \gamma^2(b - a)\}^{1/2}.$$

If we apply this to (2-9) and recall the definition (2-4) of γ, we obtain

$$\| S(\Pi'; f; z) - S(\Pi''; f; z) \| < \epsilon.$$

By Theorem II(4-1), $S(\Pi; f; z)$ converges in L_2-norm (hence in probability) to a r.v., which by definition is the integral of f with respect to z.

(2-11) COROLLARY *If the hypotheses of Theorem* (2-3) *are satisfied, with the C of Lemma* (2-2)

$$\left\| \int_a^b f(t)\, dz(t) \right\|_2 \leqq 2K \int_a^b \| f(t) \|_2\, dt + \left\{ K \int_a^b \| f(t) \|_2^2\, dt \right\}^{1/2},$$

$$\left\| \int_a^b f(t)\, dz(t) \right\|_2 \leqq C\left\{ \int_a^b \| f(t) \|_2^2\, dt \right\}^{1/2}.$$

The right members exist as belated integrals by Theorem (2-3) itself, with $z(t, \omega) = t$ and $f(t, \omega)$ replaced by $\| f(t) \|_2$; we can choose $\mathfrak{F}(\tau) = \{\phi, \Omega)$ for all τ. But in Section 6 we shall see that they can also be understood to be Riemann integrals. By Lemmas (1-1) and (2-2), for all belated partitions Π of $[a, b]$

$$\| S(\Pi; f; z) \|_2 \leqq 2K \sum_{j=1}^{m} \| f(\tau_j) \|_2 \Delta_j t + \left\{ K \sum_{j=1}^{m} \| f(\tau_j) \|_2^2 \Delta_j t \right\}^{1/2},$$

$$\| S(\Pi; f; z) \|_2 \leqq C \left\{ \sum_{j=1}^{m} \| f(\tau_j) \|_2^2 \Delta_j t \right\}^{1/2}.$$

As mesh Π tends to 0, these sums converge in L_2-distance to the integrals in the corollary, by Theorem (2-3).

3 Second existence theorem

For the needs of this section only, it is convenient to introduce the following expression. Let $a = t_1^* < t_2^* < \cdots < t_l^* = b$ be any finite subset of $[a, b]$. A partition $\Pi = (t_1, \ldots, t_{m+1}; \tau_1, \ldots, \tau_m)$ is **broken at** $t_1^*, \ldots, t_l^*$ if each interval $[t_j, t_{j+1}]$ of Π is completely contained in some one interval $[t_h^*, t_{h+1}^*]$, and the corresponding τ_j is also in that same interval $[t_h^*, t_{h+1}^*]$.

The following lemma is merely a convenience in proving Theorem (3-2).

(3-1) *LEMMA* *Let the standing hypotheses* (2-1) *hold, and let f be* L_1-*bounded. Let* $a = t_1^* < t_2^* < \cdots < t_l^* = b$ *be a subset of* $[a, b]$. *Then in defining the existence and the value of the integral*

$$\int_a^b f(t, \omega)\ (dz(t, \omega))^2$$

it is sufficient to consider only belated partitions Π *that are broken at* $t_1^*, \ldots, t_l^*$.

Let M be an upper bound for $\| f(\tau) \|_1$ on T. If some t_k^* is not among the t_j, it satisfies $t_j < t_k^* < t_{j+1}$ for some j. We form a new partition by including t_k^* among the partition points. The single interval $[t_j, t_{j+1}]$ is then replaced by two intervals $[t_j, t_k^*]$, $[t_k^*, t_{j+1}]$; we assign τ_j as evaluation point for both. In the Riemann sum $S(\Pi; f; z)$ the single term

$$f(\tau_j)[z(t_{j+1}) - z(t_j)]^2$$

is replaced by the sum of two terms

$$f(\tau_j)[z(t_k^*) - z(t_j)]^2 + f(\tau_j)[z(t_{j+1}) - z(t_k^*)]^2.$$

The Riemann sum changes by the difference, whose L_1-norm is

$$\begin{aligned} &E(| -2f(\tau_j)[z(t_k^*) - z(t_j)][z(t_{j+1}) - z(t_k^*)] |) \\ &\quad = 2E(| f(\tau_j) | E(| [z(t_k^*) - z(t_j)][z(t_{j+1}) - z(t_k^*)] | \,\big|\, \mathfrak{F}(t_j))) \\ &\quad \leqq E(| f(\tau_j) | E([z(t_j^*) - z(t_j)]^2 + [z(t_{j+1}) - z(t_k^*)]^2 \mid \mathfrak{F}(t_j))) \\ &\quad \leqq MK(t_{j+1} - t_j). \end{aligned}$$

If we insert all the t_k^* one by one, the change in the Riemann sum is at most MKl mesh Π. This tends to 0 with mesh Π, so without alteration in meaning or value of the integral we may restrict our attention to partitions in which all the t_k^* are division points. But we still have to consider the evaluation points.

Let $\Pi = (t_1, \ldots, t_{m+1}; \tau_1, \ldots, \tau_m)$ be a belated partition of $[a, b]$, all the t_k^* being division points of Π. We form a new partition Π', with the same division points as Π, and with evaluation points $\tau_1', \ldots, \tau_m'$ defined thus. Each $[t_j, t_{j+1}]$ is contained in an interval $[t_k^*, t_{k+1}^*]$. If τ_j is also in $[t_k^*, t_{k+1}^*]$, we define $\tau_j' = \tau_j$; otherwise $\tau_j < t_k^*$, and then we define $\tau_j' = t_k^*$. Obviously mesh $\Pi' \leqq$ mesh Π, and Π' is broken at $t_1^*, \ldots, t_l^*$.

For every j, such that $f(\tau_j') \neq f(\tau_j)$, if $[t_k^*, t_{k+1}^*]$ contains $[t_j, t_{j+1}]$ then $\tau_j < t_k^*$ and $t_{j+1} \leqq \tau_j +$ mesh $\Pi < t_k^* +$ mesh Π. So the sum of the lengths of all those $[t_j, t_{j+1}]$ for which $f(\tau_j') \neq f(\tau_j)$ is less than l mesh Π, and

$$\begin{aligned} \| S(\Pi'; f; z, z) - S(\Pi; f; z, z) \|_1 &\leqq \sum_{j=1}^{m} E(| f(\tau_j') - f(\tau_j) | (\Delta_j z)^2) \\ &= \sum_{j=1}^{m} E(| f(\tau_j') - f(\tau_j) | E((\Delta_j z)^2 \mid \mathfrak{F}(t_j))) \\ &\leqq \sum_{j=1}^{m} \| f(\tau_j') - f(\tau_j) \|_1 K\Delta_j t \\ &\leqq 2MKl \text{ mesh } \Pi. \end{aligned}$$

This tends to 0 with mesh Π. So if there is a r.v. J with the property that $\| S(\Pi; f; z, z) - J \|_1$ tends to 0 as mesh $\Pi \to 0$, Π being belated and broken at $t_1^*, \ldots, t_l^*$, then $\| S(\Pi; f; z, z) - J \|_1$ also tends to 0 as mesh $\Pi \to 0$, Π ranging over all belated partitions of $[a, b]$.

We can now prove another fundamental existence theorem.

(3-2) *THEOREM* *Let the standing hypotheses* (2-1) *be satisfied. Let* f *be* L_1*-bounded, and be* L_1*-continuous at almost all points of* $[a, b]$. *Then the integral*

$$\int_a^b f(t)\, dz^1(t)\, dz^2(t)$$

exists, and the Riemann sums $S(\Pi; f; z^1, z^2)$ *converge in* L_1*-norm to the integral as the mesh of the belated partition* Π *of* $[a, b]$ *tends to* 0.

Let M be an upper bound for $\| f(\tau) \|_1$ on T. Let ϵ be positive, and define

$$\gamma = \epsilon/4K(2M + b - a). \tag{3-3}$$

For notational simplicity, we first consider the special case $z^1 = z^2$, and define $z = z^1 - z^1(a, \omega)$. The set N at which f is not L_1-continuous has measure 0, so there exist open intervals $(\alpha_1, \beta_1), (\alpha_2, \beta_2), \ldots$ whose union G contains N and for which $\sum (\beta_i - \alpha_i) < \gamma$. The proof of the statement

(3-4) *there is a positive* δ *such that if* $t_0 \in [a, b]\backslash G$ *and* $\tau \in T$ *and* $| t_0 - \tau | < \delta$, *then* $\| f(\tau) - f(t_0) \|_1 < \gamma/3$

is practically identical with the proof of (2-6).

We choose and fix points $t_1^* = a < t_2^* < \cdots < t_l^* = b$ such that $t_{k+1}^* - t_k^* < \delta$ $(k = 1, \ldots, l - 1)$. Each $f(t_h^*, \omega)$ has finite L_1-norm, but may fail to be bounded. But the r.v. $\text{mid}\{f(t_h^*, \omega), V, -V\} = \max\{-V, \min\{V, f(t_h^*, \omega)\}\}$ is bounded, and by either the dominated convergence theorem or the monotone convergence theorem

$$\lim_{V\to\infty} \int_\Omega | f(t_h^*, \omega) - \text{mid}\{f(t_h^*, \omega), V, -V\} | \, P(d\omega) = 0.$$

So we can fix a V such that this integral is less than $\gamma/3$ for $h = 1, \ldots, l$. The corresponding r.v. $\text{mid}\{f(t_h^*, \omega), V, -V\}$ we name v_h; then v_h is a bounded r.v., and

$$\| v_h - f(t_h^*) \|_1 < \gamma/3 \qquad (h = 1, \ldots, l). \tag{3-5}$$

We define a process $s(\mathbf{t}, \omega)$ by setting $s(t, \omega) = v_h(\omega)$ when $t_h^* \leqq t < t_{h+1}^*$. Then, with the same notation, for all t in $[a, b)$

$$\| s(t) \|_1 = \| v_h \|_1 \leqq \| f(t_h^*) \|_1 \leqq M. \tag{3-6}$$

If $\Pi = (t_1, \ldots, t_{m+1}; \tau_1, \ldots, \tau_m)$ is any belated partition of $[a, b]$ broken at $t_1^*, \ldots, t_l^*$, we again denote by $\sum_G$ the sum over all those j for which $[t_j, t_{j+1}] \subset G$, and by $\sum_{\text{not}}$ the sum over all j for which $[t_j, t_{j+1}]$ is not contained in G. Then

$$(3\text{-}7)\qquad \| S(\Pi; f - s; z, z) \|_1$$
$$= E\left(\left| \sum_G + \sum_{\text{not}} [f(\tau_j) - s(\tau_j)] (\Delta_j z)^2 \right| \right)$$
$$\leqq \sum_G + \sum_{\text{not}} E(| f(\tau_j) - s(\tau_j) \mid E([\Delta_j z]^2 \mid \mathfrak{F}(t_j)))$$
$$\leqq \sum_G + \sum_{\text{not}} \| f(\tau_j) - s(\tau_j) \|_1 K \, \Delta_j t.$$

For all j, we have by (3-6)

$$\| f(\tau_j) - s(\tau_j) \|_1 \leqq \| f(\tau_j) \|_1 + \| s(\tau_j) \|_1 \leqq 2M.$$

If $[t_j, t_{j+1}]$ is not contained in G, it contains a point t_0 of $[a, b]\backslash G$. Since Π is broken at $t_1^*, \ldots, t_l^*$, there is an interval $[t_h^*, t_{h+1}^*]$ that contains $[t_j, t_{j+1}]$. It also contains τ_j, and since it has length less than δ, both $| t_0 - \tau_j |$ and $| t_0 - t_h^* |$ are less than δ. By (3-4) and (3-5),

$$\| f(\tau_j) - s(\tau_j) \|_1 \leqq \| f(\tau_j) - f(t_0) \|_1 + \| f(t_0) - f(t_h^*) \|_1$$
$$+ \| f(t_h^*) - v_h \|_1 < \gamma.$$

Applying these estimates in (3-7) yields

$$(3\text{-}8)\qquad \| S(\Pi; f - s; z, z) \|_1 \leqq 2MK \sum_G \Delta_j t + \sum_{\text{not}} \gamma K \, \Delta_j t$$
$$\leqq 2MK\gamma + \gamma K(b - a)$$
$$= \epsilon/4.$$

The point t_2^* is one of the t_j, say $t_2^* = t_{n+1}$. Then

$$(3\text{-}9)\qquad \sum_{j=1}^{n} s(\tau_j) \, (\Delta_j z)^2 = \sum_{j=1}^{n} v_1 \, \Delta_j(z^2) - 2 \sum_{j=1}^{n} v_1 z(t_j) \, \Delta_j z$$
$$= v_1(z(t_2^*)^2 - z(t_1^*)^2) - 2v_1 \sum_{j=1}^{n} z(t_j) \, \Delta_j z.$$

The points $(t_1, \ldots, t_{n+1}; t_1, \ldots, t_n)$ form a belated partition of $[t_1^*, t_2^*]$, with mesh not greater than mesh Π. The process $z(\mathbf{t}, \omega)$ is L_2-bounded and L_2-continuous on $[a, b]$, as follows at once from (2-1,iii). So by Theorem (2-3), as mesh Π tends to 0 the last sum in (3-9) converges in L_2-norm to a limit (which happens to be $\int z \, dz$). Since v_1 is bounded, and the other terms in the right member of (3-9) are independent of Π, as mesh $\Pi \to 0$ the right member of (3-9) converges in L_2-norm (hence in L_1-norm) to a limit. The same applies to the part of Π in $[t_k^*, t_{k+1}^*]$ for $k = 2, 3, \ldots, l-1$; so as mesh $\Pi \to 0$ the sum

$$S(\Pi; s; z, z) = \sum_{j=1}^{m} s(\tau_j)\,(\Delta_j z)^2$$

converges in L_1-norm to some r.v. J. Therefore there is a positive δ_1, which we can choose less than the δ of (3-4), such that

(3-10) if mesh $\Pi < \delta_1$ and Π is belated and broken at $t_1^*, \ldots, t_l^*$, then $\| S(\Pi; s; z, z) - J \|_1 < \epsilon/4$.

Let Π', Π'' be two partitions of mesh less than δ_1 and broken at $t_1^*, \ldots, t_l^*$. By (3-8) and (3-10),

$$\begin{aligned} &\| S(\Pi'; f; z, z) - S(\Pi''; f; z, z) \|_1 \\ &\quad \leqq \| S(\Pi'; f - s; z, z) \|_1 + \| S(\Pi'; s; z, z) - J \|_1 \\ &\qquad + \| J - S(\Pi''; s; z, z) \|_1 + \| S(\Pi''; s - f; z, z) \|_1 \\ &\quad < \epsilon. \end{aligned}$$

By the Cauchy criterion, Theorem II(4-1), $S(\Pi; f; z, z)$ converges in L_1-norm to some limit as the mesh of the partition Π, belated and broken at $t_1^*, \ldots, t_l^*$, tends to 0. By Lemma (3-1), the conclusion of Theorem (3-2) is valid when $z^1 = z^2$.

When this is not the case, we need only note that if z^1 and z^2 satisfy the standing hypotheses (2-1), so do $z^3 = (z^1 + z^2)/2$ and $z^4 = (z^1 - z^2)/2$; then

$$S(\Pi; f; z^1, z^2) = S(\Pi; f; z^3, z^3) - S(\Pi; f; z^4, z^4).$$

By the part of the proof already completed, as mesh Π tends to 0 both terms in the right member converges in L_1-norm to a limit. So therefore does the left member, and the proof is complete.

For the second-order integral we can easily establish an estimate analogous to that in Corollary (2-11).

(3-11) *COROLLARY If the hypotheses of Theorem* (3-2) *are satisfied,*

$$\left\| \int_a^b f(t)\, dz^1(t)\, dz^2(t) \right\|_1 \leqq K \int_a^b \| f(t) \|_1\, dt.$$

Let $\Pi = (t_1, \ldots, t_{m+1}; \tau_1, \ldots, \tau_m)$ be a belated partition of $[a, b]$. Then

$$E\left(\left|\sum_{j=1}^{m} f(\tau_j)\,\Delta_j z^1\,\Delta_j z^2\right|\right) \leqq \sum_{j=1}^{m} E[E(|f(\tau_j)|\,|\Delta_j z^1\,\Delta_j z^2|\,|\,\mathfrak{F}[t_j])]$$

$$\leqq \tfrac{1}{2}\sum_{j=1}^{m} E[|f(\tau_j)|\,E([(\Delta_j z^1)^2 + (\Delta_j z^2)^2]\,|\,\mathfrak{F}(t_j))]$$

$$\leqq \sum_{j=1}^{m} E(|f(\tau_j)|\cdot K\,\Delta_j t)$$

$$= KS(\Pi; \|f\|_1; \mathbf{t}).$$

As mesh $\Pi \to 0$, by Theorem (3-2) the left member of this inequality converges to the left member of the conclusion. Since $\|f(\tau)\|_1$ is bounded and a.e. continuous, and $\mathbf{t}$ satisfies the hypotheses (2-1) concerning noise processes, by Theorem (2-3) the right member converges to the right member of the conclusion.

The proof of the next statement is trivial.

(3-12) *COROLLARY If f is L_1-bounded and is L_1-continuous a.e. on $[a, b]$,*

$$\left\|\int_a^b f(t, \omega)\,dt\right\|_1 \leqq \int_a^b \|f(t, \omega)\|_1\,dt.$$

4 Third and fourth existence theorems

We can prove the existence of the belated integral under hypotheses weaker than those in Theorems (2-3) and (3-2), but under these weaker hypotheses we can no longer establish the useful estimates in Corollaries (2-11) and (3-11).

(4-1) *THEOREM Let the standing hypotheses* (2-1) *be satisfied. Let $f(\tau, \omega)$ be a process having the form $b(\tau, \omega)f_1(\tau, \omega)$, where $b(\tau, \omega)$ and $f_1(\tau, \omega)$ are adapted to $\mathfrak{F}_\tau$, and b is a.s. sample-function bounded and is continuous in probability a.e. in $[a, b]$, and f_1 is L_2-bounded and is*

L_2-continuous a.e. in $[a, b]$. Then the integral

$$\int_a^b f(t)\,dz(t)$$

exists.

Let $B(\omega)$ be a r.v. such that a.s. $|b(\tau, \omega)| \leqq B(\omega)$ for all τ in T. Let ϵ be positive. There exists a number V such that the set

$$\Omega_V = \{\omega \in \Omega : B(\omega) > V\}$$

has $P(\Omega_V) < \epsilon/4$. For all (τ, ω) in $T \times \Omega$, we define

$$b_V(\tau, \omega) = \text{mid}\{b(\tau, \omega), V, -V\}.$$

Then

$$b_V(\tau, \omega) = b(\tau, \omega) \quad \text{if} \quad \omega \in \Omega \backslash \Omega_V. \tag{4-2}$$

Let N be the subset of $[a, b]$, with Lebesgue measure 0, on which either b fails to be continuous in probability or f_1 fails to be L_2-continuous. If t_0 is in $[a, b]\backslash N$, as the point τ of T tends to t_0, $b(\tau, \omega)$ tends in probability to $b(t_0, \omega)$. Since

$$|b_V(\tau, \omega) - b_V(t_0, \omega)| \leqq |b(\tau, \omega) - b(t_0, \omega)|, \tag{4-3}$$

the left member converges in probability to 0. Clearly,

$$\begin{aligned} |b_V(\tau, \omega)f_1(\tau, \omega) - b_V(t_0, \omega)f_1(t_0, \omega)| &\leqq |b_V(\tau, \omega)| \cdot |f_1(\tau, \omega) - f_1(t_0, \omega)| \\ &\quad + |b_V(\tau, \omega) - b_V(t_0, \omega)| \cdot |f_1(t_0, \omega)|. \end{aligned} \tag{4-4}$$

The first term in the right member is at most $V|f_1(\tau, \omega) - f_1(t_0, \omega)|$, and therefore tends to 0 in L_2-norm as $\tau \to t_0$. For the second term, let ϵ be positive. There is a number $K > 1$ such that the set

$$\Omega_K = \{\omega \in \Omega : |f_1(t_0, \omega)| \geqq 2K\}$$

has $P(\Omega_K) < \epsilon/2$. Since the left member of (4-3) converges to 0 in probability, there is a positive δ such that for all τ in T with $|t_0 - \tau| < \delta$, the set

$$\Omega_\epsilon = \{\omega \in \Omega : |b_V(\tau, \omega) - b_V(t_0, \omega)| \geqq \epsilon/2K\}$$

has $P(\Omega_\epsilon) < \epsilon/2K < \epsilon/2$. Therefore except on $\Omega_K \cup \Omega_\epsilon$, whose P-measure is less than ϵ, we have

$$|b_V(\tau, \omega) - b_V(t_0, \omega)| \cdot |f_1(t_0, \omega)| < \epsilon.$$

That is, the last term in (4-4) tends to 0 in probability to 0 as $\tau \to t_0$. Since its square does not exceed

$$4V^2 \, | \, f_1(t_0, \omega) \, |^2,$$

which has finite expectation, by the dominated convergence theorem it tends to 0 in L_2-norm. Therefore

(4-5) $b_V f_1$ satisfies the hypotheses of Theorem (2-3) and is integrable with respect to z.

Let δ be a positive number such that for every belated partition Π of $[a, b]$ with mesh $\Pi < \delta$,

$$\bar{\rho}\left(S(\Pi; b_V f_1; z), \int_a^b b_V(s) f_1(s) \, dz(s) \right) < \epsilon/4.$$

If Π' and Π'' are any two such partitions, we have

$$\bar{\rho}(S(\Pi'; b_V f_1; z), \; S(\Pi''; b_V f_1; z)) < \epsilon/2.$$

But by (4-2), since $P(\Omega_V) < \epsilon/4$,

$$\bar{\rho}(S(\Pi'; b_V f_1; z), \; S(\Pi'; b f_1; z)) < \epsilon/4,$$

and likewise with Π'' in place of Π'. Hence if Π' and Π'' are belated partitions of $[a, b]$ with mesh less than δ,

$$\bar{\rho}(S(\Pi'; f; z), \; S(\Pi''; f; z)) < \epsilon.$$

By Theorem II(4-1), f is integrable with respect to z over $[a, b]$.

(4-6) *THEOREM Let the standing hypotheses* (2-1) *be satisfied. Let* $f(\tau, \omega)$ *be a process having the form* $b(\tau, \omega) f_1(\tau, \omega)$, *where* $b(\tau, \omega)$ *and* $f_1(\tau, \omega)$ *are adapted to* $\mathfrak{F}_\tau$, *and* b *is a.s. sample-function bounded and is continuous in probability a.e. in* $[a, b]$, *and* f_1 *is* L_1*-bounded and is* L_1*-continuous a.e. in* $[a, b]$. *Then the integrals*

$$\int_a^b f(t) \, dt, \qquad \int_a^b f(t) \, dz^1(t) \, dz^2(t)$$

exist.

For $\epsilon > 0$ we define V and b_V as in the proof of Theorem (4-1). Using L_1-norms instead of L_2-norms, we prove that $b_V f_1$ is L_1-bounded and is L_1-continuous a.e. in $[a, b]$. So by Theorem (3-2) and Corollary (3-12), $b_V f_1$ is integrable with respect to t and with respect to (z^1, z^2). The proof is completed like that of Theorem (4-1).

5 The vanishing of certain integrals

In I(0-3) and the sentences following it, we presented a memory rule for the calculus of stochastic processes. Expressed in terms of integrals, the basic idea was that all integrals of order greater than 2 are 0, and so are all integrals of the form $\int f\, dz\, dt$. In this section we prove these statements under suitable hypotheses. We also prove that the second-order integral $\int f\, dz^1\, dz^2$ is 0 if $z^1(t) - z^1(s)$ and $z^2(t) - z^2(s)$ are conditionally independent given $\mathfrak{F}(s)$, so that in particular the integral is 0 if z^1 and z^2 are independent Brownian motions.

(5-1) *THEOREM* *Let z^1 and z^2 satisfy the standing hypotheses* (2-1), *and let $v^1, \ldots, v^q$ $(q \geqq 1)$ be processes on $[a, b]$ with a.s. continuous sample functions. Let g be a process on T such that*

(5-2) $$|g(\tau, \omega)| \leqq B(\omega) f(\tau, \omega),$$

where $B(\boldsymbol{\omega})$ is an a.s. finite-valued r.v. and f is L_1-bounded and adapted to $\mathfrak{F}_{\boldsymbol{\tau}}$. Then (i) *as the mesh of the* (*not necessarily belated*) *partition $\Pi = (t_1, \ldots, t_{m+1}; \tau_1, \ldots, \tau_m)$ tends to* 0,

$$\sum_{j=1}^{m} |B(\boldsymbol{\omega})\, \Delta_j z^1\, \Delta_j z^2\, \Delta_j v^1 \cdots \Delta_j v^q| \to 0$$

in probability; and (ii)

$$\int_a^b g(t)\, dz^1(t)\, dz^2(t)\, dv^1(t) \cdots dv^q(t) = 0.$$

We shall prove that as the mesh of the belated partition Π tends to 0,

(5-3) $$\sum |f(\tau_j) B\, \Delta_j z^1\, \Delta_j z^2\, \Delta_j v^1 \cdots \Delta_j v^q| \to 0$$

in probability. With $f = 1$ this yields conclusion (i), in which Π does not have to be belated because the τ_j do not enter; and by (5-2) it yields (ii).

It is enough to prove (5-3) with $B(\boldsymbol{\omega}) = 1$; otherwise we replace v^1 by Bv^1. For $\delta > 0$, we define

$$\operatorname{Osc}(\delta, \omega) = \sup\left\{\prod_{i=1}^{q} |v^i(t, \omega) - v^i(s, \omega)| : a \leqq s \leqq t \leqq b, t - s < \delta\right\}.$$

For all ω in a set Ω_1 with $P(\Omega_1) = 1$, all $v^i(\boldsymbol{\tau}, \boldsymbol{\omega})$ are continuous, so the right member of this definition is unchanged if we restrict s and t to rational numbers. For each s and t, $\prod [v^i(t) - v^i(s)]$ is a r.v., so $\operatorname{Osc}(\delta, \boldsymbol{\omega})$ is also a r.v. For each ω in Ω_1, it tends to 0 with δ.

By hypothesis, there is an M such that $\| f(\tau) \|_1 \leqq M$ for all τ in T. Then for every belated partition

$$E\left(\sum_{j=1}^{m} | f(\tau_j) \; \Delta_j z^1 \; \Delta_j z^2 |\right)$$

$$= E\left(\sum_{j=1}^{m} | f(\tau_j) | \, E(| \Delta_j z^1 \; \Delta_j z^2 | \,\Big|\, \mathfrak{F}(t_j))\right)$$

$$\leqq E\left(\sum_{j=1}^{m} | f(\tau_j) | \, E(\{[\Delta_j z^1]^2 + [\Delta_j z^2]^2\}/2 \mid \mathfrak{F}(t_j))\right)$$

$$\leqq E\left(\sum_{j=1}^{m} | f(\tau_j) | \, K \, \Delta_j t\right)$$

$$\leqq MK(b - a).$$

Let ϵ be positive. The set

$$\Omega_2(\Pi) = \left\{\omega : \sum_{j=1}^{m} | f(\tau_j) \; \Delta_j z^1 \; \Delta_j z^2 | \leqq 2MK(b - a)/\epsilon\right\}$$

has measure $P(\Omega_2(\Pi)) \geqq 1 - \epsilon/2$, otherwise the preceding inequality would be false. Since $\mathrm{Osc}(\delta, \omega) \to 0$ a.s., there is a positive δ such that the set

$$\Omega_3 = \{\omega : \mathrm{Osc}(\delta, \omega) < \epsilon^2/2MK(b - a)\}$$

has $P(\Omega_3) > 1 - \epsilon/2$. Now let Π be a belated partition with mesh $\Pi < \delta$. Then for ω in $\Omega_1 \cap \Omega_2(\Pi) \cap \Omega_3$,

$$\sum | f(\tau_j, \omega) \; \Delta_j z^1 \; \Delta_j z^2 \; \Delta_j v^1 \cdots \Delta_j v^q | \leqq \sum | f(\tau_j, \omega) \; \Delta_j z^1 \; \Delta_j z^2 | \; \mathrm{Osc}(\delta, \omega) < \epsilon,$$

and $P(\Omega_1 \cap \Omega_2(\Pi) \cap \Omega_3) > 1 - \epsilon$; so (5-3) holds with $B = 1$.

(5-4) *THEOREM Let v^1 and v^2 be processes on $[a, b]$ such that the sample functions of v^2 are a.s. continuous and those of v^1 satisfy a Lipschitz condition*

$$| v^1(t, \omega) - v^1(s, \omega) | \leqq L(\omega) \qquad (s, t \text{ in } [a, b]),$$

where L is an a.s. finite-valued r.v. (in particular, v^1 may be $\mathbf{t}$*). Let g be a process on T such that*

$$| g(\tau, \omega) | \leqq B(\omega) g_1(\tau, \omega),$$

where $B(\omega)$ is an a.s. finite r.v. and g_1 is L_1-bounded. Then (i) *as the* *mesh of the* (*not necessarily belated*) *partition* Π *tends to* 0,

$$\sum | g(\tau_j)\ \Delta_j v^1\ \Delta_j v^2 | \to 0$$

in probability; and (ii)

$$\int_a^b g(t)\ dv^1(t)\ dv^2(t) = 0.$$

Conclusion (ii) evidently follows from conclusion (i). There is no loss of generality in assuming that $v^2(\mathbf{t}, \omega)$ is continuous and $B(\omega)$ and $L(\omega)$ finite and positive for all ω. We define

$$v^3(t, \omega) = v^1(t, \omega)/L(\omega),$$
$$v^4(t, \omega) = v^2(t, \omega)L(\omega)B(\omega).$$

Then for all partitions Π,

$$\sum_{j=1}^{m} | g(\tau_j)\ \Delta_j v^1\ \Delta_j v^2 | \leqq \sum_{j=1}^{m} | g_1(\tau_j)\ \Delta_j v^3\ \Delta_j v^4 |; \tag{5-5}$$

and all sample functions of v^4 are continuous, and all those of v^3 satisfy a Lipschitz condition of constant 1. There is a constant M such that $\| g_1(\tau) \|_1 \leqq M$ for all τ. Then

$$\begin{aligned} E(\sum | g_1(\tau_j)\ \Delta_j v^3 |) &\leqq E(\sum | g_1(\tau_j)\ \Delta_j t |) \\ &\leqq M(b - a). \end{aligned} \tag{5-6}$$

We define for $\delta > 0$,

$$\operatorname{Osc}(\delta, \omega) = \sup\{| v^4(t, \omega) - v^4(s, \omega) | : a \leqq s \leqq t \leqq b, t - s < \delta\}.$$

As in the preceding proof, this is a r.v. and converges to 0 a.s. as $\delta \to 0$. Let ϵ be positive. For every partition Π of $[a, b]$, the set

$$\Omega_2(\Pi) = \{\omega : \sum | g_1(\tau_j)\ \Delta_j v^3 | \leqq 2M(b - a)/\epsilon\}$$

has measure $P(\Omega_2(\Pi)) > 1 - \epsilon/2$; otherwise (5-6) would be contradicted. We choose a δ such that the set

$$\Omega_3 = \{\omega : \operatorname{Osc}(\delta, \omega) < \epsilon^2/2M(b - a)\}$$

has $P(\Omega_3) > 1 - \epsilon/2$. If Π is any partition with mesh $\Pi < \delta$, for all ω in $\Omega_2(\Pi) \cap \Omega_3$ we have

$$\sum_{j=1}^{m} | g_1(\tau_j)\ \Delta_j v^3\ \Delta_j v^4 | \leqq \epsilon.$$

Since $P(\Omega_2(\Pi) \cap \Omega_3) > 1 - \epsilon$ and (5-5) holds, conclusion (i) is established.

(5-7) *THEOREM Let the standing hypotheses* (2-1) *be satisfied, and let* $z^1(t) - z^1(s)$ *and* $z^2(t) - z^2(s)$ *be conditionally independent given* $\mathfrak{F}(s)$ *whenever* $a \leqq s \leqq t \leqq b$. *Let* f *be a process on* T *such that*

$$f(\tau, \omega) = b(\tau, \omega)f_1(\tau, \omega)$$

for all τ *in* T *and* ω *in* Ω, *where both* b *and* f_1 *are adapted to* $\mathfrak{F}_\tau$, b *has a.s. bounded sample functions and* f_1 *is* L_2-*bounded. Then*

$$\int_a^b f(t)\, dz^1(t)\, dz^2(t) = 0.$$

There is no loss of generality in assuming $\| f_1(\tau) \|_2 \leqq 1$ for all τ, since we can bring this about by dividing f_1 by a constant and multiplying b by the same constant. If $a \leqq s \leqq t \leqq b$, by the assumed conditional independence and I(2-18), a.s.

$$| E([z^1(t) - z^1(s)][z^2(t) - z^2(s)] \mid \mathfrak{F}_s) |$$

$$= | E([z^1(t) - z^1(s)] \mid \mathfrak{F}_s) | \cdot | E([z^2(t) - z^2(s)] \mid \mathfrak{F}_s) |$$

$$\leqq K^2[t - s]^2,$$

$$E([z^1(t) - z^1(s)]^2[z^2(t) - z^2(s)]^2 \mid \mathfrak{F}_s)$$

$$= E([z^1(t) - z^1(s)]^2 \mid \mathfrak{F}_s)E([z^2(t) - z^2(s)]^2 \mid \mathfrak{F}_s)$$

$$\leqq K^2[t - s]^2.$$

Let $\Pi = (t_1, \ldots, t_{m+1}; \tau_1, \ldots, \tau_m)$ be a belated partition of $[a, b]$. We choose $\Delta_j = \Delta_j z^1\, \Delta_j z^2$ and $u_j = f(\tau_j)$. The hypotheses of Lemma III(1-1) are satisfied with $C_j = D_j = K^2(\Delta_j t)^2$, so by that lemma

$$\left\| \sum_{j=1}^{m} f_1(\tau_j)\, \Delta_j z^1\, \Delta_j z^2 \right\|_2 \tag{5-8}$$

$$\leqq 2 \sum_{j=1}^{m} K^2(\Delta_j t)^2 + \left\{ \sum_{j=1}^{m} K^2(\Delta_j t)^2 \right\}^{1/2}$$

$$\leqq 2K^2(b - a) \text{ mesh } \Pi + \{K^2(b - a) \text{ mesh } \Pi\}^{1/2}.$$

This tends to 0 with mesh Π.

By hypothesis there is a finite-valued r.v. $B(\omega)$ such that a.s.

$$| b(\tau, \omega) | \leqq B(\omega) \qquad (\tau \in T).$$

Let ϵ be positive. There is a number M such that the set

$$\Omega_M = \{\omega : B(\omega) < M\}$$

has $P(\Omega_M) > 1 - \epsilon/2$. Since the left member of (5-8) converges to 0 with mesh Π, there is a positive δ such that if mesh $\Pi < \delta$,

$$\left\| \sum_{j=1}^{m} f_1(\tau_j)\, \Delta_j z^1\, \Delta_j z^2 \right\|_2^2 < \epsilon^3/2M^2.$$

This implies that the set

$$\Omega_1(\Pi) = \{\omega : |\sum f_1(\tau_j, \omega)\, \Delta_j z^1(\omega)\, \Delta_j z^2(\omega)| \leqq \epsilon/M\}$$

has measure $P(\Omega_1(\Pi)) > 1 - \epsilon/2$. Then $P(\Omega_M \cap \Omega_1(\Pi)) > 1 - \epsilon$, and on it we have

$$|\sum f(\tau_j, \omega)\, \Delta_j z^1(\omega)\, \Delta_j z^2(\omega)| < \epsilon.$$

So $S(\Pi; f; z^1, z^2)$ converges in probability to 0 as mesh $\Pi \to 0$.

The proofs of the following three statements are left as exercises.

(5-9) *Let g be a process on T and v^1, v^2 processes on $[a, b]$. Let Ω_0 be the set of ω on which $g(\boldsymbol{\tau}, \omega)$ is bounded, $v^1(\mathbf{t}, \omega)$ is Lipschitzian and $v^2(\mathbf{t}, \omega)$ is continuous. If the integral*

$$J(\omega) = \int_a^b g(t)\, dv^1(t)\, dv^2(t)$$

exists, then $J(\omega) = 0$ for all ω in Ω_0 except those in a negligible set.

(5-10) *In Theorem (5-1) remove the hypothesis that $v^1, \ldots, v^q$ have a.s. continuous sample functions, and add the hypothesis*

(5-11) *there is a positive δ and there is a function ψ on $(0, \infty)$ such that $\psi(y)/y \to 0$ as $y \to 0$ and whenever $a \leqq s \leqq t \leqq b$ and $t - s < \delta$, then a.s.*

$$E(|v^k(t) - v^k(s)|^q \mid \mathfrak{F}(s)) \leqq \psi(t - s) \qquad (k = 1, \ldots, q).$$

Then the conclusions of Theorem (5-1) hold.

(5-12) *If hypothesis (5-11) is satisfied, there is a function ψ_1 on $[0, \infty)$ such that $\psi_1(y)/y \to 0$ as $y \to 0$ and a.s.*

$$E(|v^k(t) - v^k(s)|^h \mid \mathfrak{F}(s)) \leqq \psi_1(t - s)$$

whenever $a \leqq s \leqq t \leqq b$ and $t - s < \delta$ and $3 \leqq h \leqq q$.

6 Special cases

There are some special cases, which we shall discuss shortly, in which $f(\tau, \omega)$ is not merely $\mathfrak{F}(\tau)$-measurable, but is $\mathfrak{F}(\tau')$-measurable whenever $\tau' > \tau - \delta$ for some positive δ. The only reason for using belated partitions Π was to ensure that $f(\tau_j, \omega)$ be $\mathfrak{F}(t_j)$-measurable. If the condition in the preceding sentence is satisfied, $f(\tau_j, \omega)$ is $\mathfrak{F}(t_j)$-measurable whenever mesh $\Pi < \delta$, whether Π is belated or not. We thus obtain the following theorem.

(6-1) *THEOREM Let the standing hypotheses* (2-1) *hold. Assume that there is a positive δ such that if τ and τ' are in T and $\tau' > \tau - \delta$, $f(\tau, \omega)$ is $\mathfrak{F}(\tau')$-measurable. Then as the mesh of the* (*not necessarily belated*) *partition Π of $[a, b]$ tends to* 0,

(i) *if f_2 is L_2-bounded and is L_2-continuous a.e. in $[a, b]$, then $S(\Pi; f_2; z)$ converges in L_2-distance to $\int_a^b f_2(t)\, dz(t)$;*

(ii) *if f_1 is L_1-bounded and is L_1-continuous a.e. in $[a, b]$, then $S(\Pi; f_1; z^1, z^2)$ converges in L_1-distance to $\int_a^b f_1(t)\, dz^1(t)\, dz^2(t)$;*

(iii) *if b is a.s. sample-function bounded and is continuous in probability a.e. in $[a, b]$ and f_2 and f_1 are as in* (i) *and* (ii) *respectively, then $S(\Pi; bf_2; z)$ and $S(\Pi; bf_1; z^1, z^2)$ converge in probability to the respective limits*

$$\int_a^b b(t) f_2(t)\, dz(t), \qquad \int_a^b b(t) f_1(t)\, dz^1(t)\, dz^2(t).$$

In the physically realizable case in which the $z(\mathbf{t}, \omega)$ all satisfy the same Lipschitz condition

$$| z(t, \omega) - z(s, \omega) | \leqq L\, | t - s | \qquad (s, t \text{ in } [a, b])$$

we can choose $\mathfrak{F}_\tau = \mathfrak{A}$ for all τ in T. The conditions on z in (2-1) are satisfied, and all processes f are adapted to $\mathfrak{F}_\tau$. So for such z,

if f_2 is L_2-bounded and is L_2-continuous a.e. in $[a, b]$, $S(\Pi; f_2; z)$ converges in L_2-distance to $\int f_2\, dz$;

if f_1 is L_1-bounded and is L_1-continuous a.e. in $[a, b]$, $S(\Pi; f_1; z^1, z^2)$ converges in L_1-distance to $\int f_1\, dz^1\, dz^2$;

if b is a.s. sample-function bounded and is continuous in probability a.e. in $[a, b]$, and f_1 and f_2 are as in the preceding clauses, $S(\Pi; bf_2; z)$ and $S(\Pi; bf_1; z^1, z^2)$ converge in probability to $\int bf_2\, dz$, $\int bf_1\, dz^1\, dz^2$, respectively.

Another special case that is important in applications is that in which $f(\tau, \omega)$ is a sure function, which means that it is independent of ω. It is then a function of τ alone, and by an abuse of notation we shall denote it by $f(\tau)$. If f is a sure function, we can choose $\mathfrak{F}_\tau = \{\phi, \Omega\}$ for all τ. Then f is adapted to $\mathfrak{F}_\tau$, and the conditions on z in (2-1) are merely

$$|E([z(t) - z(s)])| \leqq K(t-s),$$

$$E([z(t) - z(s)]^2) \leqq K(t-s) \qquad (a \leqq s \leqq t \leqq b).$$

The conditions that f be L_p-bounded and be L_p-continuous a.e. in $[a, b]$ ($p = 1$ or 2), and that f be a.s. sample-function bounded and be continuous in probability a.e. in $[a, b]$, all reduce to the same condition, that $f(\tau)$ be bounded on T and be continuous a.e. in $[a, b]$. If $T = [a, b]$, this is the necessary and sufficient condition for Riemann integrability of f.

If f is not merely sure, but is also of bounded variation on $[a, b]$, we can say even more. It is proved in many textbooks on real variables that if f and g are real-valued on $[a, b]$ and the ordinary Stieltjes integral $\int g\, df$ exists, so does $\int f\, dg$ and

$$\int_a^b f(t)\, dg(t) = f(b)g(b) - f(a)g(a) - \int_a^b g(t)\, df(t).$$

If f is of bounded variation, the Stieltjes integral

$$\int_a^b z(t, \omega)\, df(t)$$

exists for all ω such that $z(\mathbf{t}, \omega)$ is continuous, which we suppose true a.s. Therefore so does $\int f(t)\, dz(t, \omega)$, and

$$\text{(6-2)} \quad \int_a^b f(t)\, dz(t, \omega) = f(b)z(b, \omega) - f(a)z(a, \omega) - \int_a^b z(t, \omega)\, df(t).$$

Consequently the stochastic integral $\int f\, dz$ can be calculated pointwise, as an ordinary Stieltjes integral, for all those ω (all of Ω but a set of P-measure 0) at which $z(\mathbf{t}, \omega)$ is continuous. Moreover, the integral is highly stable under alteration of the z process. If z is replaced by a process z^* such that $|z(t, \omega) - z^*(t, \omega)|$ is uniformly small except for ω in a set of small measure, by (6-2) the integral $\int f\, dz$ will also change by only a small amount for all ω except those in a set of small measure. In this special case of sure f with bounded variation we have no reason to anticipate the "paradoxes" that occur in more general situations; and in fact this special case has been often and safely used.

Corollary (2-11) furnishes us with a bound for $\| \int f\, dz \|_2$ in this case. But if the increments of z are orthogonal, we can find the exact value of the norm of the integral. In this case, we first define

$$F(t) = E([z(t) - z(a)]^2) \qquad (a \leqq t \leqq b).$$

If $a \leqq s < t \leqq b$, because $z(s) - z(a)$ and $z(t) - z(s)$ are orthogonal we find

$$\text{(6-3)} \quad \begin{aligned} F(t) - F(s) &= E(\{z(t) - z(a)\}^2 - \{z(s) - z(a)\}^2) \\ &= E(\{z(t) - z(s)\}^2 + 2\{z(t) - z(s)\}\{z(s) - z(a)\}) \\ &= E([z(t) - z(s)]^2). \end{aligned}$$

For every real-valued function f on T (independent of ω) and every partition Π of $[a, b]$, belated or not, we have (because increments of z are orthogonal)

$$\begin{aligned} \| S(\Pi; f; z) \|_2^2 &= E\left(\left[\sum_{j=1}^{m} f(\tau_j)\ \Delta_j z\right]^2\right) \\ &= \sum_{j=1}^{m} E(f(\tau_j)^2\ (\Delta_j z)^2) + \sum_{j \neq k} f(\tau_j) f(\tau_k) E(\Delta_j z\ \Delta_k z) \\ &= \sum_{j=1}^{m} f(\tau_j)^2\ [\Delta_j F] \\ &= S(\Pi; f^2; F). \end{aligned}$$

If f satisfies any condition that guarantees that the Riemann sums $S(\Pi; f; z)$ converge in L_2-distance to a limit J (necessarily the integral of f with respect to z), then the right member of this equation also converges to a limit, and

$$\text{(6-4)} \qquad \left\| \int_a^b f(t)\ dz(t) \right\|_2 = \left\{ \int_a^b f(t)^2\ dF(t) \right\}^{1/2},$$

the right member being a Riemann–Stieltjes integral.

When f and z are both sure functions, from Theorem (6-1) we obtain the following corollary.

(6-5) COROLLARY *Let f be bounded on T and continuous a.e. in $[a, b] \subset T$. Let z be Lipschitzian on $[a, b]$. Then as the mesh of the (arbitrary,* *not necessarily belated) partition* Π *of* $[a, b]$ *tends to* 0, $S(\Pi; f; z)$ *tends to a limit*

$$\int_a^b f(t)\, dz(t).$$

In particular, this integral exists as a Riemann–Stieltjes integral.

The hypotheses in Corollary (6-5) strongly resemble the necessary and sufficient condition for the existence of the Riemann integral $\int f\, dt$. They are not necessary for the existence of the belated integral $\int f\, dz$, as the following example shows. Define $f(t) = |t|^{-1/2}$ if $t \neq 0$, $f(0) = 0$. Then the belated integral $\int_{-1}^{0} f(t)\, dt$ exists, but the similarly denoted Riemann integral does not. [If Π is a belated partition of $[-1, 0]$, $S(\Pi; f; t)$ is the integral of a step function s on $[-1, 0]$ such that $0 \leqq s \leqq f$.] However, we can state a necessary condition very close to that in Corollary (6-5).

(6-6) *If* f *has a belated integral with respect to* $\mathbf{t}$ *on* $[a, b]$, *then* f *is bounded on* $[a, c]$ *whenever* $a < c < b$, *and* f *is a.e. continuous on* $[a, b]$.

We leave the proof as an exercise, with a suggestion for the second part. If $\epsilon > 0$, there exists $\delta > 0$ such that if Π', Π'' are belated partitions of $[a, b]$ with mesh less than δ, $|S(\Pi') - S(\Pi'')| < \epsilon^2$. Choose $m > 2(b - a)/\delta$ and $t_1 = a < t_2 < \cdots < t_{m+1} = b$ with $t_{j+1} - t_j = (b - a)/m$. If the set of those intervals $[t_{j-1}, t_j]$ $(j = 2, \ldots, m)$ on which the oscillation of f is greater than ϵ had total length $> \epsilon$, we could choose τ_j' and τ_j'' in $[t_{j-1}, t_j]$ to obtain partitions Π', Π'' with $|S(\Pi') - S(\Pi'')| > \epsilon^2$.

One of the reasons that the Brownian-motion noises z are especially useful is that they are martingales. It is often useful to know that under suitable conditions, the indefinite integral of f with respect to z inherits this property.

(6-7) *THEOREM* *Let the standing hypotheses* (2-1) *hold. Assume further that* z *is a martingale with respect to* $\mathfrak{F}_{\boldsymbol{\tau}}$ *on* $[a, b]$, *and that* f *is an integrand adapted to* $\mathfrak{F}_{\boldsymbol{\tau}}$ *such that as the mesh of the belated partition* Π *of* $[a, b]$ *tends to* 0, $S(\Pi; f; z)$ *converges in* L_2*-distance. Then the process* F *on* $[a, b]$ *defined by*

$$F(t) = \int_a^t f(s)\, dz(s)$$

has a version which is also a martingale with respect to $\mathfrak{F}_{\boldsymbol{\tau}}$.

If $a \leqq s < t \leqq b$, for each positive integer n we can choose a partition Π_n of $[a, b]$ with mesh $\Pi_n < 1/n$ and with s, t among the division points of Π_n. If $k(n)$ is that integer for which $t_{k(n)} = s$, the set $\Pi_{n,s} = (t_1, \ldots, t_{k(n)};$

$\tau_1, \ldots, \tau_{k(n)-1}$) is a belated partition of $[a, s]$. Similarly from Π_n we select a belated partition $\Pi_{n,t}$ of $[a, t]$ and a belated partition $\Pi_{n,s,t}$ of $[s, t]$. This last we denote by $(t_1', \ldots, t'_{p+1}; \tau_1', \ldots, \tau_p')$. Then

$$S(\Pi_{n,s,t}; f; z) = S(\Pi_{n,t}; f; z) - S(\Pi_{n,s}; f; z);$$

so

$$\begin{aligned} &E(S(\Pi_{n,t}; f; z) - S(\Pi_{n,s}; f; z) \mid \mathfrak{F}_s) \\ &= \sum_{j=1}^{p} E(f(\tau_j') [z(t'_{j+1}) - z(t_j')] \mid \mathfrak{F}_s) \\ &= \sum_{j=1}^{p} E(E(f(\tau_j') [z(t'_{j+1}) - z(t_j')] \mid \mathfrak{F}(t_j')) \mid \mathfrak{F}_s) \\ &= \sum_{j=1}^{p} E(f(\tau_j') E([z(t'_{j+1}) - z(t_j')] \mid \mathfrak{F}(t_j')) \mid \mathfrak{F}_s) \\ &= 0. \end{aligned}$$

As $n \to \infty$, $S(\Pi_{n,t}; f; z)$ and $S(\Pi_{n,s}; f; z)$ tend in L_2-distance to $F(t)$ and $F(s)$, respectively, by Theorem II(4-3). So the left member of the above equation converges to $E(F(t) - F(s) \mid \mathfrak{F}_s)$, which must be 0. If $F(t, \omega)$ has been chosen to be $\mathfrak{F}(t)$-measurable for each t, as is possible, it is a martingale with respect to $\mathfrak{F}(\boldsymbol{\cdot})$.

7 *Examples: Brownian motions; point processes*

The two types of noise processes that have been most frequently used in models of systems are the "Gaussian white noises," in which the z are Wiener processes, and the point processes, such as the Poisson processes, in which the disturbing effect occurs impulsively at discrete times. We shall now study these as instances of our general theory.

In the case of the Wiener processes, or Brownian motion, the second-order integrals assume an especially convenient form. If z is a standard Wiener process, or Brownian motion, with respect to the $\mathfrak{F}_{\boldsymbol{\cdot}}$, it is a martingale, and by well-known computations, if $a \leqq s < t \leqq b$ we have a.s.

$$E([z(t) - z(s)]^2 \mid \mathfrak{F}_s) = t - s,$$

$$E([z(t) - z(s)]^4 \mid \mathfrak{F}_s) = 3(t - s)^2.$$

Now let f be adapted to $\mathfrak{F}_{\boldsymbol{\cdot}}$, L_2-bounded on T, and L_2-continuous a.e. in $[a, b]$. Then f is integrable with respect to (z, z), and the Riemann sums converge in L_1-distance to the integral. Let $\Pi = (t_1, \ldots, t_{m+1}; \tau_1, \ldots, \tau_m)$ be a belated partition of $[a, b]$; define

$$\Delta_j = (\Delta_j z)^2 - \Delta_j t \qquad (j = 1, \ldots, m).$$

Then by the equations above

$$\begin{aligned} E(\Delta_j \mid \mathfrak{F}(t_j)) &= 0, \\ E(\Delta_j^2 \mid \mathfrak{F}(t_j)) &= E([\Delta_j z]^4 \mid \mathfrak{F}(t_j)) - 2\Delta_j t E([\Delta_j z]^2 \mid \mathfrak{F}(t_j)) + (\Delta_j t)^2 \\ &= 2(\Delta_j t)^2. \end{aligned}$$

By Lemma (1-1), with $u_j = f(\tau_j)$,

$$\begin{aligned} \| S(\Pi; f; z, z) - S(\Pi; f; t) \|_2 &\leqq 0 + \left\{ \sum_{j=1}^{m} \| f(\tau_j) \|_2^2 2(\Delta_j t)^2 \right\}^{1/2} \\ &\leqq (\text{mesh } \Pi)^{1/2} \{2 \sup \| f(\tau) \|_2^2 (b - a)\}^{1/2}. \end{aligned}$$

So $S(\Pi; f; t)$ has the same limit in L_1-distance as $S(\Pi; f; z, z)$ has, and therefore one version of the second-order integral is given by

$$\text{(7-1)} \qquad \int_a^b f(t)\ (dz(t))^2 = \int_a^b f(t)\ dt.$$

More generally, let f be a process on T of the form

$$f(\tau, \omega) = b(\tau, \omega) f_2(\tau, \omega),$$

where b and f_2 are adapted to $\mathfrak{F}_\tau$, b is a.s. sample-function bounded and is continuous in probability a.e. in $[a, b]$, and f_2 is L_2-bounded and is L_2-continuous a.e. in $[a, b]$. There is a r.v. $B(\omega)$ such that a.s. $| b(\tau, \omega) | \leqq B(\omega)$. If $\epsilon > 0$, there is an M such that the set

$$\Omega_M = \{\omega : B(\omega) \leqq M\}$$

has $P(\Omega_M) > 1 - \epsilon/2$. If we define

$$b_M(\tau, \omega) = \text{mid}\{b(\tau, \omega), M, -M\},$$

then $b_M = b$ on $T \times \Omega_M$. As proved previously [cf. (4-5)], $b_M f_2$ is L_2-bounded and is L_2-continuous a.s. in $[a, b]$, so

$$\text{(7-2)} \qquad \int_a^b b_M f_2\ (dz)^2 = \int_a^b b_M f_2\ dt.$$

Both members of (7-1) exist; by Corollary II(2-22), on all but a negligible subset of Ω_M they coincide with the respective members of (7-2). So the two members of (7-1) coincide except on a set of measure less than ϵ, where ϵ is arbitrary, and therefore they are a.s. equal.

The version of the second-order integral given by the right member of (7-1) is not a strict version. Let $I(\omega)$ be a r.v. such that $I(\omega) = 0$ if $z(\mathbf{t}, \omega)$ is Lipschitzian and $I(\omega) = 1$ otherwise. Then a.s. $I(\omega) = 1$, so another version of the second-order integral is

$$\int_a^b f(t, \omega)\ (dz(t, \omega))^2 = \int_a^b I(\omega) f(t, \omega)\ dt. \tag{7-3}$$

This is a strict version, since it has the same value 0 as the Riemann–Stieltjes integral of f with respect to $(dz)^2$ whenever z is Lipschitzian and f is Riemann integrable.

If z^1 and z^2 are independent Brownian motions, we already know by Theorem (5-7) that

$$\int_a^b f(t)\ dz^1(t)\ dz^2(t) = 0 \tag{7-4}$$

whenever f is adapted to $\mathfrak{F}_{\boldsymbol{\tau}}$ and is either L_2-bounded or sample-function bounded. Statements (7-1) and (7-4) are sometimes summarized as

$$dz^1\, dz^2 = 0, \qquad (dz)^2 = dt,$$

which is a convenient mnemonic device if we keep in mind that there are some hypotheses on f.

A point process is one in which events, each with its own probability distribution, occur at a discrete set of random times. A simple example is "shot effect," in which electrons reach the plate of a vacuum tube at Poisson-distributed times. Another is a birth-and-death process, in which at random times a population changes by $+1$ with probability p and by -1 with probability $q = 1 - p$. The times at which events happen can be regarded as the discontinuities of a right-continuous random function $N(\boldsymbol{\tau}, \omega)$ on $(-\infty, \infty)$, each sample function $N(\boldsymbol{\tau}, \omega)$ being continuous except at points, finitely many in each bounded time interval, at which $N(\tau, \omega) - N(\tau-, \omega) = +1$. For the events, we construct a real-valued process $u(\boldsymbol{\tau}, \omega)$, defined on $(-\infty, \infty)$, such that all the r.v.'s $u(\tau, \omega)$ have the same distribution, with finite mean μ and variance σ^2, and corresponding to any finite set $\{\tau_1, \tau_2, \ldots, \tau_k\}$ the $u(\tau_k, \omega)$ are independent. We further suppose that all the $u(\tau, \omega)$ are independent of all the increments $N(t, \omega) - N(s, \omega)$. Now for each ω in Ω we define $z(\mathbf{t}, \omega)$ to be that function such that $z(0, \omega) = 0$ and $z(t, \omega) - z(s, \omega) = \sum_{s<\tau\leq t} u(\tau, \omega)[N(\tau, \omega) - N(\tau-, \omega)]$. This looks like an uncountable sum, but $N(\tau, \omega) - N(\tau-, \omega)$ is 0 except at finitely many points in $(s, t]$.

Suppose that we are interested in the magnitude at time b of an effect produced by these events. We shall suppose that the effect at time b of a disturbance u at time t is proportional to u, with a proportionality constant $\Phi(t, b)$, and that effects are additive. We also assume that events before time a ($<b$) have no effect (the system is "switched on" at time a), and of course we assume that events at times $t > b$ have no effect at time b. Then the total effect at time b is

$$D(b, \omega) = \sum_{a<\tau\leqq b} \Phi(\tau, b)[z(\tau, \omega) - z(\tau-, \omega)]. \tag{7-5}$$

If $\Phi(\tau, b)$ happens to be continuous in τ, we can readily write $D(b, \boldsymbol{\omega})$ as a stochastic integral. For each ω, there are finitely many points $t_1^*(\omega), \ldots, t_h^*(\omega)$ at which $N(\boldsymbol{\tau}, \omega)$ is discontinuous. (Here h depends on ω.) Let $\delta(\omega)$ be the least of the numbers $t_{i+1}^*(\omega) - t_i^*(\omega)$, $i = 1, 2, \ldots, h - 1$. Then if $\Pi = (t_1, \ldots, t_{k+1}; \tau_1, \ldots, \tau_k)$ is any partition (not necessarily belated) of $[a, b]$ with mesh $\Pi < \delta$, each t_i^* will lie in an interval $(t_{n(i)}, t_{n(i)+1}]$ of the partition, no two in any one interval. By definition of z,

$$\Delta_{n(i)}z = u(t_i^*, \omega);$$

so

$$S(\Pi; \Phi(\tau, b); z(t, \omega)) = \sum_{i=1}^{h} \Phi(\tau_{n(i)}, b) u(t_i^*, \omega).$$

As mesh Π tends to 0, $\tau_{n(i)}$ tends to $t_i^*(\omega)$, so $\Phi(\tau_{n(i)}, b)$ tends to $\Phi(t_i^*(\omega), b)$. So as mesh Π tends to 0,

$$S(\Pi; \Phi(\boldsymbol{\tau}, b); z(\mathbf{t}, \omega)) \to \sum_{i=1}^{h} \Phi(t_i^*, \omega) u(t_i^*, \omega) = D(b, \omega).$$

By definition

$$D(b, \omega) = \int_a^b \Phi(t, b)\, dz(t, \omega), \tag{7-6}$$

the integral existing in the strong sense that it is the pointwise limit of $S(\Pi; \Phi(\tau, b); z)$ for general partitions Π, not necessarily belated.

Suppose next that for each n and each interval $[a, b]$, the joint distribution of the discontinuities of N, conditioned on there being n discontinuities in $(a, b]$, is "continuous," that is, it is given as the integral of some density on $[a, b] \times \cdots \times [a, b]$. If $\Phi(\boldsymbol{\tau}, b)$ is finite at each τ and is continuous at almost all τ in $[a, b]$, there is probability 0 that any one of the n discontinuities of N shall coincide with a discontinuity of $\Phi(\boldsymbol{\tau}, b)$. This is true for each n, so there is a set $\Omega_1 \subset \Omega$ with $P\Omega_1 = 1$ such that if ω is in Ω_1, $\Phi(\boldsymbol{\tau}, \omega)$ is continuous at every discontinuity of $N(\boldsymbol{\tau}, \omega)$. The argument in the preceding paragraph holds for all such ω, so again (7-6) holds. However, this equation is not particularly useful unless we add some hypotheses concerning N. We assume to begin with that $E([N(b) - N(a)]^2)$ is finite. To simplify typography, if ϕ is any function on $[a, b]$ and (s, t) and (s', t') are subintervals of $[a, b]$, we define

$$\Delta\phi = \phi(t) - \phi(s), \qquad \Delta'\phi = \phi(t') - \phi(s').$$

The expectation

$$\nu(t) = E(N(t) - N(a))$$

is finite for $a \leqq t \leqq b$, and

$$E(\Delta N) = \Delta\nu.$$

The conditional distribution of Δz, under the condition $\Delta N = n$, is the distribution of the sum of n independent choices $u_1, \ldots, u_n$ of a r.v. whose distribution is that of $u(a)$. Hence

$$\begin{aligned} E(\Delta z \mid \Delta N = n) &= E(u_1 + \cdots + u_n) = n\mu, \\ E([\Delta z]^2 \mid \Delta N = n) &= E[(u_1 + \cdots + u_n)^2] \\ &= \sum_{i=1}^{n} E(u_i^2) + \sum_{i \neq j} E(u_i u_j) \\ &= n[\sigma^2 + \mu^2] + \sum_{i \neq j} E(u_i)E(u_j) \\ &= n\sigma^2 + (n\mu)^2. \end{aligned}$$

This implies

$$\begin{aligned} E(\Delta z) &= \mu \sum nP(\Delta N = n) \\ &= \mu\, \Delta\nu, \\ E([\Delta z]^2) &= \sigma^2 \sum nP(\Delta N = n) + \mu^2 \sum n^2 P(\Delta N = n) \\ &= \sigma^2\, \Delta\nu + \mu^2 E([\Delta N]^2). \end{aligned} \tag{7-7}$$

If we now assume that there is a constant K such that

$$\Delta\nu \leqq K(t - s), \qquad E([\Delta\nu]^2) \leqq K(t - s)$$

for all $(s, t) \subset [a, b]$, by Theorem (6-1) the Riemann sums $S(\Pi; \Phi(\tau, b); z)$ converge in L_2-distance to the integral in Eq. (7-6). Hence

$$E\left(\int_a^b \Phi(t, b)\, dz(t)\right) = \mu \int_a^b \Phi(t, b)\, d\nu(t). \tag{7-8}$$

We now add the strong assumption that N has independent increments, and define $\nu_2(t)$ to be the variance of $N(t) - N(a)$. Then $\Delta\nu_2$ is the variance of ΔN; and, since $E(\Delta N) = \Delta\nu$,

$$\Delta\nu_2 = E([\Delta N - \Delta\nu]^2).$$

By (7-7),

$$\text{(7-9)} \qquad \begin{aligned} E([\Delta z - \mu\Delta\nu]^2) &= E([\Delta z]^2) - 2\mu\,\Delta\nu E(\Delta z) + \mu^2(\Delta\nu)^2 \\ &= \sigma^2\,\Delta\nu + \mu^2 E([\Delta N]^2) - \mu^2[\Delta\nu]^2 \\ &= \sigma^2\,\Delta\nu + \mu^2\,\Delta\nu_2. \end{aligned}$$

If $a \leqq s \leqq t \leqq s' \leqq t' \leqq b$, the conditional distribution of $\Delta z\,\Delta' z$ under the condition $\Delta N = n$, $\Delta' N = n'$ is the same as the distribution of $(u_1 + \cdots + u_n)(v_1 + \cdots + v_{n'})$ when $u_1, \ldots, v_{n'}$ are independent choices of the value of an r.v. whose distribution is that of $u(a)$. So

$$E(\Delta z\,\Delta' z \mid \Delta N = n, \Delta' N = n') = nn'\mu^2,$$

whence, by multiplying by $P(\Delta N = n \text{ and } \Delta' N = n')$ and summing, we obtain

$$\begin{aligned} E(\Delta z\,\Delta' z) &= \mu^2 E(\Delta N\,\Delta' N) \\ &= \mu^2\,\Delta\nu\,\Delta'\nu. \end{aligned}$$

Therefore

$$E([\Delta z - \mu\,\Delta\nu][\Delta' z - \mu\,\Delta'\nu]) = 0,$$

and the process $z - \mu\nu$ has orthogonal increments. By (6-4), (7-8), and (7-9)

$$\text{(7-10)} \qquad \begin{aligned} \operatorname{var} \int_a^b \Phi(t, b)\,dz(t) &= E\left(\int_a^b \Phi(t, b)\,d[z(t) - \mu\nu(t)]\right)^2 \\ &= \int_a^b \Phi(t, b)^2\,d[\sigma^2\nu + \mu^2\nu_2] \\ &= \sigma^2 \int_a^b \Phi(t, b)^2\,d\nu(t) + \mu^2 \int_a^b \Phi(t, b)^2\,d\nu_2(t). \end{aligned}$$

An especially important special case is that in which N is a Poisson process, satisfying with some positive λ

$$P(\Delta N = k) = e^{-\lambda\,\Delta t}(\lambda\,\Delta t)^k/k!.$$

In this case

$$\nu(t) = \lambda(t - a), \qquad \nu_2(t) = \lambda(t - a).$$

If the events u consist of the arrival of an electron at the plate of a vacuum tube, carrying charge e, the quantity $\Delta z = e\,\Delta N$ is the total charge arriving between times s and t. If unit charge arriving at time t produces at time b an effect $\Phi(t, b)$ at some point in a linear circuit, the total effect at time b is given by Eq. (7-6). So by Eqs. (7-8) and (7-10),

$$E(D(b, \omega)) = e\lambda \int_a^b \Phi(t, b)\, dt,$$

$$\operatorname{var} D(b, \omega) = e^2\lambda \int_a^b \Phi(t, b)^2\, dt.$$

This result (for Φ a function of $b - t$) was obtained by H. Campbell in 1909; it is known as Campbell's theorem.

For numerous other examples of point processes, we refer the reader to Blanc-Lapierre and Fortet [2, in particular Vol. 1, Chap. 5] and to Srinavasan and Vasudevan [12, in particular Chap. 2].

8 Extension to the Itô-belated integral

All the existence theorems for the belated integral that are in the preceding sections have direct analogs for the Itô-belated integral. The purpose of this section is to present these analogs. But first we prove a theorem that is needed in later proofs and that also has some interest in its own right. It gives an indication of the degree to which the Itô-belated integral generalizes the belated integral.

We have seen in (6-5) and (6-6) that in the deterministic case the belated integral $\int f\, dt$ is a very slight extension of the Riemann integral. We shall now show that in the deterministic case the Itô-belated integral generalizes the Lebesgue integral.

(8-1) *THEOREM Let f be extended-real-valued and Lebesgue integrable on $[a, b]$. Then to each positive ϵ there corresponds a $\delta(\mathbf{t})$ a.e. positive on $[a, b)$ and a positive number δ^* such that*

for every partition $\Pi = \{(A_1, \tau_1), \ldots, (A_k, \tau_k)\}$ in $[a, b]$ such that $A_j \subset (\tau_j - \delta(\tau_j), \tau_j + \delta(\tau_j))$ and $\sum_j mA_j > (b - a) - \delta^*$, it is true that

$$\left| S(\Pi; f; \mathbf{t}) - \int_a^b f(t)\, dt \right| < \epsilon,$$

the integral being the Lebesgue integral.

Let us denote by L the Lebesgue integral of f from a to b. It is a well-known property of the Lebesgue integral that for each positive ϵ there exists a lower semicontinuous function g on $[a, b]$ such that $g \geqq f$ and

$$\int_a^b g(t)\,dt < L + \epsilon/4.$$

We define

$$f_2(t) = g(t) + \epsilon/4(b-a) \qquad (a \leqq t \leqq b).$$

Then

(A) f_2 is lower semicontinuous on $[a, b]$; $f_2(t) > f(t)$ for all t in $[a, b]$ such that $f(t)$ is finite and

$$\int_a^b f_2(t)\,dt < L + \epsilon/2.$$

By applying the same procedure to $-f$ and then changing signs, we find that there exists a function f_1 on $[a, b]$ such that

(B) f_1 is upper semicontinuous on $[a, b]$; $f_1(t) < f(t)$ for all t in $[a, b]$ such that $f(t)$ is finite and

$$\int_a^b f_1(t)\,dt > L - \epsilon/2.$$

If $f(\bar{t})$ is finite, by (A) $f_2(\bar{t}) > f(\bar{t})$ and f_2 is lower semicontinuous at $\bar{t}$, so there exists a positive number $\delta_2(\bar{t})$ such that

(C) if t is in $[a, b]$ and $\bar{t} - \delta_2(\bar{t}) < t < \bar{t} + \delta_2(\bar{t})$, then $f_2(t) > f(\bar{t})$.

Likewise, from (B) we deduce that for such $\bar{t}$ there exists a positive $\delta_1(\bar{t})$ such that

(D) if t is in $[a, b]$ and $\bar{t} - \delta_1(\bar{t}) < t < \bar{t} + \delta_1(\bar{t})$, then $f_1(t) < f(\bar{t})$.

We define

$$\delta(\bar{t}) = \min\{\delta_1(\bar{t}), \delta_2(\bar{t})\} \quad \text{if} \quad |f(\bar{t})| < \infty; \qquad \delta(\bar{t}) = 0 \quad \text{otherwise}.$$

Because f_1 and f_2 are integrable over $[a, b]$, there is a positive number δ^* such that if E is any subset of $[a, b]$ with $mE < \delta^*$,

$$\left|\int_E f_1(t)\,dt\right| < \epsilon/2 \qquad \text{and} \qquad \left|\int_E f_2(t)\,dt\right| < \epsilon/2.$$

Now we shall show that the $\delta(t)$ and δ^* thus defined have the properties described in the theorem.

Let Π be any partition in $[a, b]$ with the properties required in the statement of the theorem; for Π we use the same notation as in the theorem. Since each $A_j \neq \phi$, $\delta(\bar{t}_j) > 0$; so $|f(\bar{t}_j)| < \infty$, and

$$f_1(t) < f(\bar{t}_j) < f_2(t) \qquad (t \text{ in } A_j).$$

Integrating over A_j and then summing over $j = 1, \ldots, k$ yields

$$\sum_{j=1}^{k} \int_{A_j} f_1(t)\, dt \leqq S(\Pi; f; \mathbf{t}) \leqq \sum_{j=1}^{k} \int_{A_j} f_2(t)\, dt,$$

the integrals being Lebesgue integrals. The set $D = [a, b] \backslash [\bigcup_{j=1}^{k} A_j]$ has measure less than δ^*, so by the choice of δ^*

$$\int_D f_1(t)\, dt - \epsilon/2 < 0, \qquad \int_D f_2(t)\, dt + \epsilon/2 > 0.$$

From these two and the preceding inequality we obtain

$$\int_a^b f_1(t)\, dt - \epsilon/2 < S(\Pi; f; \mathbf{t}) < \int_a^b f_2(t)\, dt + \epsilon/2.$$

From this, with (A) and (B),

$$L - \epsilon < S(\Pi; f; \mathbf{t}) < L + \epsilon,$$

which completes the proof.

As a corollary, if f is Lebesgue integrable over $[a, b]$, it has an Itô-belated integral with respect to $\mathbf{t}$ over $[a, b]$, the two integrals being equal. By a theorem in McShane [10, p. 45], the limit of $S(\Pi; f; \mathbf{t})$ exists in the sense of Theorem (8-1) if and only if the Lebesgue integral exists. However, we shall not make any use of this fact. It is not known if the existence of the Itô-belated integral $\int f\, dt$ implies that the Lebesgue integral exists.

Given the algebras $\mathfrak{F}_{\boldsymbol{\tau}}$ as in the standing hypotheses (2-1), we have defined a **simple process** $s(\mathbf{t}, \omega)$ **adapted to** $\mathfrak{F}_{\boldsymbol{\tau}}$ thus: There are points $t_1^* = a < t_2^* < \cdots < t_{l+1}^* = b$ and r.v.'s $v_1, \ldots, v_l$ such that for $j = 1, \ldots, l$, v_j is $\mathfrak{F}(t_j^*)$-measurable and $s(t, \omega) = v_j$ $(t_j^* \leqq t < t_{j+1}^*)$.

Simple functions have integrals of the obvious form, as we now prove.

(8-2) *If $s(\mathbf{t}, \omega)$ is a simple process on $[a, b]$ adapted to $\mathfrak{F}_{\boldsymbol{\tau}}$, with the notation of the preceding sentence, and $E(v_j^2) < \infty$ for each j, then s has an Itô-belated integral with respect to z, and*

$$\int_a^b s(t)\, dz(t) = \sum_{j=1}^{l} v_j[z(t_{j+1}^*) - z(t_j^*)].$$

Moreover, the Riemann sums $S(\Pi; s; z)$ converge in L_2-distance to the integral as (δ, δ^*) shrinks.

By Theorem II(5-4) it is enough to prove this for each interval $[t_j^*, t_{j+1}^*]$ separately. We simplify the typography by writing a, b, v for t_j^*, t_{j+1}^*, v_j, respectively. Let ϵ be positive; define

$$\delta^* = [\epsilon/C \| v \|]^2,$$

where C is the constant defined in Lemma (2-2). Let $\Pi = \{(A_1, \tau_1), \ldots, (A_k, \tau_k)\}$ be a (δ, δ^*) partition of $[a, b)$ (it does not even have to be belated). The set difference $[a, b)\backslash\bigcup_1^k A_i$ consists of intervals $[\alpha_1, \beta_1), \ldots, [\alpha_h, \beta_h)$ of total length less than δ^*. Hence by Lemma (2-2)

$$\begin{aligned} \| v[z(b) - z(a)] - S(\Pi; s; z) \|_2 &= \left\| v \sum_{i=1}^{h} [z(\beta_i) - z(\alpha_i)] \right\| \\ &\leqq C\left\{\sum_{i=1}^{l} \| v \|^2(\beta_i - \alpha_i)\right\}^{1/2} \\ &< \epsilon. \end{aligned}$$

We can now prove an extension of Theorems (2-3) and (4-1).

(8-3) *THEOREM Let the standing hypotheses* (2-1) *be satisfied. Let f be a measurable process adapted to $\mathfrak{F}_{\boldsymbol{\tau}}$. If $| f(\mathbf{t}, \omega) |^2$ is a.s. Lebesgue integrable from a to b, the Itô-belated integral of f with respect to z exists. Moreover, if $\| f(\mathbf{t}) \|^2$ is Lebesgue integrable from a to b, the Riemann sums $S(\Pi_b; f; z)$ for belated partitions Π_b in $[a, b)$ converge in L_2-distance to the integral as (δ, δ^*) shrinks.*

We consider first the case in which $\| f \|^2$ is integrable from a to b. Let ϵ be positive, and define

$$\gamma = \epsilon^2/32C^2,$$

where C is defined in Lemma (2-2). By Theorem I(3-9), there is a bounded simple function $s(\mathbf{t}, \boldsymbol{\omega})$ adapted to $\mathfrak{F}_{\boldsymbol{\tau}}$ such that

$$\int_a^b \| f(t) - s(t) \|^2 \, dt < \gamma,$$

the integral being the Lebesgue integral. By Theorem (8-1), there exist a $\delta_1(\mathbf{t})$ positive a.e. on $[a, b)$ and a positive δ_1^* such that if Π is any (δ_1, δ_1^*) belated partition in $[a, b)$,

$$\left| S(\Pi; \| f(\mathbf{t}) - s(\mathbf{t}) \|^2; \mathbf{t}) - \int_a^b \| f(\mathbf{t}) - s(\mathbf{t}) \|^2 \, dt \right| < \gamma,$$

whence

$$\left| S(\Pi; \| f(\mathbf{t}) - s(\mathbf{t}) \|^2; \mathbf{t}) \right| < 2\gamma.$$

By Lemma (2-2),

$$\| S(\Pi; f - s; z) \| \leqq C\{S(\Pi; \| f(\mathbf{t}) - s(\mathbf{t}) \|^2; \mathbf{t})\}^{1/2} < \epsilon/4.$$

By (8-2), there exist a $\delta_2(\mathbf{t})$ positive a.e. on $[a, b)$ and a positive δ_2^* such that for every (δ_2, δ_2^*) belated partition Π in $[a, b)$,

$$\left\| S(\Pi; s; z) - \int_a^b s(t)\, dz(t) \right\|_2 < \epsilon/4.$$

Now define $\delta(t) = \min(\delta_1(t), \delta_2(t))$ $(a \leqq t < b)$, $\delta^* = \min(\delta_1^*, \delta_2^*)$. If Π' and Π'' are any two (δ, δ^*) belated partitions in $[a, b)$, both the preceding estimates hold for both partitions, so

$$\begin{aligned} &\| S(\Pi'; f; z) - S(\Pi''; f; z) \| \\ &\qquad \leqq \| S(\Pi'; f - s; z) \| \\ &\qquad + \| S(\Pi'; s; z) - S(\Pi''; s; z) \| + \| S(\Pi''; f - s; z) \| \\ &\qquad < \epsilon. \end{aligned}$$

By Theorem II(4-1), as extended to Itô-belated integrals, as (δ, δ^*) shrinks the Riemann sums $S(\Pi; f; z)$ converge in L_2-distance to a limit, which is the integral.

Suppose now that $| f(\mathbf{t}, \omega) |^2$ is a.s. integrable. We define $f_n(t, \omega) = \operatorname{mid}(f(t, \omega), n, -n)$ $(n = 1, 2, 3, \ldots)$. Then by the part of the proof already completed the Itô-belated integral

$$F_n(t) = \int_a^t f_n(s)^2\, ds$$

exists for each t in $[a, b]$. It is the limit in L_2-distance of sums $S(\Pi; f_n^2; \mathbf{s})$ for (δ, δ^*) belated partitions Π in $[a, t)$, and all these sums are $\mathfrak{F}_t$-measurable, so F_n is adapted to $\bar{\mathfrak{F}}_{\boldsymbol{\tau}}$ [cf. I(3-12)]. With probability 1, as $n \to \infty$, $F_n(t, \omega)$ converges to

$$F(t, \omega) = \int_a^t f(s, \omega)^2\, ds,$$

so F is adapted to $\bar{\mathfrak{F}}_{\boldsymbol{\tau}}$. Then so is the function $I_n(\mathbf{t}, \boldsymbol{\omega})$ which for each t in $[a, b]$ is the indicator function of the set $\Omega_{t,n} = \{\omega : F(t, \omega) \leqq n\}$ $(n = 1, 2,$

3, . . .). By hypothesis, $F(b, \omega)$ is a.s. finite. So if $\epsilon > 0$, we can and do choose n so that

$$P(\Omega_{b,n}) > 1 - \epsilon/4.$$

Then

$$I_n(t, \omega)f(t, \omega) = f(t, \omega) \qquad (a \leqq t \leqq b; \quad \omega \in \Omega_{b,n}),$$

so for all partitions Π in $[a, b)$

$$\bar{\rho}(S(\Pi; I_n\cdot f; z), S(\Pi; f; z)) < \epsilon/4. \tag{8-4}$$

The function $I_n\cdot f$ satisfies the hypotheses of the theorem. For each ω in Ω, let $t(\omega)$ be the supremum of numbers t such that $\omega \in \Omega_{t,n}$. Since F is continuous in t, ω is in $\Omega_{t(\omega),n}$, and

$$\int_a^b [I_n(s, \omega)f(s, \omega)]^2\, ds = \int_a^{t(\omega)} [f(s, \omega)]^2\, ds \leqq n.$$

Hence

$$\int_a^b \| I_n\cdot f \|^2\, ds = E\left(\int_a^b [I_n(s)f(s)]^2\, ds\right) \leqq n.$$

So by the part of the proof already completed there exist a $\delta(\mathbf{t}) > 0$ a.e. on $[a, b)$ and a $\delta^* > 0$ such that for every (δ, δ^*) belated partition Π in $[a, b)$,

$$\bar{\rho}\left(S(\Pi; I_n\cdot f; z), \int_a^b I_n\cdot f\, dz\right) < \epsilon/4.$$

This and (8-4) imply that for any two such partitions Π', Π'' we have

$$\bar{\rho}(S(\Pi'; f; z), \qquad S(\Pi''; f; z)) < \epsilon.$$

By the extension of Theorem II(4.1) to the Itô-belated integral, f is integrable with respect to z.

In proving an existence theorem for second-order integrals we need the following analog of statement (8-2).

(8-5) *If $s(\mathbf{t}, \boldsymbol{\omega})$ is a bounded simple process on $[a, b]$ adapted to $\mathfrak{F}_{\boldsymbol{\tau}}$, and z is measurable and $z(a, \omega) = 0$ a.s., s has an Itô-belated integral with respect to (z, z) over $[a, b)$, and as (δ, δ^*) shrinks the Riemann sums $S(\Pi; s; z, z)$ for belated partitions Π in $[a, b)$ converge in L_1-distance to the integral.*

As in proving (8-2), it is enough to prove this when $s(t, \omega)$ is constantly equal to a bounded $\mathfrak{F}_a$-measurable r.v., v. Let M_v be an upper bound for $\{| v(\omega) | : \omega \in \Omega\}$, and let M_z be an upper bound for $\| z(t) \|_2$ $(a \leqq t \leqq b)$; this exists, by (2-1,iii). By Theorem (8-3), there exist a $\delta(\mathbf{t})$ positive a.e. on

$[a, b)$ and a positive δ_1^* such that for every (δ, δ_1^*) belated partition Π in $[a, b)$,

$$\left\| S(\Pi; vz; z) - \int_a^b vz(t)\, dz(t) \right\|_2 < \epsilon/6.$$

We define

$$\delta^* = \min\{\delta_1^*, \epsilon/3M_vK, [\epsilon/6CM_vM_z]^2\}.$$

Let $\Pi = \{(A_1, \tau_1), \ldots, (A_k, \tau_k)\}$ be a (δ, δ^*) belated partition in $[a, b)$. We denote the set of all initial and final points of the A_i by $t_1 = a < t_2 < \cdots < t_{h+1} = b$. For all j in a certain subset J of $\{1, \ldots, h\}$, t_j is the initial point of one of the intervals $A_1, \ldots, A_k$; for all j in $J^* = \{1, \ldots, h\}\backslash J$, $[t_j, t_{j+1})$ is not one of the A_i. Then

$$\begin{aligned} S(\Pi; s; z, z) &= \sum_{j\in J} v[z(t_{j+1}) - z(t_j)]^2 \\ &= \sum_{j=1}^{h} v[z(t_{j+1})^2 - z(t_j)^2] \\ &\quad - 2\sum_{j=1}^{h} vz(t_j)[z(t_{j+1}) - z(t_j)] \\ &\quad - \sum_{j\in J^*} v[z(t_{j+1}) - z(t_j)]^2 \\ &= v[z(b)^2 - z(a)^2] - 2\int_a^b vz(t)\, dz(t) \\ &\quad - 2\left[S(\Pi; vz; z) - \int_a^b vz(t)\, dz(t) \right] \\ &\quad + 2\sum_{j\in J^*} vz(t_j)[z(t_{j+1}) - z(t_j)] \\ &\quad - \sum_{j\in J^*} v[z(t_{j+1}) - z(t_j)]^2. \end{aligned}$$

The third term in the right member has L_1-norm less than $\epsilon/3$, by choice of δ and δ_1^*. By Lemma (2-2), the L_2-norm (and therefore the L_1-norm) of the fourth term is at most

$$2C\left\{ \sum_{j\in J^*} \| vz(t_j) \|^2 (t_{j+1} - t_j) \right\}^{1/2} \leqq 2CM_vM_z(\delta^*)^{1/2} < \epsilon/3.$$

The L_1-norm of the fifth term is at most

$$E\left(\sum_{j\in J^*} |v| \cdot [z(t_{j+1}) - z(t_j)]^2\right) \leqq M_v \sum_{j\in J^*} K(t_{j+1} - t_j)$$
$$< M_v K\delta^*$$
$$< \epsilon/3.$$

So

$$\left|\left| S(\Pi; s; z, z) - v[z(b)^2 - z(a)^2] - 2\int_a^b vz(t)\, dz(t) \right|\right|_1 < \epsilon.$$

Now we can prove an existence theorem for second-order Itô-belated integrals.

(8-6) *THEOREM* *Let the standing hypotheses* (2-1) *be satisfied. Let* f, z^1, z^2 *be measurable processes adapted to* $\mathfrak{F}_{\boldsymbol{\tau}}$. *If* $f(\boldsymbol{\tau}; \boldsymbol{\omega})$ *is a.s. Lebesgue integrable from* a *to* b, *the Itô-belated integral of* f *with respect to* (z^1, z^2) *exists. Moreover, if* $E(|f(\mathbf{t})|)$ *is integrable from* a *to* b, *the Riemann sums* $S(\Pi_b; f; z^1, z^2)$ *converge in* L_1*-distance to the integral as* (δ, δ^*) *shrinks.*

We consider first the special case $z^1 = z^2$, and we define $z(\mathbf{t}) = z^1(\mathbf{t}) - z^1(a)$. Suppose that $E(|f|)$ is integrable; let ϵ be positive. If we define

$$f_n(t, \omega) = \text{mid}(f(t, \omega), n, -n), \tag{8-7}$$

we can fix an n such that

$$\int_a^b E(|f_n - f|)\, dt < \epsilon/16K.$$

Then, by Theorem I(3-8), we can find a bounded simple process $g(\mathbf{t}, \boldsymbol{\omega})$ adapted to $\mathfrak{F}_{\boldsymbol{\tau}}$ such that

$$\int_a^b E(|f_n - g|^2)\, ds < \epsilon^2/256(b - a)K^2,$$

whence by the Cauchy–Buniakowski–Schwarz inequality

$$\int_a^b E(1 \cdot |f_n - g|)\, ds < \epsilon/16K.$$

This and the preceding inequality imply

$$\int_a^b E(|f - g|)\, dt < \epsilon/8K.$$

By Theorem (8-1), there exist a $\delta_1(\mathbf{t}) > 0$ a.e. on $[a, b)$ and a $\delta_1^* > 0$ such that for every (δ_1, δ_1^*) belated partition Π in $[a, b]$

$$\left| S(\Pi; E(|f - g|); \mathbf{t}) - \int_a^b E(|f - g|)\, dt \right| < \epsilon/8K.$$

Hence for all such

$$| S(\Pi; E(|f - g|); \mathbf{t}) | < \epsilon/4K.$$

Let $\{(A_1, \tau_1), \ldots, (A_k, \tau_k)\}$ be a (δ_1, δ_1^*)-belated partition in $[a, b]$; we denote A_j by $[\alpha_j, \beta_j)$. Then

$$\begin{aligned}
E(| S(\Pi; f - g; z, z) |) &\leqq \sum_{j=1}^{k} E(| f(\tau_j) - g(\tau_j) | (\Delta_j z)^2) \\
&\leqq \sum_{j=1}^{k} E(E(| f(\tau_j) - g(\tau_j) | (\Delta_j z)^2 \mid \mathfrak{F}(\alpha_j))) \\
&= \sum_{j=1}^{k} E(| f(\tau_j) - g(\tau_j) | E([\Delta_j z]^2 \mid \mathfrak{F}(\alpha_j))) \\
&\leqq \sum_{j=1}^{k} E(| f(\tau_j) - g(\tau_j) | \cdot K\, \Delta_j t) \\
&= K S(\Pi; | f - g |; \mathbf{t}) \\
&< \epsilon/4.
\end{aligned}$$

This can be written in the form

$$\text{(8-8)} \qquad \| S(\Pi; f; z, z) - S(\Pi; g; z, z) \|_1 < \epsilon/4.$$

By statement (8-5), there exist $\delta_2(\mathbf{t}) > 0$ a.e. on $[a, b)$ and a $\delta_2^* > 0$ such that if Π is any (δ_2, δ_2^*) belated partition in $[a, b]$,

$$\text{(8-9)} \qquad \left\| S(\Pi; g; z, z) - \int_a^b g(t)\, dz(t)\, dz(t) \right\|_1 < \epsilon/4.$$

Define $\delta(\tau) = \min\{\delta_1(\tau), \delta_2(\tau)\}$ for all τ in $[a, b)$, and $\delta^* = \min\{\delta_1^*, \delta_2^*\}$. If Π' and Π'' are both (δ, δ^*) belated partitions in $[a, b]$, by the two preceding inequalities

$$\| S(\Pi'; f; z, z) - S(\Pi''; f; z, z) \|_1 < \epsilon.$$

So by Theorem II(4-1) the sums $S(\Pi; f; z, z)$ converge in L_1-distance to a limit as (δ, δ^*) shrinks.

We now abandon the extra hypothesis that $\| f(\mathbf{t}) \|_1$ is integrable. If we define f_n by (8-7), as in the proof of Theorem (8-3) we find that the process defined by

$$F_n(t) = \int_a^t | f_n(s) | \, ds$$

is adapted to $\mathfrak{F}_{\boldsymbol{\tau}}$. Hence so is $F(\mathbf{t})$, where

$$F(t) = \int_a^t | f(s) | \, ds.$$

By a trivial modification of the proof in Theorem (8-3), f is integrable with respect to (z, z).

If z^1 and z^2 satisfy the standing hypotheses (2-1) but are not equal, we define

$$z'(t, \omega) = [z^1(t, \omega) + z^2(t, \omega)]/2,$$

$$z''(t, \omega) = [z^1(t, \omega) - z^2(t, \omega)]/2.$$

These also satisfy hypotheses (2-1), and for every partition Π in $[a, b]$ we have

$$S(\Pi; f; z^1, z^2) = S(\Pi; f; z', z') - S(\Pi' f; z'', z'').$$

If f satisfies the hypotheses of the theorem, by the preceding proof both sums in the right member have limits in probability as (δ, δ^*) shrinks, and therefore so does the left member. Moreover, if $\| f(\mathbf{t}) \|_1$ is integrable, both sums in the right member converge in L_1-distance, hence so does the left member.

The theorems of Section 5, that enable us to get rid of all integrals of third or higher order and of some second-order integrals, can be extended to the Itô-belated integral, as we now prove.

(8-10) *THEOREM* *Let z^1 and z^2 satisfy the standing hypotheses* (2-1), *and let $v^1, \ldots, v^q$ be processes on $[a, b]$ with a.s. continuous sample functions. If z^1, z^2 and $f(\boldsymbol{\tau}, \boldsymbol{\omega})$ are measurable and adapted to $\mathfrak{F}_{\boldsymbol{\tau}}$ and $f(\boldsymbol{\tau}, \omega)$ is a.s. Lebesgue integrable from a to b, then*

$$\int_a^b f(t) \, dz^1(t) \, dz^2(t) \, dv^1(t) \cdots dv^q(t) = 0.$$

It is enough to prove this for nonnegative f. Suppose first that $z^1 = z^2$;

define $z = z^1$. By Theorem (8-6) the integral

$$J(\omega) = \int_a^b f(t)\ (dz(t))^2$$

exists. Let ϵ be positive. Since $J(\omega)$ is a.s. finite, there exists an M such that the set

$$\Omega_M = \{\omega \in \Omega : J(\omega) \geqq M\}$$

has $P(\Omega_M) < \epsilon/3$. Since $S(\Pi_b; f; z, z)$ converges to J in probability as (δ, δ^*) shrinks, there is a $\delta_2(\mathbf{t})$ positive a.e. on $[a, b)$ and a positive number δ^* such that for every (δ_2, δ^*) belated partition Π in $[a, b]$, the set

$$\Omega_\Pi = \{\omega \in \Omega : S(\Pi; f; z, z)(\omega) \geqq J(\omega) + 1\}$$

has $P(\Omega_\Pi) < \epsilon/3$.

We define $\text{Osc}(\delta, \omega)$ as in the proof of Theorem (5-1), and recall that it is an r.v. that converges a.s. to 0 as $\delta \to 0$. There is therefore a positive δ_1 such that the set

$$\Omega_3 = \{\omega : \text{Osc}(\delta_1, \omega) \geqq \epsilon/(M + 1)\}$$

has $P(\Omega_3) < \epsilon/3$. Now we define $\delta(t) = \min\{\delta_2(t), \delta_1\}$ for all t in $[a, b)$. If Π is a (δ, δ^*) belated partition in $[a, b)$, the set $\Omega_M \cup \Omega_\Pi \cup \Omega_3$ has P-measure less than ϵ, and for all ω not in that set

$$\begin{aligned} | S(\Pi; f; z^1, z^2, v^1, \ldots, v^q)(\omega) | &\leqq \sum_{j=1}^{k} f(\tau_j)(\Delta_j z)^2 \,\text{Osc}(\delta, \omega) \\ &< \epsilon. \end{aligned}$$

So the conclusion of the theorem holds if $z^1 = z^2$. If this is not satisfied, we use the device in the last paragraph of the proof of Theorem (3-2) to complete the proof.

(8-11) *THEOREM Let v^1 and v^2 be processes on $[a, b]$ such that the sample functions of v^2 are a.s. continuous and those of v^1 satisfy a Lipschitz condition*

$$| v^1(t) - v^1(s) | \leqq L(\omega) | t - s | \qquad (s, t \text{ in } [a, b]),$$

where L is an a.s. finite-valued r.v. [In particular, $v^1(\mathbf{t})$ may be $\mathbf{t}$.] Let g be a process on T such that $g(\boldsymbol{\tau}, \omega)$ is a.s. Lebesgue integrable over $[a, b]$. Then

$$\int_a^b g(t)\ dv^1(t)\ dv^2(t) = 0.$$

Without loss of generality we may assume $v^2(\mathbf{t}, \omega)$ continuous and $L(\omega)$ finite for all ω. We define

$$v^3(t, \omega) = v^1(t, \omega)/L(\omega),$$

$$v^4(t, \omega) = v^2(t, \omega)L(\omega).$$

Then $v^3(\mathbf{t}, \omega)$ satisfies a Lipschitz condition of constant 1, and for all partitions Π

$$\sum | g(\tau_j)\ \Delta_j v^1\ \Delta_j v^2 | \leqq \sum | g(\tau_j)\ \Delta_j t\ \Delta_j v^4 |.$$

As (δ, δ^*) shrinks, $\sum | g(\tau_j)\ \Delta_j t |$ converges in probability to the r.v.

$$J(\boldsymbol{\omega}) = \int_a^b | g(t) |\, dt.$$

We define $\operatorname{Osc}(\delta, \omega)$ as in the proof of Theorem (5-4); it is a r.v. that a.s. converges to 0 with δ. The rest of the proof is a trivial modification of that of Theorem (8-10), with $\Delta_j t$ replacing $(\Delta_j z)^2$.

(8-12) *THEOREM Let the standing hypotheses* (2-1) *be satisfied, and let* $z^1(t) - z^1(s)$ *and* $z^2(t) - z^2(s)$ *be conditionally independent given* $\mathfrak{F}_s$ *whenever* $a \leqq s \leqq t \leqq b$. *Let f be a measurable process on T adapted to* $\mathfrak{F}_{\boldsymbol{\tau}}$ *such that* $| f(\mathbf{t}, \omega) |^2$ *is a.s. Lebesgue integrable over* $[a, b]$. *Then*

$$\int_a^b f(t)\, dz^1(t)\, dz^2(t) = 0.$$

We consider first the case in which $\| f(\mathbf{t}) \|_2^2$ is integrable over $[a, b]$.

If δ^* is positive and $0 < \delta_1 < 1$, and $\delta(\mathbf{t}) > 0$ a.e. on $[a, b)$ and $\delta(\mathbf{t}) < \delta_1$ on $[a, b)$, as in the first part of the proof of Theorem (5-7) we find that for every (δ, δ^*) belated partition Π in $[a, b)$

$$\begin{aligned} \left\| \sum_{j=1}^m f(\tau_j)\ \Delta_j z^1\ \Delta_j z^2 \right\|_2 &\leqq 2 \sum_{j=1}^m \| f(\tau_j) \|_2 K^2\ (\Delta_j t)^2 \\ &\quad + \left\{ \sum_{j\ 1}^m \| f(\tau_j) \|_2^2 K^2\ (\Delta_j t)^2 \right\}^{1/2} \\ &< 2\delta_1 K^2 \sum_{j=1}^m \| f(\tau_j) \|_2\ \Delta_j t \\ &\quad + K\delta_1^{1/2} \left\{ \sum_{j=1}^m \| f(\tau_j) \|_2^2\ \Delta_j t \right\}^{1/2}. \end{aligned}$$

As (δ, δ^*) shrinks, the sums converge to finite numbers, while δ_1 is arbitrarily small, so $S(\Pi; f; z^1, z^2)$ converges in L_2-distance to 0.

If $f(\boldsymbol{\tau}, \omega)^2$ is a.s. integrable, as in the proof of Theorem (8-3), for each positive ϵ there is a function $I_n \cdot f$ such that $E([I_n \cdot f]^2)$ is integrable and $I_n(t, \omega) f(t, \omega) = f(t, \omega)$ $(a \leqq t \leqq b)$ on a set $\Omega_{b,n}$ with $P(\Omega_{b,n}) > 1 - \epsilon/4$. So for all partitions Π in $[a, b]$

$$\bar{\rho}(S(\Pi; I_n \cdot f; z^1, z^2), \qquad S(\Pi; f; z^1, z^2)) < \epsilon/4.$$

By the part of the proof already completed, ultimately

$$\bar{\rho}(S(\Pi; I_n \cdot f; z^1, z^2), 0) < \epsilon/4,$$

whence

$$\bar{\rho}(S(\Pi; f; z^1, z^2), 0) < \epsilon/2.$$

The extension of Theorem (6-7) to Itô-belated integrals is easy; the proof needs little change. Likewise Eq. (7-1) holds for Itô-belated integrals, the proof being almost unchanged.

The estimates of the integrals in Corollaries (2-11) and (3-11) are valid for Itô-belated integrals also; the proofs need only notational changes.

It is not true that for every f and z, the existence of the belated integral of f with respect to z implies the existence of the Itô-belated integral. If $z(\mathbf{t}, \boldsymbol{\omega})$ is independent of ω, continuous, and nowhere differentiable on $[a, b]$, the belated integral $\int 1\, dz$ exists and the Itô-belated integral does not. There is therefore some interest in finding additional hypotheses on f and $z^1, \ldots, z^q$ which, together with the existence of the belated integral, imply the existence of the Itô-belated integral. We leave the proof of the following statement as an exercise.

(8-13) *Let the standing hypotheses* (2-1) *be satisfied.*

(i) *If the belated integral*

$$(*) \qquad \int_a^b f(t)\, dz(t)$$

exists and f is L_2-bounded, the Itô-belated integral exists.

(ii) *If the belated integral*

$$(**) \qquad \int_a^b f(t)\, dz^1(t)\, dz^2(t)$$

exists and f is L_1-bounded, the Itô-belated integral exists.

(iii) *If f is a.s. sample-function bounded, and either of the belated integrals* (*), (**) *exists, so does the corresponding Itô-belated integral.*

IV Continuity, Chain Rule, and Substitution

1 Continuity of sample functions

Two of the most important formulas of elementary calculus are the "chain rule" for differentiating composite functions, and the substitution theorem, which allows us to change the independent variable in a definite integral. Another important theorem asserts that an indefinite integral, over an interval $[a, t]$, is a continuous function of the upper limit t. In this chapter all three will be extended to belated integrals; in the last section, recommended only for readers interested in generality, they are extended to the Itô-belated integral.

In the theorems in Chapter III we always assumed the standing hypotheses III(2-1), but often added the hypothesis that the noise-processes z had a.s. continuous sample functions. From this point on this continuity will be assumed in all theorems, so we restate the standing hypotheses to include it.

(1-1) *STANDING HYPOTHESES AND NOTATION*

(i) *$(\Omega, \mathcal{A}, P)$ is a probability triple; T is a set of real numbers; and $[a, b]$ is an interval contained in T.*

(ii) *$\{\mathfrak{F}_\tau : \tau \in T\}$ is a family of σ-subalgebras of $\mathcal{A}$ such that if τ and τ' are in T and $\tau < \tau'$, then $\mathfrak{F}_\tau \subset \mathfrak{F}_{\tau'}$. When convenient we shall use $\mathfrak{F}[\tau]$ to denote $\mathfrak{F}_\tau$.*

(iii) *Every process denoted by z, with or without affix, will be a real-valued process adapted to $\mathfrak{F}_{\boldsymbol{\tau}}$, having a.s. continuous sample functions $z(\mathbf{t}, \omega)$, and satisfying with some positive K the inequalities*

$$| E(z(t) - z(s) \mid \mathfrak{F}_s) | \leqq K(t - s),$$

$$E([z(t) - z(s)]^2 \mid \mathfrak{F}_s) \leqq K(t - s)$$

a.s. whenever $a \leqq s \leqq t \leqq b$.

(iv) *$f(\boldsymbol{\tau}, \omega)$ is a measurable real-valued process on T adapted to $\mathfrak{F}_{\boldsymbol{\tau}}$.*

Let $f(\boldsymbol{\tau}, \omega)$ be integrable with respect to $z^1(\mathbf{t}, \omega), \ldots, z^q(\mathbf{t}, \omega)$ over $[a, b]$. Then the integral

$$J(t, \omega) = \int_a^t f(s, \omega)\, dz^1(s, \omega) \cdots dz^q(s, \omega) \tag{1-2}$$

exists for each t in $[a, b]$. This process J we call an **indefinite integral** of f with respect to $(z^1, \ldots, z^q)$. Such an indefinite integral is a special case of a finite set of sums of indefinite integrals of the form

$$x^i(t, \omega) = x^i(a, \omega) + \int_a^t f^i(s, \omega)\, ds + \sum_{\rho=1}^{r} \int_a^t g_\rho{}^i(s, \omega)\, dz^\rho(s, \omega) \tag{1-3}$$

$$+ \sum_{\rho,\sigma=1}^{r} \int_a^t g^i_{\rho,\sigma}(s, \omega)\, dz^\rho(s, \omega)\, dz^\sigma(s, \omega)$$

$$(i = 1, \ldots, n).$$

We know that for each t it is possible to choose a strict version for each integral (if the integral exists). However, this kind of strictness is more than is needed to attain the consistency property described in II(1-3). We use instead a weaker, but adequate and more convenient, concept of strictness for systems (1-3), and in particular for indefinite integrals (1-2).

(1-4) *DEFINITION A vector process* $(x^1(\mathbf{t}, \omega), \ldots, x^n(\mathbf{t}, \omega))$ *on* $[a, b]$ *is a* **strict** *version of the right member of the equation*

$$x^i(t, \omega) = x^i(a, \omega) + \int_a^t f^i(s, \omega)\, ds$$

$$+ \sum_{\rho=1}^{r} \int_a^t g_\rho{}^i(s, \omega)\, dz^\rho(s, \omega)$$

$$+ \sum_{\rho,\sigma=1}^{r} \int_a^t g^i_{\rho,\sigma}(s, \omega)\, dz^\rho(s, \omega)\, dz^\sigma(s, \omega),$$

or satisfies the equation **strictly**, *if for each t in $[a, b]$ the equation holds a.s. and for each ω_0 such that all $z^\rho(\mathbf{t}, \omega_0)$ are Lipschitzian and each integrand $f^i(\mathbf{t}, \omega_0)$, $g_\rho{}^i(\mathbf{t}, \omega_0)$, $g^i_{\rho,\sigma}(\mathbf{t}, \omega_0)$ is Riemann integrable over $[a, b]$, the equation is valid for all t in $[a, b]$ if ω is replaced by ω_0 and all integrals interpreted as ordinary Riemann–Stieltjes integrals.*

It is easy to see that for each of the ω_0 specified in this definition all the Riemann–Stieltjes integrals exist, and the second-order integrals have the value 0. We can prove that the first-order integrals are continuous functions of t for such ω_0. In fact, we shall prove a stronger statement.

(1-5) *For each ω_0 such that $z^1(\mathbf{t}, \omega_0), \ldots, z^q(\mathbf{t}, \omega_0)$ are Lipschitzian and the integral in* (1-2) *exists as a deterministic (belated) integral, $J(\mathbf{t}, \omega_0)$ is continuous.*

If we define Ω_0 to consist of the single point ω_0, with probability measure 1, the standing hypotheses (1-1) still hold with Ω_0 in place of Ω. We omit the ω_0 from the notation. There is a positive δ such that if Π is any belated partition of $[a, b]$ with mesh $\Pi < \delta$,

$$| S(\Pi; f; z^1, \ldots, z^q) - J(b) | < \epsilon.$$

We choose points $t_0 = a < t_1 < \cdots < t_h = b$ with all $t_j - t_{j-1} < \delta$. For each t in $[a, b]$, there is a j for which $t_{j-1} < t \leqq t_j$. We construct a Cauchy partition

(1-6) $$\Pi(t) = (t_0, t_1, \ldots, t_{j-1}, t; t_0, t_1, \ldots, t_{j-1});$$

this has mesh $\Pi(t) < \delta$. So by Theorem II(4-3),

$$| S(\Pi(t); f; z^1, \ldots, z^q) - J(t) | \leqq \epsilon.$$

Clearly the sum

$$S(\Pi(t);f;z^1,\ldots,z^q) = \sum_{i=0}^{j-2} f(t_i)\prod_{k=1}^{q}[z^k(t_{i+1}) - z^k(t_i)] + f(t_{j-1})\prod_{k=1}^{q}[z^k(t) - z^k(t_{j-1})]$$

is continuous on $[a, b]$. So $J(t)$ can be uniformly approximated to within an arbitrary positive ϵ by continuous functions, and (1-5) is established.

The object of this section is to prove that if the f and z^k satisfy the hypotheses which in Chapter III were shown to guarantee the existence of the integral in (1-2), we can choose a strict version of the indefinite integral in which *every* sample function $F(\mathbf{t}, \omega)$ is continuous, not merely those corresponding to the ω_0 of Definition (1-4). We prove this separately for $q = 1$ and for $q = 2$. For $q > 2$, we do not need to prove it, since the integral is then 0 in all cases of interest to us.

(1-7) *THEOREM Let the standing hypotheses* (1-1) *be satisfied. Let* $f(\boldsymbol{\tau}, \omega)$ *be a process that can be represented in the form* $b(\boldsymbol{\tau}, \omega)f_1(\boldsymbol{\tau}, \omega)$, *where* b *and* f_1 *are adapted to* $\mathfrak{F}_{\boldsymbol{\tau}}$ *and are measurable, and* b *is a.s. sample-function bounded and is continuous in probability a.e. in* $[a, b]$, *and* f_1 *is* L_2*-bounded and is* L_2*-continuous a.e. in* $[a, b]$. *Then there is a process* $F(\mathbf{t}, \omega)$ *on* $[a, b]$ *that has continuous sample functions and is a strict version of the indefinite integral*

(1-8) $$\int_a^t f(s, \omega)\, dz(s, \omega) \qquad (a \leqq t \leqq b).$$

We first prove this under the supplementary hypothesis

(1-9) the process b is bounded.

In this case $f(\boldsymbol{\tau}, \omega)$ is L_2-bounded and is L_2-continuous a.e. in $[a, b]$, as we showed in the proof of Theorem III(4-1).

For each t in $[a, b]$, we choose a version $F_1(t, \omega)$ of the integral (1-8). If M is an upper bound for $\| f(\tau) \|_2$ on T, by Corollary III(2-11) we have for $a \leqq t' \leqq t'' \leqq b$

$$\| F_1(t'', \omega) - F_1(t', \omega) \|_2 \leqq C\left\{\int_{t'}^{t''} M^2\, ds\right\}^{1/2}.$$

The right member tends to 0 with $t'' - t'$. So F_1 is L_2-continuous, which implies that $F_1(t, \omega)$ is continuous in probability on $[a, b]$.

For each positive integer n, we choose a bounded simple process $g_n(\mathbf{t}, \omega)$ on $[a, b]$ such that

(1-10) $$\int_a^b E([f - g_n]^2)\,dt < 2^{-4n}/(C + K[b-a]^{1/2})^2;$$

this is possible by Theorem I(3-8). We define

(1-11) $$h_n(t, \omega) = f(t, \omega) - g_n(t, \omega).$$

If $a = t_0 < t_1 < \cdots < t_h = b$, and g_n has the constant value v_j on $[t_j, t_{j+1})$ for $j = 0, \ldots, h-1$, and $t_{j-1} < t \leqq t_j$, the r.v.

$$G_n(t, \omega) = \sum_{i=0}^{j-2} v_i(\omega)[z(t_{i+1}, \omega) - z(t_i, \omega)] + v_{j-1}(\omega)[z(t, \omega) - z(t_{j-1}, \omega)]$$

is a version of the integral

$$G_n(t, \omega) = \int_a^t g_n(s, \omega)\,dz(s, \omega),$$

as stated in II(4-7). Since almost all sample functions of z are continuous, we have the following statement.

(1-12) For all ω except those in a negligible set N_1, all the functions $G_n(\mathbf{t}, \omega)$ $(n = 1, 2, 3, \ldots)$ are continuous on $[a, b]$.

Let $H_n(t, \omega)$ be a version of the indefinite integral

(1-13) $$H_n(t, \omega) = \int_a^t h_n(s, \omega)\,dz(s, \omega).$$

For each t in $[a, b]$, by (1-11) both $F_1(t, \omega)$ and $G_n(t, \omega) + H_n(t, \omega)$ are versions of the integral (1-8), so they differ only on a negligible subset of Ω. Now let D be a denumerable subset of $[a, b]$ that contains a and b and is dense in $[a, b]$. Then by the preceding sentence,

(1-14) for all ω except those in a negligible set N_2, the equation

$$G_n(t, \omega) = F_1(t, \omega) - H_n(t, \omega)$$

holds for all t in D and all positive integers n.

We now prove an auxiliary statement.

(1-15) For each positive integer n, and for $\sigma = +1$ or -1, the process $\psi_{n,\sigma}$ defined by

$$K\int_a^t |\, h_n(s, \omega) \,|\, ds + \sigma H_n(t, \omega)$$

is a submartingale.

The first integral in this sum exists by Theorem III(2-3); for each t we fix on any version of it. Suppose that $\phi(\mathbf{t}, \omega)$ is a bounded simple process with values $v_0(\omega), \ldots, v_{h-1}(\omega)$ on the respective intervals $[t_0, t_1), \ldots, [t_{h-1}, t_h)$, where $t_0 = a$ and $t_h = b$. For each t, we define $\Phi(t, \omega)$ to be a version of

$$K \int_a^t |\phi(s, \omega)| \, ds + \sigma \int_a^t \phi(s, \omega) \, dz(s, \omega).$$

Then if $a \leqq t < t' \leqq b$, the points t, t' are contained in intervals $[t_p, t_{p+1})$, $[t_q, t_{q+1}]$ respectively. If $[\alpha, \beta)$ is any one of the intervals

$$(1\text{-}16) \qquad [t, t_{p+1}), \qquad [t_{p+1}, t_{p+2}), \qquad \ldots, \qquad [t_q, t'),$$

the simple process ϕ has a constant value $v(\omega)$ on $[\alpha, \beta)$. Then

$$\begin{aligned} E(\Phi(\beta) - \Phi(\alpha) \mid \mathfrak{F}_t) &= E(E(\Phi(\beta) - \Phi(\alpha) \mid \mathfrak{F}_\alpha) \mid \mathfrak{F}_t) \\ &= E(E(K|v|(\beta - \alpha) + \sigma v[z(\beta) - z(\alpha)] \mid \mathfrak{F}_\alpha) \mid \mathfrak{F}_t) \\ &\geqq K(\beta - \alpha) E(|v| \mid \mathfrak{F}_t) \\ &\quad - E(|v| \cdot | E(z(\beta) - z(\alpha) \mid \mathfrak{F}_\alpha)| \mid \mathfrak{F}_t) \\ &\geqq 0 \quad \text{a.s.} \end{aligned}$$

If we add these inequalities for all $[\alpha, \beta)$ in the set (1-16), we obtain

$$E(\Phi(t') - \Phi(t) \mid \mathfrak{F}_t) \geqq 0,$$

and Φ is a submartingale.

Given n and σ, by Theorem I(3-8), there is a sequence $\phi_1, \phi_2, \ldots$ of bounded simple processes such that

$$\lim_{j \to \infty} \int_a^b E(|\phi_j - h_n|^2) \, dt = 0.$$

For each j, we define

$$\Phi_j(t, \omega) = K \int_a^t |\phi_j(s, \omega)| \, ds + \sigma \int_a^t \phi_j(s, \omega) \, dz(s, \omega).$$

This is a submartingale, by the preceding proof. By Corollary III(2-11),

$$(1\text{-}17)\qquad \left\|\sigma\int_a^t \phi_j(s,\omega)\,dz(s,\omega) - \sigma\int_a^t h_n(s,\omega)\,dz(s,\omega)\right\|_2^2$$

$$\leqq C^2\left\{\int_a^t E(|\phi_j - h_n|^2\,ds)\right\},$$

which tends to 0 as $j\to\infty$. Also, by the Cauchy–Buniakowski–Schwarz inequality,

$$(1\text{-}18)\qquad \left\|\int_a^t |\phi_j(s,\omega)|\,ds - \int_a^t |h_n(s,\omega)|\,ds\right\|_2^2$$

$$\leqq E\left(\left[\int_a^t |\phi_j(s,\omega) - h_n(s,\omega)|\,ds\right]^2\right)$$

$$\leqq E\left(\left\{\int_a^t |\phi_j(s,\omega) - h_n(s,\omega)|^2\,ds\right\}\left\{\int_a^t 1\,ds\right\}\right),$$

which tends to 0 as $j\to\infty$. So at each t in $[a, b]$, the r.v.'s $\Phi_j(t,\omega)$ converge to $\psi_{n,\sigma}(t,\omega)$ in L_2-norm. This implies that $\psi_{n,\sigma}$ is also a submartingale, and (1-15) is established.

Just as we proved (1-17) and (1-18), we prove

$$\left\|\int_a^t h_n(s,\omega)\,dz(s,\omega)\right\|_2 \leqq C\left\{\int_a^t E([h_n(s,\omega)]^2)\,ds\right\}^{1/2},$$

$$\left\|\sigma\int_a^t |h_n(s,\omega)|\,ds\right\|_2 \leqq (b-a)^{1/2}\left\{\int_a^t E([h_n(s,\omega)]^2)\,ds\right\}^{1/2}.$$

By (1-11) and (1-10), these yield

$$E(\psi_{n,\sigma}^+) \leqq E\left(\left|K\int_a^t |h_n|\,ds + \sigma\int_a^t h_n\,dz\right|\right)$$

$$\leqq \left\|K\int_a^t |h_n|\,ds + \sigma\int_a^t h_n\,dz\right\|_2$$

$$\leqq 2^{-4n}.$$

If we define

$$A_{n,\sigma} = \{\omega\in\Omega : \sup\{\psi_{n,\sigma}(t,\omega) : t\in D\} > 2^{-n}\},$$

by Theorem I(3-11) $P(A_{n,\sigma}) \leqq 2^{-n}$. The set

$$A_n = \{\omega \in \Omega : \sup\{|H_n(t, \omega)| : t \in D\} > 2^{-n}\}$$

is contained in the union of $A_{n,1}$ and $A_{n,-1}$ so

$$P(A_n) \leqq 2^{1-n}.$$

For each positive integer k, the union of the A_n with $n > k$ has P-measure at most 2^{1-k}, and except on this union we have

$$|H_n(t, \omega)| \leqq 2^{-n} \qquad (t \in D; \quad n = k + 1, k + 2, \ldots),$$

which implies the uniform convergence on D of $H_n(t, \omega)$ to 0 for all ω not in that union. Therefore

(1-19) for all ω except those in a negligible set N_3, $H_n(\mathbf{t}, \omega)$ converges to 0 uniformly on D as $n \to \infty$.

Statements (1-12), (1-14), and (1-19) imply that for all ω except those in the negligible set $N = N_1 \cup N_2 \cup N_3$, the $G_n(\mathbf{t}, \omega)$ are continuous on $[a, b]$, and $G_n(t, \omega)$ converges to $F_1(t, \omega)$ uniformly on D. Let ω be in $\Omega \backslash N$ and let ϵ be positive. We fix n such that $|G_n(t, \omega) - F_1(t, \omega)| < \epsilon/3$ on D. Because $G_n(\mathbf{t}, \omega)$ is continuous on the closed interval $[a, b]$, there is a $\delta > 0$ such that if s, t are in D and $|s - t| < \delta$, then $|G_n(s, \omega) - G_n(t, \omega)| < \epsilon/3$. This and the preceding inequality imply $|F_1(s, \omega) - F_1(t, \omega)| < \epsilon$. So $F_1(\mathbf{t}, \omega)$ is uniformly continuous on D. It is therefore the restriction to D of a function $F_2(\mathbf{t}, \omega)$ continuous on $[a, b]$. For ω in N, we define $F_2(\mathbf{t}, \omega) = 0$.

Let t be any point of $[a, b]$, and let $t_1, t_2, \ldots$ be a sequence of points of D tending to t. For each t_j, $F_1(t_j, \omega)$ and $F_2(t_j, \omega)$ are equal for all ω in $\Omega \backslash N$. Since F_1 is continuous in probability on $[a, b]$, $F_2(t_j, \boldsymbol{\omega})$ converges to $F_1(t, \boldsymbol{\omega})$ in probability. But since the sample functions of F_2 are continuous, $F_2(t_j, \omega)$ converges to $F_2(t, \omega)$ for each ω. Hence $F_1(t, \omega) = F_2(t, \omega)$ except on a negligible set, and $F_2(\mathbf{t}, \boldsymbol{\omega})$ is a version of the integral (1-8) with continuous sample paths.

Let Ω_0 be the set of all ω_0 in Ω such that $z(\mathbf{t}, \omega_0)$ is Lipschitzian and the integral

$$\int_a^b f(t, \omega_0)\, dz(t, \omega_0)$$

exists. For each ω_0 in Ω_0, define

$$J(t, \omega_0) = \int_a^t f(s, \omega_0)\, dz(s, \omega_0) \qquad (a \leqq t \leqq b).$$

By (1-5), this is a continuous function of t on $[a, b]$. For each t in D, if

$\Pi_1, \Pi_2, \ldots$ is a sequence of Cauchy partitions of $[a, t]$ with mesh Π_n tending to 0, $S(\Pi_n; f; z)$ tends to $F_2(t, \boldsymbol{\omega})$ in probability because $F_2(t, \boldsymbol{\omega})$ is a version of the integral, and tends to $J(t, \omega_0)$ at each ω_0 in Ω_0. By Lemma II(2-20),

$$F_2(t, \omega_0) = J(t, \omega_0) \tag{1-20}$$

except on a negligible subset $N(t)$ of Ω_0. The union $N(D)$ of the $N(t)$ for all t in D is a negligible set. If $\omega \in \Omega_0 \backslash N(D)$, (1-20) holds for all t in D, and since both members are continuous it holds at all t in $[a, b]$. Hence if for all t in $[a, b]$ we define

$$\begin{aligned} F(t, \omega) &= J(t, \omega) \qquad (\omega \in \Omega_0) \\ &= F_2(t, \omega) \qquad (\omega \in \Omega \backslash \Omega_0), \end{aligned}$$

the processes F and F_2 differ only on a negligible set, so F is a version of the integral (1-8). It has continuous sample functions and coincides with J on Ω_0, so it has the properties asserted in the theorem. The theorem is proved under the supplementary hypothesis (1-9).

If the hypotheses of the theorem hold, there is an a.s. finite-valued r.v. $B(\boldsymbol{\omega})$ such that for all ω in Ω,

$$| b(\tau, \omega) | \leqq B(\omega) \qquad (\tau \in T).$$

For each positive integer M, we define

$$\Omega_M = \{\omega \in \Omega : B(\omega) \leqq M\},$$

and we define

$$\begin{aligned} b_M(\tau, \omega) &= \operatorname{mid}\{b(\tau, \omega), M, -M\}, \\ f_M(\tau, \omega) &= b_M(\tau, \omega) f_1(\tau, \omega). \end{aligned}$$

Then by the part of the proof already completed, there is a process $F_M(\mathbf{t}, \boldsymbol{\omega})$ with continuous sample functions that is a strict version of

$$\int_a^t f_M(s, \boldsymbol{\omega})\, dz(s, \boldsymbol{\omega}) \qquad (a \leqq t \leqq b).$$

By Corollary II(2-22), $F_{M+1}(t, \omega)$ and $F_M(t, \omega)$ are identical on all but a negligible subset of Ω_M. We change F_{M+1} to agree with F_M on this subset; it is still a strict version of the integral of f_M with continuous simple functions. At every point ω of $\bigcup \Omega_M$, all F_M with sufficiently large M are defined and have the same value. This we take as $F(t, \omega)$. On the negligible set $\Omega \backslash \bigcup \Omega_M$, we choose $F(t, \omega)$ to be defined by the Riemann–Stieltjes integral (1-8) if $z(\mathbf{t}, \omega)$ is Lipschitzian and f is integrable, and to be 0 otherwise. This F has the required properties.

We now turn to second-order integrals.

(1-21) *THEOREM Let the standing hypotheses (1-1) be satisfied. Let $f(\boldsymbol{\tau}, \omega)$ be a process that can be represented in the form $b(\boldsymbol{\tau}, \omega) f_1(\boldsymbol{\tau}, \omega)$, where b and f_1 are measurable and are adapted to $\mathfrak{F}_{\boldsymbol{\tau}}$, and b has a.s. bounded sample functions and is continuous in probability a.e. in $[a, b]$, and f_1 is L_1-bounded and is L_1-continuous a.e. in $[a, b]$. Then there exists a process $F(\mathbf{t}, \omega)$ with continuous sample functions that is a strict version of the integral*

$$\int_a^t f(s, \omega)\, dz^1(s, \omega)\, dz^2(s, \omega) \qquad (a \leqq t \leqq b). \tag{1-22}$$

We prove this first under three additional hypotheses:

$$b \text{ is bounded}, \tag{1-23}$$

$$f \geqq 0, \tag{1-24}$$

$$z^1 = z^2. \tag{1-25}$$

We let z denote the common value of z^1 and z^2.

For each t in $[a, b]$ we choose a version $F_1(t, \omega)$ of the integral (1-22). If M is an upper bound for $\| f(t) \|_1$, by Corollary III(3-11) we have for $a \leqq t' \leqq t'' \leqq b$

$$\| F_1(t'', \omega) - F_1(t', \omega) \|_1 \leqq K \int_{t'}^{t''} M\, dt.$$

The right member tends to 0 with $t'' - t'$. So F_1 is L_1-continuous, which implies that $F_1(\mathbf{t}, \omega)$ is continuous in probability on $[a, b]$.

For each positive integer n, we choose a bounded simple process $g_n(\mathbf{t}, \omega)$ such that

$$\int_a^b E(| f - g_n |)\, dt < 1/2^{2n}K; \tag{1-26}$$

this is possible by Corollary I(3-9). We again define

$$h_n(t, \omega) = f(t, \omega) - g_n(t, \omega). \tag{1-27}$$

Suppose that $a = t_0 < t_1 < \cdots < t_h = b$, and that g_n has the constant value v_j on $[t_j, t_{j+1})$ for $j = 0, 1, \ldots, h - 1$. By Theorem III(3-2), g_n is integrable with respect to (z, z) over $[a, b]$. If $t_{j-1} < t \leqq t_j$, the integral

$$G_n(t, \omega) = \int_a^t g_n(s, \omega)\, [dz(s, \omega)]^2$$

is the sum of the integrals over $[t_0, t_1], [t_1, t_2], \ldots, [t_{j-1}, t]$. So to show that a version with continuous sample functions exists, it is enough to show that this is true on each subinterval $[t_{j-1}, t_j]$. To simplify notation we rename this interval $[c, d]$, and we denote the constant value of g_n on it by v. Let t be any point in $[c, d]$, and let $\Pi = (t_1', \ldots, t'_{m+1}; t_1', \ldots, t_m')$ be any Cauchy partition of $[c, t]$. Then

$$S(\Pi; g_n; z, z) = \sum_{j=1}^{m} v[z(t'_{j+1}) - z(t_j')]^2$$

$$= \sum_{j=1}^{m} v[z(t'_{j+1})^2 - z(t_j')^2] - 2 \sum_{j=1}^{m} vz(t_j')[z(t'_{j+1}) - z(t_j')].$$

The first sum is $v[z(t)^2 - z(c)^2]$ for all Π, and is a.s. continuous. By Theorem III(2-3), $vz(\mathbf{t}, \omega)$ is integrable with respect to z over $[c, t]$. So as mesh Π tends to 0, the second sum in the right member converges in probability to any version of

$$\int_a^t v(\omega)z(s, \omega)\, dz(s, \omega).$$

By Theorem (1-7), we can choose a version of this integral with continuous sample functions. Therefore:

(1-28) There is a process $G_n(\mathbf{t}, \omega)$ with a.s. continuous sample functions such that for each t in $[a, b]$, $G_n(t, \omega)$ is a version of

$$\int_a^t g_n(t, \omega)\, [dz(t, \omega)]^2.$$

Define

$$H_n(t, \omega) = \int_a^t h_n(s, \omega)\, [dz(s, \omega)]^2, \qquad \text{(1-29)}$$

choosing and fixing any version of the integral.

We also fix on any version $H^*(\omega)$ of the integral

$$\int_a^b | h_n(t, \omega) |\, [dz(t, \omega)]^2.$$

By Corollary III(3-11) and (1-26),

$$|| H^*(\omega) ||_1 \leqq 2^{-2n}.$$

Let A_n be the subset of Ω on which

$$| H^*(\omega) | \geqq 2^{-n};$$

then by the preceding inequality, $P(A_n) \leqq 2^{-n}$. By a familiar argument,

$$\lim_{n\to\infty} \int_a^b | h_n(t, \omega) | [dz(t, \omega)]^2 = 0 \tag{1-30}$$

for all ω except those in a negligible subset of Ω. For each fixed t and n, let Π_j be a belated partition of $[a, b]$ with t as a division point, and let $\Pi_j(t)$ be the partition of $[a, t]$ obtained by using those division points of Π_j up to and including t, together with the associated evaluation points. Then

$$| S(\Pi_j(t); h_n; z, z) | \leqq S(\Pi_j; | h_n |; z, z).$$

We can choose a subsequence of the Π_j such that the two members of this inequality converge a.s. to the respective members of the inequality

$$| H_n(t, \omega) | \leqq \int_a^b | h_n(t, \omega) | [dz(t, \omega)]^2. \tag{1-31}$$

So (1-31) holds except on a negligible set. Therefore if D is a denumerable set dense in $[a, b]$, there is a negligible set such that for each ω not in that set (1-31) is valid for all t in D and all positive integers n. From this and (1-30),

(1-32) for all ω except those in a negligible set N_3, $H_n(\mathbf{t}, \omega)$ converges to 0 uniformly on D.

Statement (1-14) holds for the present G_n, F_1, and H_n, with the same proof. Statements (1-28) and (1-32) correspond to (1-12) and (1-19). We can now repeat the proof of Theorem (1-7), beginning immediately after (1-19), to establish the conclusion of Theorem (1-21), subject to the additional hypotheses (1-23), (1-24), and (1-25).

Under the hypotheses of Theorem (1-21) and the supplementary hypothesis (1-23), we define

$$z' = [z^1 + z^2]/2, \qquad z'' = [z^1 - z^2]/2.$$

Then for every t in $[a, b]$ and belated partition Π of $[a, t]$ we have

$$\begin{aligned} S(\Pi; f; z^1, z^2) &= S(\Pi; f^+; z', z') - S(\Pi; f^+; z'', z'') \\ &\quad - S(\Pi; f^-; z', z') + S(\Pi; f^-; z'', z''). \end{aligned} \tag{1-33}$$

By the part of the proof already concluded, there are four processes, $F_3, \ldots, F_6$ on $[a, b]$ with continuous sample functions such that for each t, the right members of (1-33) converge in probability to the values of these four processes at t. Hence the left member of (1-33) converges in probability to

$$F_2(t, \omega) = F_3(t, \omega) - F_4(t, \omega) - F_5(t, \omega) + F_6(t, \omega),$$

and F_2 has continuous sample functions. The amendment to obtain a process with continuous sample functions that is a strict version of the indefinite integral (1-22), and the removal of the boundedness hypothesis (1-23), are exactly as in the last part of the proof of Theorem (1-7).

(1-34) *THEOREM* *Under the hypotheses of Theorem* (1-21), *there exists a process* $F(\mathbf{t}, \omega)$ *with continuous sample functions that is a strict version of the integral*

$$\int_a^t f(s, \omega)\, ds \qquad (a \leqq t \leqq b).$$

The proof is a simplification of the first part of the proof of Theorem (1-21), wherein $z^1 = z^2$.

We leave the proof of the following statement as an exercise.

(1-35) *If the hypotheses of Theorem* (1-7) *are weakened by dropping the requirement that* z *has a.s. continuous sample functions, the conclusions hold in the weakened form in which the sample functions of* F *are a.s. bounded (instead of a.s. continuous). In particular, the sample functions of* z *are a.s. bounded. Likewise, if the hypotheses of Theorem* (1-21) *are weakened by dropping the requirement that* z^1 *and* z^2 *have a.s. continuous sample functions, the conclusion holds in the weakened form in which* F *is a.s. sample-function bounded.*

2 Differentiation of a composite function

As we have mentioned before, the "stochastic calculus" we are developing is an integral calculus. We have no corresponding theory of derivation, to produce the interplay of theories that gives such power and beauty to the ordinary differential and integral calculus. Perhaps future problems will motivate a differential stochastic calculus shaped by those problems. Until then, it is reasonable to content ourselves with a fragmentary differential calculus, ancillary to the integral calculus instead of coordinate with it, and to develop some formulas that have proved themselves useful. Outstanding among these is the differentiation lemma of Itô [4].

It is convenient to introduce the notation of differentials, even though their meanings are assigned by means of integrals. Suppose that $z^1, \ldots, z^r$ is a set of processes on $[a, b]$, and that $f^i(\boldsymbol{\tau}, \omega)$, $g_\rho{}^i(\boldsymbol{\tau}, \omega)$, and $h^i_{\rho,\sigma}(\boldsymbol{\tau}, \omega)$ $(i = 1, \ldots, n;\ \rho, \sigma$ in $\{1, \ldots, r\})$ are real-valued processes on T. The statement that a set of processes $x^1(\mathbf{t}, \omega), \ldots, x^n(\mathbf{t}, \omega)$ satisfies the system of equations

$$dx^i = f^i(t, \omega)\, dt + \sum_{\rho=1}^{r} g_\rho{}^i(t, \omega)\, dz^\rho(t, \omega) + \sum_{\rho,\sigma=1}^{r} h^i_{\rho,\sigma}(t, \omega)\, dz^\rho(t, \omega)\, dz^\sigma(t, \omega) \tag{2-1}$$

with initial values

$$x^i(a, \omega) = x_0{}^i(\omega) \tag{2-2}$$

shall mean merely that for each t in $[a, b]$ the equations

$$x^i(t, \omega) = x_0{}^i(\omega) + \int_a^t f^i(s, \omega)\, ds + \sum_{\rho=1}^{r} \int_a^t g_\rho{}^i(s, \omega)\, dz(s, \omega) + \sum_{\rho,\sigma=1}^{r} \int_a^t h^i_{\rho,\sigma}(s, \omega)\, dz^\rho(s, \omega)\, dz^\sigma(s, \omega) \tag{2-3}$$

are satisfied; that is, a sum of versions of the terms in the right member is a.s. equal to the left member. Also, we say that the set $(x^1, \ldots, x^n)$ satisfies (2-1) strictly if it satisfies (2-3) strictly, as defined in Definition (1-4).

Suppose that $F(t, x)$ is defined for all t in $[a, b]$ and all x in R^n. If $x^1, \ldots, x^n$ satisfy Eqs. (2-1), a purely formalistic procedure for finding dF suggests itself at once. Expand $F(t + dt, x + dx)$ by Taylor's formula. For each factor dx^i substitute the right member of (2-1). Expand, and discard all terms that contain three or more factors in the set $dt, dz^1, \ldots, dz^r$ and all that contain a factor dt and at least one other factor from the set. [For these we know by Theorems III(5-1) and III(5-4) that the corresponding integrals in (2-3) vanish.] This formalism yields an equation. But to accept this equation with no further study is to exhibit a naïve faith in the indestructibility of formulas, even when ripped out of context. We have proved nothing like Taylor's formula for stochastic processes, and since (2-1) has no meaning except through the interpretation (2-3) we have no right to substitute the right member of (2-1) for dx^i in the (nonexistent) Taylor's formula. And yet, the result we obtain by this heedless faith in formalism is correct. This is the content of the next theorem, whose importance is outstanding.

(2-4) *THEOREM* (Itô's differentiation formula) *Let hypotheses* (1-1) *be satisfied. Let* f^i, $g_\rho{}^i$, $h^i_{\rho,\sigma}$ $(i = 1, \ldots, n;\ \rho, \sigma = 1, \ldots, r)$ *be processes on T that have factorizations*

$$f^i(\tau, \omega) = b^i(\tau, \omega) f^{i*}(\tau, \omega),$$

$$g_\rho{}^i(\tau, \omega) = b_\rho{}^i(\tau, \omega) g_\rho^{i*}(\tau, \omega),$$

$$h^i_{\rho,\sigma}(\tau, \omega) = b^i_{\rho,\sigma}(\tau, \omega) h^{i*}_{\rho,\sigma}(\tau, \omega)$$

in which b^i, $b_\rho{}^i$, *and* $b^i_{\rho,\sigma}$ *are measurable processes adapted to* $\mathfrak{F}_{\boldsymbol{\tau}}$, *continuous in probability a.e. in* $[a, b]$ *and having a.s. bounded sample functions; the* g_ρ^{i*} *are* L_2*-bounded processes adapted to* $\mathfrak{F}_{\boldsymbol{\tau}}$ *and are* L_2*-continuous a.e. in* $[a, b]$; *and the* f^{i*} *and* $h^{i*}_{\rho,\sigma}$ *are* L_1*-bounded processes adapted to* $\mathfrak{F}_{\boldsymbol{\tau}}$ *and are* L_1*-continuous a.e. in* $[a, b]$. *Assume also*

(i) $x^1, \ldots, x^n$ *are processes on* $[a, b]$ *with continuous sample functions, and they satisfy*

(2-5)
$$dx^i = f^i(t)\, dt + \sum_{\rho=1}^{r} g_\rho{}^i(t)\, dz^\rho(t) + \sum_{\rho,\sigma=1}^{r} h^i_{\rho,\sigma}(t)\, dz^\rho(t)\, dz^\sigma(t)$$

strictly, and

(ii) $F(\mathbf{t}, \mathbf{x})$ *is a function on* $[a, b] \times R^n$ *that is continuous together with its partial derivatives*

$$F_0 = \partial F/\partial t, \qquad F_i = \partial F/\partial x^i, \qquad F_{i,j} = \partial^2 F/\partial x^i x^j \qquad (i, j = 1, \ldots, n)$$

for all (t, x) *in* $[a, b] \times R^n$.

Then the composite function

$$F(\mathbf{t}, x(\mathbf{t}, \boldsymbol{\omega})) = F(\mathbf{t}, x^1(\mathbf{t}, \boldsymbol{\omega}), \ldots, x^n(\mathbf{t}, \boldsymbol{\omega}))$$

strictly satisfies the differential equation

(2-6)
$$dF = F_0\, dt + \sum_{i=1}^{n} F_i f^i\, dt + \sum_{i=1}^{n} \sum_{\rho=1}^{r} F_i g_\rho{}^i\, dz^\rho + \sum_{i=1}^{n} \sum_{\rho,\sigma=1}^{r} \left\{ F_i h^i_{\rho,\sigma} + \frac{1}{2} \sum_{j=1}^{n} F_{i,j} g_\rho{}^i g_\sigma{}^j \right\} dz^\rho\, dz^\sigma,$$

wherein the F_0, F_i, *and* $F_{i,j}$ *are evaluated at* $(t, x^1(t, \omega), \ldots, x^n(t, \omega))$.

We shall first prove this under three additional hypotheses.

(2-7) The processes b^i, $b_\rho{}^i$, $b^i_{\rho,\sigma}$ are bounded.

(2-8) All partial derivatives of F of order $\leqq 3$ are defined and continuous, and their absolute values do not exceed a constant B_1 for all (t, x) in $[a, b] \times R^n$.

(2-9) There exists a bound B_2 for all $|f^i(\tau, \omega)|$, $|g_\rho{}^i(\tau, \omega)|$, and $|h^i_{\rho,\sigma}(\tau, \omega)|$ for (τ, ω) in $T \times \Omega$.

If (2-7) is satisfied, the processes f^i, $g_\rho{}^i$, and $h^i_{\rho,\sigma}$ have the same properties of boundedness and continuity in L_1-norm or L_2-norm as the f^{i*}, etc. have, so we can simplify notation by assuming that b^i, $b_\rho{}^i$, and $b^i_{\rho,\sigma}$ are all identically 1.

Let p be a positive integer and t a point in $[a, b]$. We first subdivide $[a, b]$ into p equal subintervals by points $t_1 = a, t_2, \ldots, t_{p+1} = b$. Let $h(t)$ be the greatest integer such that $t_{h(t)} < t$. Then, if $a < t \leqq b$,

(2-10) $$\Pi(t) = (t_1, \ldots, t_{h(t)}, t; t_1, \ldots, t_{h(t)})$$

is a Cauchy partition of $[a, t]$ with mesh $\Pi(t) \leqq (b - a)/p$. We define some symbols (omitting the variable ω):

(2-11) $$\Delta_h t = t_{h+1} - t_h = (b - a)/p \qquad (h = 1, \ldots, h(t) - 1),$$

$$\Delta_{h(t)} t = t - t_{h(t)},$$

$$\Delta_h z^\rho = z^\rho(t_{h+1}) - z^\rho(t_h) \qquad (h = 1, \ldots, h(t) - 1;\ \rho = 1, \ldots, r),$$

$$\Delta_{h(t)} z^\rho = z^\rho(t) - z^\rho(t_{h(t)}) \qquad (\rho = 1, \ldots, r),$$

$$T_h{}^0 = \Delta_h t \qquad (h = 1, \ldots, h(t)),$$

$$T_h{}^i = f^i(t_h)\,\Delta_h t + \sum_{\rho=1}^{r} g_\rho{}^i(t_h)\,\Delta_h z^\rho + \sum_{\rho,\sigma=1}^{r} h^i_{\rho,\sigma}(t_h)\,\Delta_h z^\rho\,\Delta_h z^\sigma,$$

$$y^i(t) = x_0{}^i + S(\Pi(t); f^i; t) + \sum_{\rho=1}^{r} S(\Pi(t); g_\rho{}^i; z^\rho)$$

$$+ \sum_{\rho,\sigma=1}^{r} S(\Pi(t); h^i_{\rho,\sigma}; z^\rho, z^\sigma).$$

By Theorems III(2-3) and III(3-2), with an obvious estimate of the sums involving $\Delta_h t$, $y^i(t, \omega)$ converges to $x^i(t, \omega)$ in L_1-norm, and by Theorem II(4-3) and the remark following it the convergence is uniform.

We define

$$F_{i,j,k} = \partial^3 F/\partial x^i\,\partial x^j\,\partial x^k,$$

the subscript 0 indicating differentiation with respect to t. Then by hypothesis (2-8), as $p \to \infty$, $F(t, y(t, \omega))$ converges in L_1-norm to $F(t, x(t, \omega))$ uniformly on $[a, b]$, and likewise with F_i or F_{ij} in place of F. But the F, F_i, and F_{ij} have been supposed uniformly bounded, so the L_2-norms of $F(t, y(t, \omega)) - F(t, x(t, \omega))$, etc., remain below a constant multiple of their L_1-norms. Therefore

(2-12) as $p \to \infty$, $F(t, y(t, \omega))$ converges in L_2-norm to $F(t, x(t, \omega))$ uniformly on $[a, b]$, and likewise for the derivatives F_i and $F_{i,j}$.

Since $y^i(t_1) = x_0{}^i$,

$$\text{(2-13)}\quad F(t, y(t)) = F(a, x_0) + \sum_{h=1}^{h(t)-1} [F(t_{h+1}, y(t_{h+1})) - F(t_h, y(t_h))] + [F(t, y(t)) - F(t_{h(t)}, y(t_{h(t)}))].$$

By the theorem of mean value, on the line segments joining the consecutive points $(a, y(a, \omega))$, $(t_2, y(t_2, \omega), \ldots, (t, y(t, \omega))$ there are respective points $(\bar{t}_1, \bar{y}_1), \ldots, (\bar{t}_{h(t)}, \bar{y}_{h(t)})$ (dependent in general on ω) such that for $h = 1, \ldots, h(t) - 1$

$$\begin{aligned}
\text{(2-14)}\quad & F(t_{h+1}, y(t_{h+1})) - F(t_h, y(t_h)) \\
&= F(t_h + T_h{}^0, y^1(t_h) + T_h{}^1, \ldots, y^n(t_h) + T_h{}^n) - F(t_h, y(t_h)) \\
&= \sum_{i=0}^{n} F_i(t_h, y(t_h)) T_h{}^i + \frac{1}{2} \sum_{i,j=0}^{n} F_{i,j}(t_h, y(t_h)) T_h{}^i T_h{}^j \\
&\quad + \frac{1}{6} \sum_{i,j,k=0}^{n} F_{i,j,k}(\bar{t}_h, \bar{y}_h) T_h{}^i T_h{}^j T_h{}^k;
\end{aligned}$$

a similar equation holds for $h = h(t)$ except that $(t_{h+1}, y(t_{h+1}))$ is replaced by $(t, y(t))$ in the left member. If we substitute the expressions for $T_h{}^i$ from (2-11) and expand, we obtain for each h a collection of terms each containing at least one and at most six factors from the set $\{\Delta_h t, \Delta_h z^1, \ldots, \Delta_h z^r\}$. We add the right members of the equations (2-14) with $h = 1, 2, \ldots, h(t)$, and bracket together all those terms in which the $\Delta_h t$ and $\Delta_h z^\rho$ factors are alike except for the subscripts h. As $p \to \infty$, all bracketed sums containing three or more Δ factors will converge in probability to 0, by Theorem III(5-1), and all those with one Δt factor and at least one other Δ factor will converge in probability to 0, by Theorem III(5-4). The remaining terms in the sum will have one of the forms

(2-15) $$\sum_{h=1}^{h(t)} F_0(t_h, y(t_h))\ \Delta_h t,$$

(2-16) $$\sum_{h=1}^{h(t)} F_i(t_h, y(t_h)) g_\rho{}^i(t_h)\ \Delta_h z^\rho,$$

(2-17) $$\sum_{k=1}^{h(t)} F_i(t_k, y(t_k)) h^i_{\rho,\sigma}(t_k)\ \Delta_k z^\rho\ \Delta_k z^\sigma,$$

(2-18) $$\frac{1}{2}\sum_{k=1}^{h(t)} F_{i,j}(t_k, y(t_k)) g_\rho{}^i(t_k) g_\sigma{}^i(t_k)\ \Delta_k z^\rho\ \Delta_k z^\sigma.$$

Because of (2-12) and (2-9), we obtain from Lemma III(2-2)

$$\lim_{\rho\to\infty} \left\| \sum_{h=1}^{h(t)} [F_i(t_h, x(t_h)) - F_i(t_h, y(t_h))] g_\rho{}^i(t_h)\ \Delta_h z^\rho \right\| = 0.$$

Hence if in the sums (2-16) we replace $(t_h, y(t_h))$ by $(t_h, x(t_h))$, the limit in L_2-norm (hence in probability) is unchanged. A similar proof applies to (2-17) and (2-18). But after this change is made, the sums (2-15), (2-16), (2-17), and (2-18) are transformed into the Riemann sums, for partition $\Pi(t)$, of the functions F_0, $F_i g_\rho{}^i$, etc., and by Theorems III(2-3) and III(3-2) these sums converge in probability to the corresponding belated integrals. The meaning of (2-6) is that a.s.

(2-19) $$\begin{aligned} F(t, x(t, \omega)) &= F(a, x_0(\omega)) + \int_a^t F_0\, ds + \sum_{i=1}^n \int_a^t F_i f^i(s, \omega)\, ds \\ &+ \sum_{i=1}^n \sum_{\rho=1}^r \int_a^t F_i g_\rho{}^i(s, \omega)\, dz^\rho(s, \omega) \\ &+ \sum_{i=1}^n \sum_{\rho,\sigma=1}^r \int_a^t F_i h^i_{\rho,\sigma}(s, \omega)\, dz^\rho(s, \omega)\, dz^\sigma(s, \omega) \\ &+ \frac{1}{2} \sum_{i,j=1}^n \sum_{\rho,\sigma=1}^r \int_a^t F_{i,j} g_\rho{}^i(s, \omega) g_\sigma{}^j(s, \omega)\, dz^\rho(s, \omega)\, dz^\sigma(s, \omega), \end{aligned}$$

the F_0, F_i, and $F_{i,j}$ being evaluated at $(s, x(s, \omega))$. The Riemann sums in the transformed equations (2-16), (2-17), and (2-18) converge to these respective limits, and Eq. (2-6) is established under the supplementary hypotheses (2-7), (2-8), and (2-9). The rest of the proof will consist of a succession of devices to allow us to eliminate these extra hypotheses.

First we retain hypotheses (2-7) and (2-8), but abandon (2-9).

For each positive M and all i, ρ, σ, τ, and ω in their respective ranges we define

$$f_M{}^i(\tau, \omega) = \text{mid}\{ f^i(\tau, \omega), M, -M\}, \tag{2-20}$$

and we define $g^i_{\rho,M}$ and $h^i_{\rho,\sigma,M}$ analogously. We denote by $x_M{}^i$ the solutions of Eqs. (2-5) with initial value $x_0{}^i$ and with $f_M{}^i$, $g^i_{\rho,M}$, $h^i_{\rho,\sigma,M}$ in place of f^i, $g_\rho{}^i$, $g^i_{\rho,\sigma}$, respectively. Then (omitting ω from the notation)

$$\begin{aligned} x^i(t) - x_M{}^i(t) = &\int_a^t [f^i(s) - f_M{}^i(s)]\, ds \\ &+ \sum_{\rho=1}^r \int_a^t [g_\rho{}^i(s) - g^i_{\rho,M}(s)]\, dz^\rho(s) \\ &+ \sum_{\rho,\sigma=1}^r \int_a^t [h^i_{\rho,\sigma}(s) - h^i_{\rho,\sigma,M}(s)]\, dz^\rho(s)\, dz^\sigma(s). \end{aligned} \tag{2-21}$$

Now, by hypothesis

(2-22) if h is any one of the functions f^i or $h^i_{\rho,\sigma}$ $(\rho, \sigma = 1, \ldots, r; i = 1, \ldots, n)$, then for each τ in T

$$\lim_{M\to\infty} E(|\, h(\tau, \omega) - h_M(\tau, \omega)\, |) = 0;$$

(2-23) if h is any one of the functions $g_\rho{}^i$ $(i = 1, \ldots, n;\ \rho = 1, \ldots, r)$, then

$$\lim_{M\to\infty} E([h(\tau, \omega) - h_M(\tau, \omega)]^2) = 0.$$

The integrals corresponding to the terms in the right member of (2-6) all exist, by Theorems III(2-3), III(3-2), III(4-1), and III(4-6), both as written and with $f_M{}^i$, $g^i_{\rho,M}$, $h^i_{\rho,\sigma,M}$ in place of f^i, $g_\rho{}^i$, $h^i_{\rho,\sigma}$. Moreover, the expectations written in (2-22) and (2-23) are bounded and almost everywhere continuous on $[a, b]$, so they are Lebesgue integrable; and by (2-22) and (2-23) their integrals tend to 0 as $M \to \infty$. So by Corollaries III(3-12),

III(2-11), and III(3-11) we have for $a \leqq t \leqq b$

(2-24) $$\lim_{M\to\infty} \left\| \int_a^t [f^i(s) - f_M{}^i(s)]\, ds \right\|_1 = 0,$$

(2-25) $$\lim_{M\to\infty} \left\| \int_a^t [g_\rho{}^i(s) - g^i_{\rho,M}(s)]\, dz^\rho(s) \right\|_1 = 0$$

$(\rho = r+1, \ldots, r)$;

(2-26) $$\lim_{M\to\infty} \left\| \int_a^t [h^i_{\rho,\sigma}(s) - h^i_{\rho,\sigma,M}(s)]\, dz^\rho(s)\, dz^\sigma(s) \right\|_1 = 0$$

$(\rho, \sigma = 1, \ldots, r)$,

uniformly on $[a, b]$. So by (2-21)

(2-27) $x_M{}^i(t)$ converges to $x^i(t)$ in L_1-norm, uniformly on $[a, b]$, as $M \to \infty$.

By hypothesis (2-8), (2-27) implies

(2-28) if G is any one of the functions

$F(t, x), \quad F_i(t, x), \quad F_{i,j}(t, x)$

$(i, j = 0, 1, \ldots, n; \quad a \leqq t \leqq b; \quad x \in \bar{R}^n)$,

$G(t, x_M(t, \omega))$ converges to $G(t, x(t, \omega))$ uniformly in L_1-norm on $[a, b]$.

If $\rho = 1, \ldots, r$, we have by Corollary III(2-11), for $a \leqq t \leqq b$,

(2-29) $$\left\| \int_a^t \{F_i(s, x_M(s)) g^i_{\rho,M}(s) - F_i(s, x(s)) g_\rho{}^i(s)\}\, dz^\rho \right\|_2^2$$

$$\leqq C^2 \int_a^t E([F_i(s, x_M(s)) g^i_{\rho,M}(s) - F_i(s, x(s)) g_\rho{}^i(s)]^2\, ds.$$

But if $s \in [a, b]$

(2-30) $$E([F_i(s, x_M(s)) g^i_{\rho,M}(s) - F_i(s, x(s)) g_\rho{}^i(s)]^2$$
$$\leqq 2E(F_i(s, x_M(s))^2[g_{\rho,M}(s) - g_\rho{}^i(s)]^2)$$
$$+ 2E([F_i(s, x_M(s)) - F_i(s, x(s))]^2 g_\rho{}^i(s)^2].$$

In each term in the right member the expression whose expectation is taken remains below a bounded multiple of $(g_\rho{}^i(s, \omega))^2$, whose expectation is finite, and each tends to 0 in probability, so both terms in the right member of (2-30) tend to 0 as $M \to \infty$. Therefore by (2-29)

$$(2\text{-}31)\quad \lim_{M\to\infty} \left\| \int_a^t \{F_i(s, x_M(s))g^i_{\rho,M}(s) - F_i(s, x(s)g_\rho{}^i(s)\}\, dz^\rho(s) \right\|_2 = 0.$$

For $\rho, \sigma = 1, \ldots, r$,

$$(2\text{-}32)\quad E(| F_i(s, x_M(s))h^i_{\rho,\sigma,M}(s) - F_i(s, x(s))h^i_{\rho,\sigma}(s) |)$$
$$\leqq E(| F_i(s, x_M(s)) |\cdot| h^i_{\rho,\sigma,M}(s) - h^i_{\rho,\sigma}(s) |)$$
$$+ E(| F_i(s, x_M(s)) - F_i(s, x(s)) |\cdot| h^i_{\rho,\sigma}(s) |).$$

The expressions following the symbol E in the right member remain below a bounded multiple of $| h^i_{\rho,\sigma}(s) |$, whose expectation is finite; and they converge in probability to 0. So the right member of (2-32) tends to 0 as $M \to \infty$. By Corollary III(3-11)

$$(2\text{-}33)\quad \left\| \int_a^t [F_i(s, x_M(s))h^i_{\rho,\sigma,M}(s) - F_i(s, x(s))h^i_{\rho,\sigma}(s)]\, dz^\rho(s)\, dz^\sigma(s)) \right\|_1$$
$$\leqq K \int_a^t E(| F_i(s, x_M)h^i_{\rho,\sigma,M}(s) - F_i(x, s)h^i_{\rho,\sigma}(s) |)\, ds,$$

so the left member of (2-33) tends to 0 as $M \to \infty$. A similar argument proves

$$(2\text{-}34)\quad \lim_{M\to\infty} \left\| \int_a^t [F_{i,j}(s, x_M(s))g^i_{\rho,M}(s)g^j_{\sigma,M}(s) - F_{i,j}(s, x(s))g_\rho{}^i(s)g_\sigma{}^j(s)]\, dz^\rho(s)\, dz^\sigma(s) \right\|_1 = 0,$$

$$(2\text{-}35)\quad \lim_{M\to\infty} \left\| \int_a^t [F_i(s, x_M(s))f_M{}^i(s) - F_i(s, x(s))f^i(s)]\, ds \right\|_1 = 0,$$

$$(2\text{-}36)\quad \lim_{M\to\infty} \left\| \int_a^t [F_0(s, x_M(s)) - F_0(s, x(s))]\, ds \right\|_1 = 0.$$

By the part of the proof already completed, the analog of (2-19) with $x_M{}^i, f_M{}^i, g^i_{\rho,M}, h^i_{\rho,\sigma,M}$ in place of $x^i, f^i, g_\rho{}^i, h^i_{\rho,\sigma}$ is valid. By the limit relations just proved, with (2-28), as $M \to \infty$ each term of this analog tends in L_1-distance to the corresponding term of (2-19) as written. So (2-19) is valid if hypotheses (2-7) and (2-8) are satisfied.

Now we abandon (2-8), but use instead the supplementary hypothesis

(2-37) there is a B_3 such that $F(t, x) = 0$ whenever $|x| \geqq B_3$.

For every $\epsilon > 0$ we can first define F_ϵ on $[a - \epsilon, b + \epsilon]$ by setting

$$F_\epsilon(t, x) = F(a + [t - a + \epsilon](b - a)/(b - a + 2\epsilon), x)$$
$$(a - \epsilon \leqq t \leqq b + \epsilon, \quad x \text{ in } R^n).$$

Then as $\epsilon \to 0$, F_ϵ converges uniformly on $[a, b] \times R^n$ to F, and likewise for any continuous partial derivative of F. Next, let ϕ be a nonnegative infinitely differentiable function vanishing outside the unit sphere in R^n and having $\int \phi\, dx = 1$, and for each positive δ define

$$\phi_\delta(x) = \delta^{-n}\phi(\delta x^1, \ldots, \delta x^n).$$

Then if $\delta < \epsilon$, the convolution

$$F_\epsilon * \phi_\delta(x) = \int_{R^n} F_\epsilon(x - y)\phi_\delta(y)\, dy$$

is defined and infinitely differentiable on $[a, b] \times R^n$, and it converges uniformly to F_ϵ as $\delta \to 0$. Likewise, each derivative of F_ϵ that is continuous is the uniform limit as $\delta \to 0$ of the corresponding derivative of $F_\epsilon * \phi_\delta$. So, by hypothesis (ii),

(2-38) there exists a sequence $F^{(1)}, F^{(2)}, \ldots$ of functions infinitely differentiable on $[a, b] \times R^n$, with all derivatives bounded, such that each derivative listed in hypothesis (ii) is the uniform limit of the corresponding derivative of $F^{(n)}$.

Under assumptions (2-7) and (2-37), we have shown that (2-19) holds with $F^{(n)}$ in place of F. The limit passage is trivial, so under assumptions (2-7) and (2-37), Eq. (2-19) holds. We now show that hypothesis (2-37) is superfluous. By hypothesis, the x^i have continuous sample functions. Therefore there exists a finite-valued r.v. $B(\omega)$ such that

(2-39) $$|x^i(t, \omega)| \leqq B(\omega) \qquad (i = 1, \ldots, n; \quad a \leqq t \leqq b)$$

except on a set with P-measure 0.

Now for each positive integer M we define ϕ_M to be an infinitely differentiable function on R^n with values in $[0, 1]$, such that $\phi_M(x) = 1$ if $|x| \leqq M$ and $\phi_M(x) = 0$ if $|x| \geqq M + 1$. By (2-39) there is an increasing sequence $A_1, A_2, \ldots$ of measurable subsets of Ω with $P(A_M) \to 1$ such that if $\omega \in A_M$, then $|x(t, \omega)| \leqq M$ for $a \leqq t \leqq b$.

Let t be in $[a, b]$, and let ϵ be positive. We can and do choose an M for which $PA_M > 1 - \epsilon/2$. By the part of the proof already completed,

(2-40) $$F(t, x(t, \omega))\phi_M(x(t, \omega)) = F(a, x_0(\omega))\phi_M(x_0(\omega)) + \int_a^t [F\phi_M]_0\, ds + \sum_{i=1}^n \int_a^t [F\phi_M]_i f^i\, ds + \sum_{i=1}^n \sum_{\rho=1}^r \int_a^t [F\phi_M]_i g_\rho{}^i\, dz^\rho + \sum_{i=1}^n \sum_{\rho,\sigma=1}^r \int_a^t [F\phi_M]_i h^i_{\rho,\sigma}\, dz^\rho\, dz^\sigma + \frac{1}{2}\sum_{i,j=1}^n \sum_{\rho,\sigma=1}^r \int_a^t [F\phi_M]_{i,j} g_\rho{}^i g_\sigma{}^j\, dz^\rho\, dz^\sigma.$$

If $\omega \in A_M$, the integrands in (2-19) and (2-40) are equal, so by Corollary II(2-22), for each t the right members of (2-19) and (2-40) are equal for all ω in $A_M \backslash N(t)$, where $N(t)$ is a negligible set. Let N be the union of $N(t)$ for all rational t in $[a, b]$. Then N is negligible, and by continuity (2-40) holds for all ω in $\Omega_M \backslash N$. So the right members of (2-19) and (2-40) have $\bar{\rho}$-distance less than $\epsilon/2$. Likewise the left members of (2-19) and (2-40) coincide on A_M, so they have $\bar{\rho}$-distance less than $\epsilon/2$. Since (2-40) is valid, the right and left members of (2-19) have $\bar{\rho}$-distance less than ϵ. But ϵ is an arbitrary positive number, so the two members of (2-19) are a.s. equal, and (2-19) holds.

It remains to remove supplementary hypothesis (2-7). By the hypotheses of the theorem, there exists an a.s. finite-valued r.v. $B(\omega)$ such that for all (τ, ω) in $T \times \Omega$, all i in $\{1, \ldots, n\}$ and all ρ and σ in $\{1, \ldots, r\}$

$$|b^i(\tau, \omega)| \leqq B(\omega), \qquad |b_\rho{}^i(\tau, \omega)| \leqq B(\omega), \qquad |b^i_{\rho,\sigma}(\tau, \omega)| \leqq B(\omega).$$

For each positive M, we define

$$\Omega_M = \{\omega \in \Omega : B(\omega) \leqq M\}.$$

Then for each positive ϵ, there is an M such that $P(\Omega_M) > 1 - \epsilon$. We define

$$\begin{aligned}
b_M{}^i(\tau, \omega) &= \operatorname{mid}\{b^i(\tau, \omega), M, -M\},\\
b^i_{\rho,M}(\tau, \omega) &= \operatorname{mid}\{b_\rho{}^i(\tau, \omega), M, -M\},\\
b^i_{\rho,\sigma,M}(\tau, \omega) &= \operatorname{mid}\{b^i_{\rho,\sigma}(\tau, \omega), M, -M\},\\
f_M{}^i(\tau, \omega) &= b_M{}^i(\tau, \omega) f^{i*}(\tau, \omega),\\
g^i_{\rho,M}(\tau, \omega) &= b^i_{\rho,M}(\tau, \omega) g^{i*}_\rho(\tau, \omega),\\
h^i_{\rho,\sigma,M}(\tau, \omega) &= b^i_{\rho,\sigma,M}(\tau, \omega) h^{i*}_{\rho,\sigma}(\tau, \omega),
\end{aligned}$$

and we define $x_M{}^i$ by (2-5) with f^i, etc., replaced by $f_M{}^i$, etc. Then (2-19) holds with $x_M{}^i, f_M{}^i$, etc. in place of x^i, f^i, etc. For each t, we have by Corollary II(2-22)

$$x_M{}^i(t, \omega) = x^i(t, \omega)$$

for all ω in $\Omega_M \backslash N(t)$, where $P(N(t)) = 0$. So for all such ω, Eq. (2-19) holds. Since $P(\Omega_M)$ can be chosen arbitrarily close to 1, (2-19) holds except on a negligible set.

So far we have not considered the property of strictness. Let Ω_0 be the set of all points ω_0 in Ω such that all $z^\rho(\mathbf{t}, \omega_0)$ are Lipschitzian and all $f^i(\mathbf{t}, \omega_0)$, $g_\rho{}^i(\mathbf{t}, \omega_0)$, $h^i_{\rho,\sigma}(\mathbf{t}, \omega_0)$ are Riemann integrable over $[a, b]$. Then $F(\mathbf{t}, x^1(\mathbf{t}, \omega_0), \ldots, x^n(\mathbf{t}, \omega_0))$ is also Riemann integrable. Let ω_0 be a point of Ω_0. For every subset A of Ω, we define $P_0(A)$ to be 1 if $\omega_0 \in A$ and 0 if $\omega_0 \notin A$. The hypotheses of Theorem (2-4) are satisfied with this new measure, so for each t in $[a, b]$, Eq. (2-19) holds on a set of P_0-measure 1; that is, it holds at ω_0. In (2-19) the integrals are belated integrals. But the Riemann integrals also exist, and the two kinds of integrals have the same value. So (2-29) holds with ω_0 in place of ω and with the integrals interpreted as Riemann–Stieltjes integrals, and (2-6) is strictly satisfied.

(2-41) *COROLLARY If for each indefinite integral in the right member of* (2-19) *we choose a strict version with continuous sample functions, then* (2-19) *holds identically in t on* $[a, b]$ *for all* ω *except those in a negligible set, and in particular for all* ω *such that all* $z^\rho(\mathbf{t}, \omega)$ *are Lipschitzian and all integrands in* (2-19) *are Riemann integrable over* $[a, b]$.

Let Ω_0 be the set of ω such that all $z(\mathbf{t}, \omega)$ are Lipschitzian and all integrands in (2-19) Riemann integrable. For each t in $[a, b]$, the sum of the chosen versions of the integrals in (2-19) is equal to the left member except on a negligible set $N(t)$ disjoint from Ω_0. Let N be the union of $N(t)$ for all rational t in $[a, b]$. This is negligible and disjoint from Ω_0. For all ω in $\Omega \backslash N$, the right and left members of (2-19) are equal for all rational t in $[a, b]$; being continuous, they are equal for all t in $[a, b]$.

3 *Applications of Itô's differentiation formula*

When all z^ρ are identically 0, Theorem (2-4) is the customary chain rule for differentiation, except that it has a hypothesis, superfluous in this case, that the partial derivatives F_{ij} are continuous.

Suppose next that $z^1, \ldots, z^r$ are independent Brownian motions. Let $I_\rho(\omega)$ be 0 if $z^\rho(\mathbf{t}, \omega)$ is Lipschitzian and be 1 otherwise. As we saw in III(7-4) and III(7-3), the second-order integral

$$\int_a^t g^i_{\rho,\sigma}(s, \omega)\, dz^\rho(s, \omega)\, dz^\sigma(s, \omega)$$

has a strict version 0 if $\rho \neq \sigma$, and if $\rho = \sigma$ it has a strict version

$$\int_a^t g^i_{\rho,\rho}(s, \omega) I_\rho(\omega)\, dt.$$

In this case the conclusion of Theorem (2-4) becomes

$$\text{(3-1)}\quad dF = \left\{F_0 + \sum_{i=1}^n F_i f^i + \sum_{i=1}^n \sum_{\rho=1}^r \left[F_i h^i_{\rho,\rho} + \frac{1}{2} \sum_{j=1}^n F_{i,j} g_\rho{}^i g_\rho{}^j\right] I_\rho\right\} dt$$

$$+ \sum_{i=1}^n \sum_{\rho=1}^r F_i g_\rho{}^i\, dz^\rho,$$

the equation being satisfied strictly. If we replace I_ρ by 1 we obtain the usual form of the Itô differentiation formula, except that the integrals are belated integrals instead of Itô integrals. However, the extension to Itô-belated integrals will be presented in the last section of this chapter. The version with I_ρ replaced by 1 is no longer strict.

As a simple example, with $n = r = 1$, we define

$$dx = 1\, dz, \qquad F(t, x) = x^2.$$

Then

$$\text{(3-2)}\qquad dF = F_1\, dz + \tfrac{1}{2} F_{11}\, (dz)^2$$
$$= 2x\, dz + (dz)^2,$$

or in the integrated form (1-29)

$$\text{(3-3)}\qquad \int_a^t z(s)\, dz(s) = [z(t)^2 - z(a)^2]/2 - \frac{1}{2}\int_a^t (dz)^2.$$

Since $x = z$ is a strict version of $\int 1\, dz$, by Theorem (2-4) this equation holds strictly. In particular, if z is Brownian motion and $I(\omega)$ is defined as before, we obtain

$$\text{(3-4)}\qquad \int_a^t z(s)\, dz(s) = \tfrac{1}{2}[z(t)^2 - z(a)^2 - I(\omega)(t - a)],$$

in agreement with II(2-10) and with Doob [3, p. 444]. In the latter reference

$I(\omega)$ is replaced by 1. The right member of (3-4) is changed only on a negligible set, so the right member is still a version of the left; but it is no longer strict, and lacks both consistency and stability (cf. Chapter II, Section 1).

If m is a positive integer, and $n = r = 1$, we define

$$dx = 1\, dz, \qquad x(a) = z(a), \qquad F(t, x) = x^m.$$

Then

$$\begin{aligned} dF &= F_1\, dz + \tfrac{1}{2}F_{11}\, (dz)^2 \\ &= mx^{m-1}\, dz + [m(m-1)/2]x^{m-2}\, (dz)^2, \end{aligned}$$

or

$$\text{(3-5)} \quad \int_a^t z^{m-1}\, dz$$

$$= [z(t)^m - z(a)^m]/m - [(m-1)/2] \int_a^t z(s)^{m-2}\, (dz(s))^2.$$

In particular, if z is Brownian motion and $I(\omega)$ is defined as before,

$$\text{(3-6)} \quad \int_a^t z^{m-1}\, dz = [z(t)^m - z(a)^m]/m - [(m-1)/2]I(\omega) \int_a^t z(s)^{m-2}\, ds.$$

This is not as simple as (3-4), but at least the integral in the right member is of a simple type, being an ordinary Riemann integral for each ω.

Suppose next that $n = 2$, and that

$$F(t, x^1, x^2) = x^1x^2.$$

If x^1 and x^2 satisfy (2-5), by the memory device I(0-2′) we have

$$\text{(3-7)} \qquad d(x^1x^2) = x^1\, dx^2 + x^2\, dx^1 + dx^1\, dx^2,$$

which means

$$\text{(3-8)} \qquad d(x^1x^2) = [x^1f^2 + x^2f^1]\, dt + \sum_{\rho=1}^{r} [x^1g_\rho{}^2 + x^2g_\rho{}^1]\, dz^\rho$$

$$+ \sum_{\rho,\sigma=1}^{r} [x^1h^2_{\rho,\sigma} + x^2h^1_{\rho,\sigma} + g_\rho{}^1g_\sigma{}^2]\, dz^\rho\, dz^\sigma.$$

The fact that this equation holds yields the information that if each of two processes is a finite sum of indefinite integrals, so also is their product. But Theorem (2-4) already contained the more general information that if each of the processes $x^1, \ldots, x^n$ is a sum of indefinite integrals, any twice-continuously-differentiable function of them is also a sum of indefinite integrals.

From (3-8) we can deduce the following statement, whose proof we leave as an exercise:

If $g_1, \ldots, g_r$ are L_2-bounded and are L_2-continuous a.e. in $[a, b]$, and $z^1, \ldots, z^r$ are independent standard Brownian motions, and

$$x = \sum_{\rho=1}^{r} \int_a^b g_\rho(t)\, dz^\rho(t),$$

then the variance of x is

$$E(x^2) = \sum_{\rho=1}^{r} \int_a^b E(g_\rho(t)^2)\, dt.$$

By operating formally on (3-7) we would obtain the formula for integration by parts

$$\text{(3-9)} \qquad \int_a^b x^1(s)\, dx^2(s) = x^1(b)x^2(b) - x^1(a)x^2(a) - \int_a^b x^2(s)\, dx^1(s) - \int_a^b dx^1(s)\, dx^2(s).$$

However, the integrals in (3-9) have the appearance of belated integrals, but those belated integrals have not yet been shown to exist. Equation (3-9) is to be interpreted via (3-8), to mean that if x^1 and x^2 satisfy (2-5), then

$$\text{(3-10)} \qquad \int_a^b x^1(s)[f^2(s)\, ds] + \sum_\rho \int_a^b x^1(s) g_\rho{}^2(s)\, dz^\rho(s) + \sum_{\rho,\sigma} \int_a^b x^1(s) h^2_{\rho,\sigma}(s)\, dz^\rho(s)\, dz^\sigma(s)$$

$$= x^1(b)x^2(b) - x^1(a)x^2(a) - \int_a^b x^2(s) f^1(s)\, ds - \sum_\rho \int_a^b x^2(s) g_\rho{}^1(s)\, dz^\rho(s) - \sum_{\rho,\sigma} \int_a^b x^2(s) h^1_{\rho,\sigma}(s)\, dz^\rho(s)\, dz^\sigma(s) - \sum_{\rho,\sigma} \int_a^b g_\rho{}^1(s) g_\sigma{}^2(s)\, dz^\rho(s)\, dz^\sigma(s).$$

If all z^ρ vanish and Ω consists of a single point, Eqs. (2-5) reduce to

$$x^i(t) = x^i(a) + \int_a^t f^i(t)\,dt \qquad (i = 1, 2),$$

and (3-10) is the standard formula

$$\int_a^b x^1 f^2\,dt = x^1(b)x^2(b) - x^1(a)x^2(a) - \int_a^b x^2 f^1\,dt$$

of ordinary calculus.

If we choose $n = 3$ and $F = x^1x^2x^3$, we obtain a formula (which we leave as an exercise for the reader) that can be regarded as an integration by parts formula for second-order integrals, but does not seem to be useful.

As another exercise, we leave it to the reader to prove by means of the identity

$$\Delta_j(x^1x^2) = x^1(t_j)\,\Delta_j x^2 + x^2(t_j)\,\Delta_j x^1 + \Delta_j x^1\,\Delta_j x^2$$

that if any two of the three belated integrals in (3-9) exist, so does the third, and (3-9) is valid. But it should be pointed out that the mere existence of the belated integrals in (3-9) does not permit us to deduce (3-10) because we have as yet no theory of substitution in integrals.

In the special case

$$x^1 = z, \qquad x^2 = z$$

Eq. (3-9) reduces to

$$\int_a^b z(s)\,dz(s) = z(b)^2 - z(a)^2 - \int_a^b z(s)\,dz(s) - \int_a^b (dz(s))^2,$$

which yields (3-3) once more.

We shall study stochastic differential equations in the next chapter; but the special case of a single linear homogeneous equation can be treated easily by means of Theorem (2-4). We seek to solve an equation

$$\text{(3-11)} \qquad d\phi = \phi f\,dt + \sum_{\rho=1}^{r} \phi g_\rho\,dz^\rho + \frac{1}{2}\sum_{\rho,\sigma=1}^{r} \phi g_{\rho,\sigma}\,dz^\rho\,dz^\sigma$$

with initial conditions

$$\text{(3-12)} \qquad \phi(a, \omega) = \phi_0(\omega),$$

where $\phi_0(\omega)$ is an $\mathfrak{F}_a$-measurable r.v. Here and for the rest of this section we assume that every function g_ρ (or later, $g_\rho{}^*$) satisfies the hypotheses of existence theorem III(4-1), and every f and $h_{\rho,\sigma}$ (or f^* and $h^*_{\rho,\sigma}$) satisfies the hypotheses of Theorem III(4-6). To solve (3-11) we define

$$h_{\rho,\sigma}(\tau,\omega) = g_{\rho,\sigma}(\tau,\omega) - g_\rho(\tau,\omega)g_\sigma(\tau,\omega), \tag{3-13}$$

$$x(t) = \int_a^t f(s)\,ds + \sum_{\rho=1}^{r}\int_a^t g_\rho(s)\,dz^\rho(s) + \frac{1}{2}\sum_{\rho,\sigma=1}^{r}\int_a^t h_{\rho,\sigma}(s)\,dz^\rho(s)\,dz^\sigma(s). \tag{3-14}$$

Clearly, $h_{\rho,\sigma}$ satisfies the hypotheses of Theorem III(4-6), so all the integrals in (3-14) exist. Here and henceforth we assume, as is permissible, that for $x(t)$ or any process similarly defined we choose a strict version with continuous sample functions. Let

$$F(t,x) = \exp x \qquad (a \leqq t \leqq b,\quad x \in R).$$

Then by Corollary (2-41),

$$dF = F\cdot f\,dt + \sum_{\rho=1}^{r} F\cdot g_\rho\,dz^\rho + \sum_{\rho,\sigma=1}^{r} \tfrac{1}{2}F\{h_{\rho,\sigma} + g_\rho g_\sigma\}\,dz^\rho\,dz^\sigma,$$

so the process $\phi(\mathbf{t},\boldsymbol{\omega}) = \phi_0(\boldsymbol{\omega})F(\mathbf{t}, x(\mathbf{t},\boldsymbol{\omega}))$, satisfies (3-11) and (3-12).

We shall now show that the solutions of (3-11) and (3-12) have the uniqueness property that if ϕ_1 and ϕ_2 are both solutions with continuous sample functions such that (3-11) holds for all ω in the set Ω_0 of points for which the $z^\rho(\mathbf{t},\omega)$ are Lipschitzian and the f, g_ρ and $g_{\rho,\sigma}$ (or equivalently the $h_{\rho,\sigma}$) are Riemann integrable over $[a,b]$, then $\phi_1(\mathbf{t},\omega) = \phi_2(\mathbf{t},\omega)$ except at most on a negligible set disjoint from Ω_0. Given two such ϕ_1, ϕ_2, we define $\phi = \phi_2 - \phi_1$. This satisfies (3-11) and has

$$\phi(a,\omega) = 0 \qquad (\omega \in \Omega). \tag{3-15}$$

Let y be a strict version of

$$y(t) = -\int_a^t f(s)\,ds - \sum_{\rho=1}^{r}\int_a^t g_\rho(s)\,dz^\rho(s) - \frac{1}{2}\sum_{\rho,\sigma=1}^{r}\int_a^t h_{\rho,\sigma}(s)\,dz^\rho(s)\,dz^\rho(s), \tag{3-16}$$

with continuous sample functions, and define

$$\psi(t,\omega) = \exp y(t,\omega). \tag{3-17}$$

Then

$$d\psi = -\psi f\,dt - \sum_{\rho=1}^{r}\psi g_\rho\,dz^\rho - \frac{1}{2}\sum_{\rho,\sigma=1}^{r}\psi\{h_{\rho,\sigma} - g_\rho g_\sigma\}\,dz^\rho\,dz^\sigma \tag{3-18}$$

and by (3-8) we deduce $d(\phi\psi) = 0$. This implies that $\phi\psi$ is independent of t for all ω except those in a negligible set disjoint from Ω_0. By (3-15), $\phi\psi$ is 0 for all such ω; and since $\psi \neq 0$, we must have $\phi = 0$, and so $\phi_1 = \phi_2$, for all such ω.

In the special case in which the z^ρ are independent Brownian motions and the $g_{\rho,\sigma}$ are 0, a solution of (3-11) is given by

$$\phi(t, \omega) = \phi_0(\omega) \exp x(t, \omega),$$

where

$$x(t, \omega) = \int_a^t \left[f(s) - \sum_{\rho=1}^{r} g_\rho(s)^2/2 \right] ds + \sum_{\rho=1}^{r} \int_a^t g_\rho(s)\, dz^\rho(s).$$

But this solution is not strict, not consistent, and not stable.

In the elementary theory of differential equations there is a well-known formula for the solution of a (single) linear equation. This can be generalized to the case of a linear nonhomogeneous stochastic differential equation. We seek to solve the equation

$$\text{(3-19)} \qquad d\phi = [\phi f + f^*]\, dt + \sum_{\rho=1}^{r} [\phi g_\rho + g_\rho{}^*]\, dz^\rho + \frac{1}{2} \sum_{\rho,\sigma=1}^{r} [\phi g_{\rho,\sigma} + g^*_{\rho,\sigma}]\, dz^\rho\, dz^\sigma,$$

with the initial conditions

$$\text{(3-20)} \qquad \phi(a, \omega) = \phi_0(\omega),$$

$\phi_0(\omega)$ being an $\mathfrak{F}(a)$-measurable r.v. The uniqueness of the solution is proved exactly as before.

We again define y by (3-16) and ψ by (3-17). By (3-8)

$$d(\phi\psi) = \psi f^*\, dt + \sum_{\rho=1}^{r} \psi g_\rho{}^*\, dz^\rho + \sum_{\rho,\sigma=1}^{r} \psi[\tfrac{1}{2} g^*_{\rho,\sigma} - g_\rho{}^* g_\sigma]\, dz^\rho\, dz^\sigma.$$

We define

$$\text{(3-21)} \qquad \Psi(t, \omega) = \int_a^t \psi(s) f^*(s)\, ds + \sum_{\rho=1}^{r} \int_a^t \psi(s) g_\rho{}^*(s)\, dz^\rho(s) + \sum_{\rho,\sigma=1}^{r} \int_a^t \psi(s)[\tfrac{1}{2} g^*_{\rho,\sigma}(s) - g_\rho{}^*(s) g_\sigma(s)]\, dz^\rho(s)\, dz^\sigma(s).$$

Except on a negligible set we have for all t in $[a, b]$

$$\phi(t)\psi(t) = \phi(a) + \Psi(t),$$

so that [recalling (3-17)]

$$\phi(t) = \exp[-y(t)]\{\phi(a) + \Psi(t)\}. \tag{3-22}$$

When all z^ρ are 0 this reduces to the familiar formula for the solution of one linear (deterministic) differential equation. When the z^ρ are independent standard Brownian motions and $g_{\rho,\sigma} = g^*_{\rho,\sigma} = 0$, it reduces to

$$\begin{aligned}\phi(t) &= \phi(a)\exp(-y(t)) \\ &\quad + \int_a^t [\exp y(s) - y(t)]\left[f^*(s) - \sum_{\rho=1}^r g_\rho{}^*(s)g_\rho(s)\right] ds \\ &\quad + \sum_{\rho=1}^r \int_a^t [\exp y(s)]g_\rho{}^*(s)\, dz^\rho(s),\end{aligned} \tag{3-23}$$

but this is neither strict, consistent, nor stable.

An interesting special case is obtained by specializing (3-19) to the form

$$d\phi = -r\phi\, dt + rc\, dz, \tag{3-24}$$

where c is a real number and r is a positive number. Then $f = -r$, $g^* = rc$, and

$$y(t) = \int_a^t r\, ds = r(t-a),$$

so

$$\psi(t) = \exp r(t-a).$$

By (3-22),

$$\phi(t) = \exp r(a-t)\left\{\phi(a) + \int_a^t [\exp r(s-a)]rc\, dz(s)\right\}. \tag{3-25}$$

Let us define

$$\Phi(t) = \int_a^t \phi(s)\, ds. \tag{3-26}$$

We leave as exercises the proofs of the next two statements.

(3-27) If z is a martingale, so is $\phi(\mathbf{t}, \omega)\exp r\mathbf{t}$.

(3-28) If z is a Markov process, ϕ and the vector process (ϕ, Φ) are Markov processes.

Since ϕ has continuous sample functions, Φ has continuously differentiable sample functions. We shall now prove that if $E(\phi(a)^2) < \infty$ and $c \neq 0$, the process

$$\Phi(t)/c + z(a)$$

converges to $z(t)$ uniformly in L_2-norm as $r \to \infty$. This gives us a means of approximating any process z satisfying (1-1) by a process with continuously differentiable sample functions. Moreover, the error in the approximation is given explicitly by a simple formula [see (3-29) below].

By (3-26) and (3-25),

$$\Phi(t) = [\phi(a)/r][1 - \exp r(a - t)] + rc \int_a^t \exp(-rs) \left\{ \int_a^s (\exp r\tau)\, dz(\tau) \right\} ds,$$

so by the integration by parts formula (3-9) or (3-10),

$$\Phi(t) = \phi(a)[1 - \exp r(a - t)]/r + c\left\{ -e^{-rt} \int_a^t e^{rs}\, dz(s) + \int_a^t e^{-rs} e^{rs}\, dz(s) \right\},$$

whence

$$\Phi(t)/c - z(t) + z(a) = \phi(a)[1 - \exp r(a - t)]/rc - \int_a^t \exp r(s - t)\, dz(s). \tag{3-29}$$

By Corollary III(2-11),

$$\| \Phi(t)/c - z(t) + z(a) \|_2 \leqq \| \phi(a) \|_2 [1 - \exp r(a - t)]/rc + C\{[1 - e^{2r(a-t)}]/2r\}^{1/2},$$

which tends to 0 uniformly on $[a, \infty)$ as $r \to \infty$.

When ϕ is normally distributed and z is a Brownian motion, the process ϕ determined by (3-24) is called an Ornstein–Uhlenbeck process. We leave as exercises the following statements concerning such processes.

(3-30) If $a \leqq s \leqq t$ and $\phi(a) = 0$, then

$$E(\phi(s)\phi(t)) = (rc^2/2) \exp r(s - t),$$

$$E(\phi(t) \mid \phi(s) = x) = x \exp r(s - t), \tag{3-31}$$

$$\text{variance of } \phi(t) = (\text{variance of } \phi(a)) \exp 2r(s - t) + (rc^2/2)[1 - \exp 2r(s - t)]. \tag{3-32}$$

4 Substitution

As an exercise in pure formalism, from (2-1) we can form a substitution rule thus. Given a process $\phi(\tau, \omega)$, multiply each term in (2-1) by $\phi(t, \omega)$ and write $\int_a^t$ before each term. If f is sufficiently well behaved, the terms of the resulting equation have meaning. But it would be truly naïve to believe that we have thus proved the truth of the equation; for (2-1) is itself mere formalism, whose meaning is given by an integral equation of the type of (2-3). The naïveté becomes even more apparent if we try to evaluate $\int \phi\, dx^1\, dx^2$ by substituting for dx^1 and dx^2 from (2-1). In fact, by similar naïve trust in formulas we would try to evaluate

$$\iint \phi(x, y)\, dx\, dy$$

with the help of a substitution $x = X(u, v)$, $y = Y(u, v)$ by replacing dx by $(X_u\, du + X_v\, dv)$ and dy by $(Y_u\, du + Y_v\, dv)$, with familiar disastrous consequences.

Nevertheless, as in the preceding section we find that under not too excessive hypotheses the equations that we reach by such blind faith in formulas are in fact correct. Of course the proof cannot follow such simple paths, and in fact it is rather tedious. We consider the first-order and the second-order integrals separately.

(4-1) *THEOREM Let the standing hypotheses (1-1) be satisfied. Let $\phi, f, g_\rho, h_{\rho,\sigma}$ $(\rho, \sigma = 1, \ldots, r)$ be processes on T adapted to $\mathfrak{F}_\tau$ with the following property:*

There are processes b_0, b_ρ, $b_{\rho,\sigma}$ and f^1, $g_\rho{}^1$, $h^1_{\rho,\sigma}$ $(\rho, \sigma = 1, \ldots, r)$ on T such that

(i) *ϕ, b_0, b_ρ, $b_{\rho,\sigma}$ are measurable and a.s. sample-function bounded and are continuous in probability a.e. in $[a, b]$;*

(ii) *f^1 and the $h^1_{\rho,\sigma}$ are L_1-bounded and a.e. L_1-continuous in $[a, b]$;*

(iii) *the $g_\rho{}^1$ are L_2-bounded and are L_2-continuous a.e. in $[a, b]$;*

(iv) *for all (τ, ω) in $T \times \Omega$,*

$$f(\tau, \omega) = b_0(\tau, \omega) f^1(\tau, \omega), \qquad g_\rho(\tau, \omega) = b_\rho(\tau, \omega) g_\rho{}^1(\tau, \omega),$$
$$h_{\rho,\sigma}(\tau, \omega) = b_{\rho,\sigma}(\tau, \omega) h^1_{\rho,\sigma}(\tau, \omega).$$

Define

(4-2)
$$x(t) = x_0 + \int_a^t f(s)\, ds + \sum_{\rho=1}^{r} \int_a^t g_\rho(s)\, dz^\rho(s)$$
$$+ \sum_{\rho,\sigma=1}^{r} \int_a^t h_{\rho,\sigma}(s)\, dz^\rho(s)\, dz^\sigma(s)$$

where x_0 is any real $\mathfrak{F}_a$-measurable r.v. Then all the belated integrals in the equation

$$\int_a^b \phi(t)\,dx(t) = \int_a^b \phi(t)f(t)\,dt + \sum_{\rho=1}^{r} \int_a^b \phi(t)g_\rho(t)\,dz^\rho(t) \tag{4-3}$$

$$+ \sum_{\rho,\sigma=1}^{r} \int_a^b \phi(t)h_{\rho,\sigma}(t)\,dz^\rho(t)\,dz^\sigma(t)$$

exist, and the equation is valid.

By Theorems III(4-1) and III(4-6), all the integrals in the right members of Eqs. (4-2) and (4-3) exist. It remains to prove that the left member of (4-3) exists and that (4-3) is valid. We prove this first under the supplementary hypothesis

(4-4) there is a number M such that

$$|\phi(\tau,\omega)| \leqq M, \qquad |b_0(\tau,\omega)| \leqq M, \qquad |b_\rho(\tau,\omega)| \leqq M,$$

$$|b_{\rho,\sigma}(\tau,\omega)| \leqq M$$

for all (τ, ω) in $T \times \Omega$ and all ρ, σ in $\{1, \ldots, r\}$.

For each belated partition $\Pi = (t_1, \ldots, t_{m+1}; \tau_1, \ldots, \tau_m)$ of $[a, b]$ we define ϕ_Π on $[a, b]$ by

$$\phi_\Pi(t,\omega) = \phi(\tau_j,\omega) \qquad (t_j \leqq t < t_{j+1};\ \ j = 1, \ldots, m),$$

$$\phi_\Pi(b,\omega) = \phi(\tau_m,\omega),$$

and we define $b_{0,\Pi}$, $b_{\rho,\Pi}$, $b_{\rho,\sigma,\Pi}$ analogously. By Theorem II(4-2)

$$S(\Pi;\phi;x) = \sum_{j=1}^{m} \phi(\tau_j) \left\{ \int_{t_j}^{t_{j+1}} f(s)\,ds + \sum_{\rho=1}^{r} \int_{t_j}^{t_{j+1}} g_\rho(s)\,dz^\rho(s) \right. \tag{4-5}$$

$$\left. + \sum_{\rho,\sigma=1}^{r} \int_{t_j}^{t_{j+1}} h_{\rho,\sigma}(s)\,dz^\rho(s)\,dz^\sigma(s) \right\}$$

$$= \int_a^b \phi_\Pi(s)f(s)\,ds + \sum_{\rho=1}^{r} \int_a^b \phi_\Pi(s)g_\rho(s)\,dz^\rho(s)$$

$$+ \sum_{\rho,\sigma=1}^{r} \int_a^b \phi_\Pi(s)h_{\rho,\sigma}(s)\,dz^\rho(s)\,dz^\sigma(s).$$

We now use some statements from the proof of Theorem III(4-1). First, by the proof of (4-5) in that theorem, the process $g_\rho = b_\rho g_\rho^1$ satisfies the hypotheses of Theorem III(2-3); so we may and shall use it in place of the f in Theorem III(4-1). Second, for all t in $[a, b]$,

$$\phi_\Pi(t, \omega) = \phi(\tau_j, \omega),$$

where $|t - \tau_j| \leqq$ mesh Π. In establishing Theorem III(4-1) we proved that a.e. in $[a, b]$, the last term in the right member of Eq. III(4-4) converges to 0 in L_2-norm. The same proof, with $\phi(\tau_j)$ in place of $b_V(\tau)$, shows that a.e. in $[a, b]$,

$$E(\{\phi_\Pi(t)g_\rho(t) - \phi(t)g_\rho(t)\}^2) \to 0$$

as mesh $\Pi \to 0$. So, by the dominated convergence theorem, as mesh $\Pi \to 0$

$$\lim \int_a^b E([\phi_\Pi(t)g_\rho(t) - \phi(t)g_\rho(t)]^2)\, dt = 0.$$

By Corollary III(2-11),

$$\lim \left\| \int_a^b \phi_\Pi(t)g_\rho(t)\, dz^\rho(t) - \int_a^b \phi(t)g_\rho(t)\, dz^\rho(t) \right\|_2 = 0,$$

whence

$$\lim \bar{p}\left(\int_a^b \phi_\Pi(t)g_\rho(t)\, dz^\rho(t),\ \int_a^b \phi(t)g_\rho(t)\, dz^\rho(t)\right) = 0.$$

If we replace the references to Theorem III(4-1) and Corollary III(2-11) by references to Theorem III(4-6) and Corollary III(3-11), only trivial changes in the preceding proof are needed to show that as mesh $\Pi \to 0$,

$$\lim \bar{p}\left(\int_a^b \phi_\Pi(t)h_{\rho,\sigma}(t)\, dz^\rho(t)\, dz^\sigma(t),\ \int_a^b \phi(t)h_{\rho,\sigma}(t)\, dz^\rho(t)\, dz^\sigma(t)\right) = 0,$$

$$\lim \bar{p}\left(\int_a^b \phi_\Pi(t)f(t)\, dt,\ \int_a^b \phi(t)f(t)\, dt\right) = 0.$$

By the last three equations, as mesh $\Pi \to 0$ the right member of Eq. (4-5) converges in probability to the right member of (4-3). So ϕ is integrable with respect to x, and Eq. (4-3) is established under the supplementary hypothesis (4-4).

If the hypotheses of the theorem are satisfied, there exists an a.s. finite-valued r.v. $B(\omega)$ such that for all τ in T and ω in Ω,

$$|\phi(\tau, \omega)| \leqq B(\omega), \qquad |b_0(\tau, \omega)| \leqq B(\omega), \qquad |b_\rho(\tau, \omega)| \leqq B(\omega),$$
$$|b_{\rho,\sigma}(\tau, \omega)| \leqq B(\omega).$$

Let ϵ be positive. There is an M such that the set

$$\Omega_M = \{\omega \in \Omega : B(\omega) > M\}$$

has $P(\Omega_M) < \epsilon/3$. We define

$$\phi_M(\tau, \omega) = \operatorname{mid}\{\phi(\tau, \omega), M, -M\},$$

and we define $b_{0,M}$, $b_{\rho,M}$, $b_{\rho,\sigma,M}$ similarly. Then conditions (4-4) are satisfied with ϕ_M, etc. in place of ϕ, etc. Also, for all ω in $\Omega \backslash \Omega_M$, and all τ in T,

$$\text{(4-6)} \qquad \phi_M(\tau, \omega) = \phi(\tau, \omega), \qquad b_{0,M}(\tau, \omega) = b_0(\tau, \omega),$$
$$b_{\rho,M}(\tau, \omega) = b_\rho(\tau, \omega), \qquad b_{\rho,\sigma,M}(\tau, \omega) = b_{\rho,\sigma}(\tau, \omega).$$

We also define

$$f_M = b_{0,M} f^1, \qquad g_{\rho,M} = b_{\rho,M} g_\rho^{\,1}, \qquad h_{\rho,\sigma,M} = b_{\rho,\sigma,M} h^1_{\rho,\sigma}.$$

Then for all ω in $\Omega \backslash \Omega_M$ and all τ in T

$$\text{(4-7)} \quad f_M(\tau, \omega) = f(\tau, \omega),\ g_{\rho,M}(\tau, \omega) = g_\rho(\tau, \omega),\ h_{\rho,\sigma,M}(\tau, \omega) = h_{\rho,\sigma}(\tau, \omega).$$

For brevity, we denote the right member of (4-2) by $R(\boldsymbol{\omega})$, and the corresponding sum with f_M, $g_{\rho,M}$, $h_{\rho,\sigma,v}$ in place of f, g_ρ, $h_{\rho,\sigma}$ respectively, by $R_M(\boldsymbol{\omega})$. Then (4-2) becomes

$$x(t) = R(\boldsymbol{\omega}).$$

We define

$$x_M(t) = R_M(\boldsymbol{\omega}).$$

Similarly we denote the right member of (4-3) by R^*; the similar sum with ϕ, f, g_ρ, $h_{\rho,\sigma}$ replaced by ϕ_M, f_M, $g_{\rho,M}$, $h_{\rho,\sigma,M}$ respectively, we denote by R_M^*. Then by the part of the proof already completed we have

$$\text{(4-8)} \qquad \int_a^b \phi_M(t)\, dx_M(t) = R_M^*.$$

By (4-6), (4-7), and Corollary II(2-22), for each t the equation

$$\text{(4-9)} \qquad x_M(t, \omega) = x(t, \omega)$$

holds for all ω in $\Omega \backslash \Omega_M$ except those in a negligible set $N(t)$. If N is the union of $N(t)$ for all rational t in $[a, b]$, N is negligible, and by continuity (4-9) holds for all t in $[a, b]$ when $\omega \in \Omega \backslash (\Omega_M \cup N)$. Again by (4-6), (4-7), (4-9), and Corollary II(2-22),

$$\text{(4-10)} \qquad R_M^*(\omega) = R^*(\omega)$$

for all ω in $\Omega \backslash \Omega_M$ except those in a negligible subset.

Because of (4-8), ultimately

$$\bar{\rho}(S(\Pi; \phi_M; x_M), R_M^*) < \epsilon/3.$$

Because (4-10) holds except on a set of P-measure less than $\epsilon/3$,

$$\bar{\rho}(R_M^*, R^*) < \epsilon/3.$$

Because of (4-6) and (4-9),

$$\bar{\rho}(S(\Pi; \phi; x), S(\Pi; \phi_M; x_M)) < \epsilon/3.$$

The last three inequalities imply that ultimately

$$\bar{\rho}(S(\Pi; \phi; x), R^*) < \epsilon,$$

so the left member of (4-3) exists and (4-3) holds.

In the previous section we pointed out that Eq. (3-9) was merely a memory device to help us remember Eq. (3-10), because the integrals in (3-9) had not been shown to exist. Now that Theorem (4-1) has been proved, the situation is changed. We now know that the integrals written in (3-9) do exist, as belated integrals, and that they have the values that lead to Eq. (3-10). This is typical of the use to which Theorem (4-1) can be put. In the literature there are many examples of integrals that look like the left member of (3-9), the process x^2 not being a martingale, so that the integral does not exist as an Itô integral. Nevertheless the value of this undefined integral is computed by Eq. (4-3). So we can regard Theorem (4-1) as an a posteriori justification of many assertions already made in published papers.

Theorem (4-1) allows us to write the second-order integral $\int \phi \, dz^\rho \, dz^\sigma$ in the form $\int \phi \, dx$, where

$$x(t) = \int_a^t 1 \, dz^\rho(s) \, dz^\sigma(s).$$

This apparent simplification is deceptive. First, it does not get rid of second-order integrals, since x is defined as a second-order integral. Second, even when z^ρ and z^σ are the same Brownian motion, so that $x(t) = t - a$, the resulting first-order integral is neither strict nor stable.

Our substitution theorem for second-order integrals has stronger hypotheses than those of Theorem (4-1). In its proof we make use of the following two lemmas.

(4-11) *LEMMA Let the standing hypotheses* (1-1) *be satisfied, and also*

(4-12) *if* $a \leqq s \leqq t \leqq b$, *then a.s.* $E([z(t) - z(s)]^4 \mid \mathfrak{F}_s) \leqq K(t - s)$.
Define $C = 2K(b - a)^{1/2} + K^{1/2}$. *Let* $[s_1, t_1), \ldots, [s_m, t_m)$ *be dis-*

joint subintervals of $[a, b)$, *and for* $j = 1, \ldots, m$ *let* u_j *be an* $\mathfrak{F}(s_j)$-*measurable r.v. with* $\| u_j \|_2 < \infty$. *Then*

$$\left\| \sum_{j=1}^{m} u_j[z^1(t_j) - z^1(s_j)][z^2(t_j) - z^2(s_j)] \right\|_2$$

$$\leqq C \left\{ \sum_{j=1}^{m} \| u_j \|_2^2 \, (t_j - s_j) \right\}^{1/2}.$$

We define Δ_j to be $[z^1(t_j) - z^1(s_j)][z^2(t_j) - z^2(s_j)]$. Since

$$| \Delta_j | \leqq [| \Delta_j z^1 |^2 + | \Delta_j z^2 |^2]/2$$

and

$$| \Delta_j |^2 \leqq [| \Delta_j z^1 |^4 + | \Delta_j z^2 |^4]/2,$$

by (1-1) and (4-12) we have

$$| E(\Delta_j \mid \mathfrak{F}(s_j) \mid \leqq K(t_j - s_j),$$

$$| E(\Delta_j^2 \mid \mathfrak{F}(s_j) \mid \leqq K(t_j - s_j).$$

Now the proof of Lemma III(2-2) applies without further change.

(4-13) *LEMMA* *Let* f *be integrable with respect to* (z^1, z^2) *over* $[a, b]$, *and let* $\| f(t) \|_2^2$ *have a Riemann or belated or Cauchy integral over* $[a, b]$. *Then, with* C *as in Lemma* (4-11),

$$(*) \qquad \left\| \int_a^b f(t) \, dz^1(t) \, dz^2(t) \right\|_2 \leqq C \left\{ \int_a^b \| f(t) \|_2^2 \, dt \right\}^{1/2}.$$

Let $\Pi(1), \Pi(2), \ldots$ be a sequence of Cauchy partitions with mesh tending to 0. By Lemma (4-11),

$$\| S(\Pi(n); f; z^1, z^2) \|_2 \leqq C\{S(\Pi(n); \| f \|_2^2; t)\}^{1/2}.$$

As $n \to \infty$, the right member tends to the right member of (*). Since $S(\Pi(n); f; z^1, z^2)$ converges in probability to the integral of f with respect to (z^1, z^2), the norm of the integral is not greater than the lower limit of $\| S(\Pi(n); f; z^1, z^2) \|_2$, and the conclusion of the lemma is valid.

(4-14) *THEOREM* *Let hypotheses* (1-1) *and* (4-12) *be satisfied. Let* $\phi(\boldsymbol{\tau}, \omega)$, $f^\beta(\boldsymbol{\tau}, \omega)$, $g_\rho{}^\beta(\boldsymbol{\tau}, \omega)$, $h_{\rho,\sigma}^{\beta}(\boldsymbol{\tau}, \omega)$ $(\beta = 1, 2; \rho, \sigma = 1, \ldots, r)$ *be measurable processes on* T *adapted to* $\mathfrak{F}_{\boldsymbol{\tau}}$, *continuous in probability a.e. in* $[a, b]$, *and a.s. sample-function bounded. Let* $x^1(\mathbf{t}, \omega)$,

$x^2(\mathbf{t}, \boldsymbol{\omega})$ be processes on $[a, b]$ that satisfy

$$dx^\beta = f^\beta(t)\, dt + \sum_{\rho=1}^{r} g_\rho{}^\beta(t)\, dz^\rho(t) \tag{4-15}$$

$$+ \sum_{\rho,\sigma=1}^{r} h^\beta_{\rho,\sigma}(t)\, dz^\rho(t)\, dz^\sigma(t) \qquad (\beta = 1, 2).$$

Then ϕ is integrable with respect to (x^1, x^2) over $[a, b)$, and

$$\int_a^b \phi(t)\, dx^1(t)\, dx^2(t) = \sum_{\rho,\sigma=1}^{r} \int_a^b \phi(t) g_\rho{}^1(t) g_\sigma{}^2(t)\, dz^\rho(t)\, dz^\sigma(t). \tag{4-16}$$

To simplify notation we define $z^0(t) = t$.

There is no loss of generality in assuming that the same z^ρ are used in defining x^1 and x^2, since the coefficients $g_\rho{}^\beta$ and $g^\beta_{\rho,\sigma}$ are permitted to vanish identically. By the linearity theorem [II(4-2)], it is sufficient to establish the conclusion when for each of the two values of β the right member of (4-15) contains only a single nonzero term. That is, we can simplify (4-15) to the form

$$dx^1 = g^1(t) \prod_{\rho \in A} dz^\rho, \qquad dx^2 = g^2(t) \prod_{\sigma \in B} dz^\sigma,$$

where A and B each have either one element or two elements. Equation (4-16) then takes the form

$$\int_a^b \phi(t)\, dx^1(t)\, dx^2(t) = \int_a^b \phi(t) g^1(t) g^2(t) \prod_{\rho \in A} dz^\rho(t) \prod_{\sigma \in B} dz^\sigma(t). \tag{4-17}$$

We first add the supplementary hypothesis

(4-18) the processes ϕ, g^1 and g^2 are bounded on $T \times \Omega$.

If this is the case, there is no further loss of generality in assuming

$$|\phi(\tau, \omega)| \leqq 1, \qquad |g^\beta(\tau, \omega)| \leqq 1 \qquad (\beta = 1, 2; \quad \tau \in T, \quad \omega \in \Omega). \tag{4-19}$$

Let ϵ be positive. With the C of III(2-2), let

$$\Pi = (t_1, \ldots, t_{m+1}; \tau_1, \ldots, \tau_m)$$

be a belated partition of $[a, b]$ with

$$\text{mesh}\, \Pi < \epsilon^2/4C^4. \tag{4-20}$$

Let q be a power of 2, and let Π_q be the Cauchy partition obtained by inserting equally spaced division points

$$t_{j,1} = t_j,\ t_{j,2},\ \ldots,\ t_{j,q+1} = t_{j+1}$$

in each interval $[t_j, t_{j+1}]$ of Π. (By definition of Cauchy partition, for each interval of Π_q the corresponding evaluation point is its left end.)

We use the abbreviation

$$\text{(4-21)} \quad \Delta'_{j,i} = \prod_{\rho \in A} [z^\rho(t_{j,i+1}) - z^\rho(t_{j,i})],$$

$$\Delta''_{j,i} = \prod_{\sigma \in B} [z^\sigma(t_{j,i+1}) - z^\sigma(t_{j,i})] \quad (j = 1, \ldots, m;\ i = 1, \ldots, q).$$

Then by Theorem III(2-3)

$$\text{(4-22)} \quad S(\Pi; \phi; x^1, x^2) = \lim_{q\to\infty} \sum_{j=1}^{m} \phi(\tau_j) \left\{\sum_{i=1}^{q} g^1(t_{j,i})\ \Delta'_{j,i}\right\}\left\{\sum_{h=1}^{q} g^2(t_{j,h})\ \Delta''_{j,h}\right\}.$$

The right member of (4-22) can be written as

$$\sum_{j=1}^{m} (V_j^1 + V_j^2 + V_j^3),$$

where

$$\text{(4-23)} \qquad V_j^1 = \sum_{i=1}^{q} \phi(\tau_j) g^1(t_{j,i}) g^2(t_{j,i})\ \Delta'_{j,i}\ \Delta''_{j,i},$$

$$\text{(4-24)} \qquad V_j^2 = \sum_{i=1}^{q} \left[\sum_{h=1}^{i-1} \phi(\tau_j) g^2(t_{j,h})\ \Delta''_{j,h}\right] g^1(t_{j,i})\ \Delta'_{j,i},$$

$$\text{(4-25)} \qquad V_j^3 = \sum_{h=1}^{q} \left[\sum_{i=1}^{h-1} \phi(\tau_j) g^1(t_{j,i})\ \Delta'_{j,i}\right] g^2(t_{j,h})\ \Delta''_{j,h}.$$

The quantity in brackets in (4-24) can be estimated by Lemma III(2-2) if B has only one element and by Lemma (4-11) if B has two elements. In either case, for each j in $\{1, \ldots, m\}$ and each i in $\{1, \ldots, q\}$, we obtain

$$\text{(4-26)} \qquad \left\| \sum_{h=1}^{i-1} \phi(\tau_j) g^2(t_{j,h})\ \Delta''_{j,h} \right\|$$

$$\leqq C\left\{\sum_{h=1}^{i-1} \| \phi(\tau_j) g^2(t_{j,h}) \|^2\ (t_{j,h+1} - t_{j,h})\right\}^{1/2}$$

$$\leqq C\{\text{mesh } \Pi\}^{1/2}$$

$$\leqq \epsilon/2C.$$

Returning with this to (4-24), we apply Lemma III(2-2) if A has only one element and Lemma (4-11) if A has two elements. We obtain for all q

$$\text{(4-27)} \qquad \left\|\sum_{j=1}^{m} V_j^2\right\| = \left\|\sum_{j=1}^{m}\sum_{i=1}^{q}\left\{\left[\sum_{h=1}^{i-1}\phi(\tau_j)g^2(t_{j,h})\,\Delta''_{j,h}\right]g^1(t_{j,i})\right\}\Delta'_{j,i}\right\|$$

$$\leqq C\left\{\sum_{j=1}^{m}\sum_{i=1}^{q}(\epsilon^2/4C^2)(t_{j,i+1}-t_{j,i})\right\}^{1/2}$$

$$< (\epsilon/2)(b-a)^{1/2}.$$

In a like manner we prove that for all q

$$\text{(4-28)} \qquad \left\|\sum_{j=1}^{m} V_j^3\right\| < (\epsilon/2)(b-a)^{1/2}.$$

If either A or B has two members, the products $\Delta'_{j,i}\,\Delta''_{j,i}$ will contain at least three factors of the form $z(t_{j,i+1}) - z(t_{j,i})$. So by Theorem III(5-1) the sum $\sum_{j=1}^{m} V_j^1$ will tend to 0 in probability as mesh $\Pi \to 0$. By (4-27) and (4-28) we know that the r.v.'s

$$\lim_{q\to\infty}\sum_{j=1}^{m} V_j^2, \qquad \lim_{q\to\infty}\sum_{j=1}^{m} V_j^3$$

tend to 0 in L_2-norm, hence in probability, as mesh $\Pi \to 0$. So unless A and B contain only one member each, the right member of (4-22) tends to 0 in probability as mesh $\Pi \to 0$, and so the left member of (4-16) is 0. So is the right member, by Theorem III(5-1), and so (4-16) holds.

A similar proof applies if A and B each contain just one member, but at least one of the z^ρ, z^σ is $z^0(t) = t$.

Suppose now that A and B each consist of exactly one member of the set $\{1, \ldots, r\}$; say A contains ρ only and B contains σ only. We then define

$$\text{(4-29)} \qquad Z(t) = \int_a^t g^1(s)g^2(s)\,dz^\rho(s)\,dz^\sigma(s).$$

By Theorem (4-1), ϕ is integrable with respect to Z, and

$$\text{(4-30)} \qquad \int_a^b \phi(t)\,dZ(t) = \int_a^b \phi(t)g^1(t)g^2(t)\,dz^\rho(t)\,dz^\sigma(t).$$

Hence, with limits in the sense of convergence in probability,

(4-31)

$$\int_a^b \phi(t)g^1(t)g^2(t)\,dz^\rho(t)\,dz^\sigma(t)$$

$$= \lim_{\text{mesh }\Pi\to 0} S(\Pi;\phi;Z)$$

$$= \lim_{\text{mesh }\Pi\to 0} \sum_{j=1}^{m} \phi(\tau_j)[Z(t_{j+1}) - Z(t_j)]$$

$$= \lim_{\text{mesh }\Pi\to 0} \sum_{j=1}^{m} \phi(\tau_j)$$

$$\times \left\{\lim_{q\to\infty} \sum_{i=1}^{q} g^1(\tau_{j,i})g^2(\tau_{j,i})[z^\rho(t_{j,i+1}) - z^\rho(t_{j,i})][z^\sigma(t_{j,i+1}) - z^\sigma(t_{j,i})]\right\}$$

$$= \lim_{\text{mesh }\Pi\to 0} \lim_{q\to\infty} \sum_{j=1}^{m} V_j^1.$$

But since the estimates (4-27) and (4-28) are independent of q, we can find a q so large that for all Π with sufficiently small mesh [specified by (4-20)], the left members of (4-22) and (4-31) differ arbitrarily little in the metric of convergence in probability, and (4-16) is established in this case also. Thus (4-16) is established under the supplementary hypothesis (4-18).

If instead of assuming (4-18) we merely assume that ϕ, g^1, and g^2 satisfy the hypotheses of the theorem we use the truncations ϕ_M, etc. as frequently before to obtain the theorem as stated. The details of the proof hardly need repetition.

5 *Extension to Itô-belated integrals*

In this section we shall deduce analogs for Itô-belated integrals of all the theorems in Sections 1 and 2, and also of Theorem (4-1).

The analog of Theorem (1-7) is the following:

(5-1) *THEOREM Let the standing hypotheses* (1-1) *be satisfied. For all ω except those in a negligible set let $f(\mathbf{t}, \omega)^2$ be (Lebesgue) integrable* *over* $[a, b]$. *Then there is a process* $F(\mathbf{t}, \omega)$ *on* $[a, b]$ *with continuous sample functions that is a strict version of the indefinite (Itô-belated) integral*

$$\int_a^t f(s, \omega)\, dz(s, \omega) \qquad (a \leqq t \leqq b). \tag{5-2}$$

We first prove the conclusion under the additional hypothesis

(5-3) $\|f(t)\|_2^2$ is Lebesgue integrable over $[a, b]$.

For each t in $[a, b]$ we choose a version $F_1(t, \omega)$ of the integral (5-2). By Corollary III(2-11), if $a \leqq t' \leqq t'' \leqq b$, then

$$\| F_1(t'', \omega) - F_1(t', \omega) \|_2 \leqq C\left\{\int_{t'}^{t''} \|f(s)\|_2^2\, ds\right\}^{1/2}.$$

Hence F_1 is L_2-continuous, and consequently continuous in probability, on $[a, b]$.

We now follow the proof of Theorem (1-7), with only obvious and trivial changes, to obtain the conclusion of Theorem (5-1) under the supplementary hypothesis (5-3).

Reverting to the hypotheses of the theorem, we define

$$F_2(t, \omega) = \int_a^t f(s, \omega)^2\, ds.$$

By hypothesis this is a.s. finite. As shown in the proof of Theorem III(8-3) (where F_2 was denoted by F), F_2 is a r.v. adapted to $\mathfrak{F}_{\tau}$. We again define $I_n(\mathbf{t}, \omega)$ to be the function such that for each t in $[a, b]$, $I_n(t, \omega)$ is the indicator function of the set

$$\Omega_{t,n} = \{\omega \in \Omega : F_2(t, \omega) \leqq n\}.$$

Then again

$$I_n(t, \omega) f(t, \omega) = f(t, \omega) \qquad (\omega \in \Omega_{b,n};\ a \leqq t \leqq b).$$

The previous proof applies with $I_n f$ in place of f. If F_n $(n = 3, 4, 5, \ldots)$ is a version of the indefinite integral of $I_n f$ with continuous sample functions, by Corollary II(2-22) we may suppose that $F_{n+1}(t, \omega) = F_n(t, \omega)$ for ω in $\Omega_{b,n}$. Except on the negligible set $\Omega \backslash \bigcup \Omega_{b,n}$, the common value of the $F_n(\mathbf{t}, \omega)$ for all large n has continuous sample functions. If $\omega \in \Omega \backslash \bigcup \Omega_{b,n}$, we define $F(t, \omega)$ as in the last two sentences of the proof of Theorem (1-7).

(5-4) THEOREM *Let the standing hypotheses* (1-1) *be satisfied. For all* ω *except those in a negligible set, let* $f(\mathbf{t}, \omega)$ *be Lebesgue integrable over* $[a, b]$. *Then there exists a process* $F(\mathbf{t}, \omega)$ *with continuous sample functions that is a strict version of the integral*

(5-5) $$\int_a^t f(s, \omega)\, dz^1(s, \omega)\, dz^2(s, \omega).$$

We prove this first under the additional hypotheses

(5-6) $E(|f(t)|)$ is Lebesgue integrable over $[a, b]$,

(5-7) $f \geqq 0$,

(5-8) $z^1 = z^2$,

and we define $z = z^1 = z^2$. For each t we choose a version $F_1(\mathbf{t}, \omega)$ of the integral (5-5). By Corollary III(3-11), F_1 is L_1-continuous (hence continuous in probability) on $[a, b]$. We can now follow the proof of Theorem (1-21) with only trivial changes to establish the conclusion of Theorem (5-4), under supplementary hypotheses (5-6), (5-7), and (5-8). The latter two are removed just as in the proof of Theorem (1-21). Hypothesis (5-6) is removed by a device similar to that used at the end of the proof of Theorem (5-1), and used earlier in the proof of Theorem III(8-6).

(5-9) *THEOREM Under the hypotheses of Theorem* (5-4), *there is a process* $F(\mathbf{t}, \omega)$ *with continuous sample functions that is a strict version of*

$$\int_a^t f(s, \omega)\, ds \qquad (a \leqq t \leqq b).$$

The proof is a simplified version of the proof of Theorem (5-4).

Next we generalize the Itô differentiation lemma to apply to Itô-belated integrals.

(5-10) *THEOREM Let hypotheses* (1-1) *be satisfied. Let* $f^i(\boldsymbol{\tau}, \omega)$, $g_\rho{}^i(\boldsymbol{\tau}, \omega)$, $h^i_{\rho,\sigma}(\boldsymbol{\tau}, \omega)$ $(i = 1, \ldots, n;\ \rho, \sigma = 1, \ldots, r)$ *be measurable processes on* T *adapted to* $\mathfrak{F}_{\boldsymbol{\tau}}$ *such that for all* ω *except those in a negligible set all the functions* $f^i(\mathbf{t}, \omega)$, $[g_\rho{}^i(\mathbf{t}, \omega)]^2$, $h^i_{\rho,\sigma}(\mathbf{t}, \omega)$ *are Lebesgue integrable over* $[a, b]$. *Assume*

(i) $x^1, \ldots, x^n$ *are processes on* $[a, b]$ *with continuous sample functions, and they satisfy*

(5-11) $$dx^i = f^i(t)\, dt + \sum_{\rho=1}^{r} g_\rho{}^i(t)\, dz^\rho(t) + \sum_{\rho,\sigma=1}^{r} h^i_{\rho,\sigma}(t)\, dz^\rho(t)\, dz^\sigma(t)$$

strictly, and

(ii) *$F(\mathbf{t}, \mathbf{x})$ is a function on $[a, b] \times R^n$ that is continuous together with its partial derivatives*

$$F_0 = \partial F/\partial t, \qquad F_i = \partial F/\partial x^i, \qquad F_{i,j} = \partial^2 F/\partial x^i\, \partial x^j \quad (i, j = 1, \ldots, n)$$

on $[a, b] \times R^n$.

Then the composite function

$$F(\mathbf{t}, x(\mathbf{t}, \boldsymbol{\omega})) = F(\mathbf{t}, x^1(\mathbf{t}, \boldsymbol{\omega}), \ldots, x^n(\mathbf{t}, \boldsymbol{\omega}))$$

strictly satisfies the differential equation

$$dF = F_0\, dt + \sum_{i=1}^{n} F_i f^i\, dt + \sum_{i=1}^{n} \sum_{\rho=1}^{r} F_i g_\rho{}^i\, dz^\rho + \sum_{i=1}^{n} \sum_{\rho,\sigma=1}^{r} \left\{ F_i h^i_{\rho,\sigma} + \frac{1}{2} \sum_{j=1}^{n} F_{i,j} g_\rho{}^i g_\sigma{}^j \right\} dz^\rho\, dz^\sigma, \tag{5-12}$$

wherein the F_0, F_i, and $F_{i,j}$ are evaluated at $(t, x^1(t, \omega), \ldots, x^n(t, \omega))$.

We first prove this under these supplementary hypotheses:

(5-13) all partial derivatives of F of order $\leqq 3$ are bounded and continuous on $[a, b] \times R^n$;

(5-14) the processes f^i, $g_\rho{}^i$, $h^i_{\rho,\sigma}$ are bounded on $T \times \Omega$ $(i = 1, \ldots, n;$ $\rho, \sigma = 1, \ldots, r)$;

(5-15) $F(\mathbf{t}, \mathbf{x})$ is independent of t. [We denote it by $F(\mathbf{x})$ for brevity.]

Let δ be a.e. > 0 on $[a, b)$, and let δ^* be positive. Given a (δ, δ^*) belated partition $\Pi = \{(A_1, \tau_1), \ldots, (A_m, \tau_m)\}$ in $[a, b)$, for each t in $[a, b)$ we define $\Pi(t)$ to be the collection of all those pairs $(A_i \cap [a, t), \tau_i)$ $(i = 1, \ldots, m)$ for which $A_i \cap [a, t)$ is not empty. Then $\Pi(t)$ is a (δ, δ^*) belated partition in $[a, t)$. We define

$$y^i(t) = x^i(a) + S(\Pi(t); f^i; t) + \sum_{\rho=1}^{r} S(\Pi(t); g_\rho{}^i; z^\rho) + \sum_{\rho,\sigma=1}^{r} S(\Pi(t); h^i_{\rho,\sigma}; z^\rho, z^\sigma) \qquad (i = 1, \ldots, n).$$

Let $[\alpha_1, \beta_1), \ldots, [\alpha_{h(t)}, \beta_{h(t)})$ be the intervals in the partition $\Pi(t)$, and for each function ϕ on $[a, b)$ define

$$\Delta_j \phi = \phi(\beta_j) - \phi(\alpha_j).$$

If t' and t'' are both in the same subinterval of $[a, b) \setminus \bigcup_{j=1}^m A_j$, then $\Pi(t')$ and $\Pi(t'')$ are the same, so $y(t') = y(t'')$. Hence

$$F(x(t)) = F(x(t)) - F(y(t)) + F(x(a)) + \sum_{j=1}^{h(t)} \Delta_j F(y(t)). \tag{5-16}$$

Since F has bounded partial derivatives, it satisfies a Lipschitz condition with a constant L; we may assume $L \geqq 1$. By the definition of the Itô-belated integral, as (δ, δ^*) shrinks, $y(t, \omega)$ tends in probability to $x(t, \omega)$ for each t in $[a, b)$. Since

$$\bar{\rho}(F(x(t)), F(y(t)) \leqq L\bar{\rho}(x(t) - y(t), 0),$$

this implies that as (δ, δ^*) shrinks,

$$\lim \bar{\rho}(F(x(t)), F(y(t)) = 0.$$

By the theorem of the mean, on the segment joining $y(\alpha_j, \omega)$ to $y(\beta_j, \omega)$ there is a point $\bar{y}_j(\omega)$ such that

$$\Delta_j F(y(t)) = \sum_{i=1}^{n} F_i(y(\alpha_j)) \Delta_j y + \frac{1}{2} \sum_{i,l=1}^{n} F_{i,l}(y(\alpha_j)) \Delta_j y^i \Delta_j y^l + \frac{1}{6} \sum_{i,l,m=1}^{n} F_{i,l,m}(\bar{y}_j) \Delta_j y^i \Delta_j y^l \Delta_j y^m.$$

In this we substitute

$$\Delta_j y^i = f^i(\tau_j) \Delta_j t + \sum_{\rho=1}^{r} g_\rho{}^i(\tau_j) \Delta_j z^\rho + \sum_{\rho,\sigma=1}^{r} h^i_{\rho,\sigma}(\tau_j) \Delta_j z^\rho \Delta_j z^\sigma$$

and sum over $j = 1, \ldots, h(t)$. The terms in the sum we group according to the superscripts ρ, σ, etc. on the factors $\Delta_j z^\rho$, etc.; each group contains $h(t)$ terms, since j ranges over $\{1, \ldots, h(t)\}$. By Theorem III(8-10), all those sums with three or more factors $\Delta_j t$ or $\Delta_j z^\rho$ will tend to 0 in probability as (δ, δ^*) shrinks. By Theorem III(8-11), the same is true of the sums with at least one factor $\Delta_j t$ and one other factor $\Delta_j t$ or $\Delta_j z^\rho$. Therefore by (5-16)

$$\text{(5-17)} \quad F(x(t)) = F(x(a)) + \sum_{j=1}^{h(t)} \Bigg\{ \sum_{i=1}^{n} F_i(y(\alpha_j)) f^i(\tau_j) \Delta_j t + \sum_{i=1}^{n} \sum_{\rho=1}^{r} F_i(y(\alpha_j)) g_\rho{}^i(\tau_j) \Delta_j z^\rho + \sum_{i=1}^{n} \sum_{\rho,\sigma=1}^{r} F_i(y(\alpha_j)) h^i_{\rho,\sigma}(\tau_j) \Delta_j z^\rho \Delta_j z^\sigma + \frac{1}{2} \sum_{i,l=1}^{n} F_{i,l}(y(\alpha_j)) g_\rho{}^i(\tau_j) g_\sigma{}^i(\tau_j) \Delta_j z^\rho \Delta_j z^\sigma \Bigg\} + R_{\text{II}},$$

where R_{II} is a r.v. that converges in probability to 0 as (δ, δ^*) shrinks.

Since F_i and $F_{i,l}$ are Lipschitzian, and $\bar{\rho}(x(t), y(t))$ converges uniformly to 0 as (δ, δ^*) shrinks, the r.v.'s $F_i(y(t, \omega))$ and $F_{i,l}(y(t, \omega))$ converge in $\bar{\rho}$-distance to $F_i(x(t, \omega))$ and $F_{i,l}(x(t, \omega))$, respectively, uniformly on $[a, b]$. By Corollaries III(2-11) and III(3-11), which are valid for the Itô-belated integral, the $x^i(\mathbf{t}, \boldsymbol{\omega})$ are L_1-continuous. Hence so are $F_i(x(\mathbf{t}, \boldsymbol{\omega}))$ and $F_{i,l}(x(\mathbf{t}, \boldsymbol{\omega}))$. Since they are bounded processes, they are also L_2-continuous. Therefore as (δ, δ^*) shrinks, $F_i(x(\alpha_j))$ tends uniformly in L_2-distance to $F_i(x(\tau_j))$, and likewise for the other functions. So if in (5-17) we replace all the $y(\alpha_j)$ by $x(\tau_j)$, as in Corollaries III(2-11) and III(3-11) the change in the right member tends to 0 in L_1-distance as (δ, δ^*) shrinks. Therefore with this replacement, (5-17) remains valid. But (5-17) as thus changed, with $\bar{\rho}(R_{\mathrm{II}}, 0) \to 0$, implies

$$
\begin{aligned}
(5\text{-}18) \quad F(x(t)) = F(x(a)) &+ \sum_{i=1}^{n} \int_a^t F_i(x(s)) f^i(s)\, ds \\
&+ \sum_{i=1}^{n} \sum_{\rho=1}^{r} \int_a^t F_i(x(s)) g_\rho{}^i(s)\, dz^\rho(s) \\
&+ \sum_{i=1}^{n} \sum_{\rho,\sigma=1}^{r} \int_a^t \Big\{ F_i(x(s)) h^i_{\rho,\sigma}(s) \\
&+ \frac{1}{2} \sum_{l=1}^{n} F_{i,l}(x(s)) g_\rho{}^i(s) g^l_\sigma(s) \Big\}\, dz^\rho(s)\, dz^\sigma(s),
\end{aligned}
$$

which is the integrated form of Eq. (5-12) under assumption (5-15).

Removal of assumption (5-15) requires nothing more than notational trickery. We define $x^0(t) = z^0(t) = t$, and rewrite $F(t, x)$ as $F(x^0, \ldots, x^n)$. Then (5-18) holds with the trivial change that ρ and σ range from 0 to n. Among the second-order integrals in the right member of (5-18) are some with ρ or σ (or both) equal to 0. These vanish, by Theorem III(8-11), and then (5-18) takes the form (5-12).

It remains to remove the supplementary hypotheses (5-13) and (5-14). We first replace (5-14) by the weaker supplementary hypothesis

(5-19) all the functions $E(|f^i|)$, $E([g_\rho{}^i]^2)$, $E(|h^i_{\rho,\sigma}|)$ are integrable over $[a, b]$.

This is done as in Section 2, beginning with the sentence containing (2-20). Removal of (5-13) can be effected just as in the proof of Theorem (2-4); and removal of (5-19) is done with the help of the functions I_n just as in the proof of Theorem (5-1). Likewise, the discussion of strictness is just as in Theorem (2-4).

The substitution theorem (4-1) can also be extended to the Itô-belated integral, as follows.

(5-20) *THEOREM Let the standing hypotheses* (1-1) *be satisfied. Let* $\phi, f, g_\rho, h_{\rho,\sigma}$ ($\rho, \sigma = 1, \ldots, r$) *be measurable processes on T adapted to* $\mathfrak{F}_\tau$ *such that the sample functions of* ϕ *are a.s. bounded and the integrals*

$$\int_a^b |f(t,\omega)|\,dt, \qquad \int_a^b |g_\rho(t,\omega)|^2\,dt, \qquad \int_a^b |h_{\rho,\sigma}(t,\omega)|\,dt$$

are a.s. finite. Define

$$x(t) = x_0 + \int_a^t f(s)\,ds + \sum_{\rho=1}^{r} \int_a^t g_\rho(s)\,dz^\rho(s) + \sum_{\rho,\sigma=1}^{r} \int_a^t h_{\rho,\sigma}(s)\,dz^\rho(s)\,dz^\sigma(s),$$

where x_0 *is any* $\mathfrak{F}_a$*-measurable r.v. Then all the Itô-belated integrals in the equation*

(5-21)
$$\int_a^b \phi(t)\,dx(t) = \int_a^b \phi(t)f(t)\,dt + \sum_{\rho=1}^{r} \int_a^b \phi(t)g_\rho(t)\,dz^\rho(t) + \sum_{\rho,\sigma=1}^{r} \int_a^b \phi(t)h_{\rho,\sigma}(t)\,dz^\rho(t)\,dz^\sigma(t)$$

exist, and the equation is valid.

Merely to simplify notation we assume $[a, b) = [0, 1)$.

By Theorems III(8-3) and III(8-6) the integrals in the right member of (5-21) exist. We first prove the theorem under the supplementary hypotheses

(5-22) the integrals

$$\int_0^1 E(|f(t)|)\,dt, \qquad \int_0^1 E(g_\rho(t)^2)\,dt, \qquad \int_0^1 E(|h_{\rho,\sigma}(t)|)\,dt$$

are all finite,

(5-23) ϕ is bounded.

If we assume (5-23), there is no further loss of generality in assuming $|\phi| \leqq 1$.

Consider first any one of the first-order integrals; for typographical convenience we omit the affix ρ. Let ϵ be positive. By Theorem I(3-8), there is a bounded simple process γ on $[a, b]$, adapted to $\mathfrak{F}_\tau$, such that

$$\int_0^1 \| g(t) - \gamma(t) \|_2^2 \, dt < (\epsilon/3C)^2. \tag{5-24}$$

Let $[t_j^*, t_{j+1}^*)$ $(j = 1, \ldots, h)$ be intervals of constancy of γ, where $t_1^* = a$ and $t_{h+1}^* = b$, and let Γ be an upper bound for $|\gamma(t, \omega)|$. By Theorem III(8-3), there exist an a.e. positive $\delta(\mathbf{t})$ and a positive number δ^* such that for every (δ, δ^*) belated partition Π in $[a, b)$,

$$\left\| S(\Pi; \phi\gamma; z) - \int_0^1 \phi(t)\gamma(t) \, dz(t) \right\|_2 < \epsilon/3. \tag{5-25}$$

We may suppose the $\delta(\tau)$ so chosen that if $t_j^* \leqq \tau < t_{j+1}^*$ then $\tau + \delta(\tau) \leqq t_{j+1}^*$; otherwise we need only to replace $\delta(\tau)$ by $\min\{\delta(\tau), t_{j+1}^* - \tau\}$. Then, with the usual notation for Π, γ is constantly equal to $\gamma(\tau_j)$ on the interval A_j. We define $\phi_\Pi(t)$ to be $\phi(\tau_j)$ on A_j and to be 0 on $[0, 1)\backslash\cup A_j$. Then

$$S(\Pi; \phi\gamma; z) = \sum_{j=1}^{h} \phi(\tau_j)\gamma(\tau_j) \, \Delta_j z = \int_0^1 \phi_\Pi(t)\gamma(t) \, dz(t). \tag{5-26}$$

Since $|\phi| \leqq 1$ we also have $|\phi_\Pi| \leqq 1$. So by (5-24)

$$\left\{\int_0^1 E[\phi(t)^2[g(t) - \gamma(t)]^2] \, dt\right\}^{1/2} \leqq \epsilon/3C,$$

$$\left\{\int_0^1 E[\phi_\Pi(t)^2[g(t) - \gamma(t)]^2] \, dt\right\}^{1/2} \leqq \epsilon/3C.$$

By Corollary III(2-11),

$$\left\| \int_0^1 \phi(t)[g(t) - \gamma(t)] \, dz(t) \right\|_2 \leqq \epsilon/3,$$

$$\left\| \int_0^1 \phi_\Pi(t)[g(t) - \gamma(t)] \, dz(t) \right\|_2 \leqq \epsilon/3.$$

From these inequalities, with (5-25) and (5-26), we obtain

$$\left\| \int_0^1 [\phi(t) - \phi_\Pi(t)]g(t) \, dz(t) \right\|_2 < \epsilon.$$

Therefore (restoring the affix ρ), as (δ, δ^*) shrinks,

$$\int_0^1 \phi_\Pi(t) g_\rho(t) \, dz^\rho(t) \to \int_0^1 \phi(t) g_\rho(t) \, dz^\rho(t) \tag{5-27}$$

in probability.

By a similar proof, using L_1-distance instead of L_2-distance, we establish that as (δ, δ^*) shrinks,

$$\int_0^1 \phi_\Pi(t) f(t)\, dt \to \int_0^1 \phi(t) f(t)\, dt \tag{5-28}$$

and

$$\int_0^1 \phi_\Pi(t) h_{\rho,\sigma}(t)\, dz^\rho(t)\, dz^\sigma(t) \to \int_0^1 \phi(t) h_{\rho,\sigma}(t)\, dz^\rho(t)\, dz^\sigma(t). \tag{5-29}$$

For every partition Π in $[0, 1)$,

$$\begin{aligned} S(\Pi; \phi; x) &= \sum_{j=1}^{h} \phi(\tau_j) \left\{ \int_{A_j} f(t)\, dt + \sum_{\rho=1}^{r} \int_{A_j} g_\rho(t)\, dz^\rho(t) \right. \\ &\qquad \left. + \sum_{\rho,\sigma=1}^{r} \int_{A_j} h_{\rho,\sigma}(t)\, dz^\rho(t)\, dz^\sigma(t) \right\} \\ &= \int_0^1 \phi_\Pi(t) f(t)\, dt + \sum_{\rho=1}^{r} \int_0^1 \phi_\Pi(t) g_\rho(t)\, dz^\rho(t) \\ &\qquad + \sum_{\rho,\sigma=1}^{r} \int_0^1 \phi_\Pi(t) h_{\rho,\sigma}(t)\, dz^\rho(t)\, dz^\sigma(t). \end{aligned} \tag{5-30}$$

By (5-27), (5-28), and (5-29), as (δ, δ^*) shrinks, the last expression for $S(\Pi; \phi; x)$ converges in probability to the right member of (5-21). This proves the existence of the integral in the left member of (5-21) and also proves the validity of the equation, subject to the supplementary hypotheses (5-22) and (5-23).

If (5-23) does not hold, for each positive M we define $\phi_M = \operatorname{mid}\{\phi, M, -M\}$. By the preceding proof, Eq. (5-21) holds with ϕ_M in place of ϕ. An argument used several times before this shows that (5-21) holds as written.

If (5-22) is not satisfied, we define and use the sets I_n as in the proof of Theorem (5-1) to complete the proof.

V Stochastic Differential Equations

1 Existence of solutions of stochastic differential equations

As before, T will denote a set of real numbers and $[a, b]$ an interval contained in T. The standard form of a system of ordinary differential equations on T is

$$(1\text{-}1)\quad x^i(t) = x^i(a) + \int_a^t f^i(s, x(s))\, ds \qquad (i = 1, \ldots, n;\ a \leqq t \leqq b).$$

The most straightforward extension of (1-1) to stochastic processes is to add one or more terms of the form

$$(1\text{-}2)\qquad \int_a^t g_\rho{}^i(s, x(s))\, dz^\rho(s)$$

to the right member, where each z^ρ is a process $z^\rho(\mathbf{t}, \omega)$ on $[a, b]$ and $(\Omega, \mathcal{A}, P)$ is a probability triple. A further generalization, and an important one, is to add one or more second-order integrals along with the first-order integrals (1-2). We do not go beyond second-order integrals, because our primary interest is in processes z^ρ with a.s. continuous sample paths, and for these all third and higher order integrals are a.s. 0.

Still a further generalization is to let the coefficients f^i, $g_\rho{}^i$, etc. depend directly on ω as well as on t and $(x^1, \ldots, x^n)$. We thus reach what for us is the standard form of a stochastic differential equation,

$$(1\text{-}3) \quad x^i(t, \omega) = x^i(a, \omega) + \sum_{\rho=1}^{r} \int_a^t g_\rho{}^i(s, x(s, \omega), \omega)\, dz^\rho(s, \omega)$$
$$+ \sum_{\rho,\sigma=1}^{r} \int_a^t h^i_{\rho,\sigma}(s, x(s, \omega), \omega)\, dz^\rho(s, \omega)\, dz^\sigma(s, \omega)$$
$$(i = 1, \ldots, n; \quad a \leqq t \leqq b).$$

Of course we cannot make any statements about solutions until we make some assumptions about the data in (1-3). At first glance a term seems lacking from (1-3), namely the integral in the right member of (1-1). It is not lacking, since we are permitted to use the identity function $\mathbf{t}: t \mapsto t$ as one of the z^ρ, say z^1, and define $g_1{}^i$ to be f^i.

In order to be able to treat Eq. (1-3) we have to make assumptions about the z^ρ slightly stronger than those in IV(1-1), and we also have to make assumptions about the $g_\rho{}^i$ and $h^i_{\rho,\sigma}$. These assumptions we now collect as follows.

(1-4) *STANDING HYPOTHESES*

(i) *$(\Omega, \mathcal{A}, P)$ is a probability triple; T is a set of real numbers; $[a, b]$ is an interval contained in T.*

(ii) *$\{\mathfrak{F}_\tau : \tau \in T\}$ is a family of σ-subalgebras of $\mathcal{A}$ such that if τ and τ' are in T and $\tau < \tau'$ then $F_\tau \subset F_{\tau'}$.*

(iii) *Every process denoted by z, with or without affixes will be a real-valued process adapted to $\mathfrak{F}_{\boldsymbol{\tau}}$, having a.s. continuous sample functions, and satisfying with some K the inequalities*

$$|\, E(z(t) - z(s) \mid \mathfrak{F}_s) \,| \leqq K(t - s),$$
$$E([z(t) - z(s)]^2 \mid \mathfrak{F}_s) \leqq K(t - s),$$
$$E([z(t) - z(s)]^4 \mid \mathfrak{F}_s) \leqq K(t - s)$$

a.s. whenever $a \leqq s \leqq t \leqq b$.

(iv) *The functions, $g_\rho{}^i$, $h^i_{\rho,\sigma}$ are defined for all t in $[a, b]$, all x in R^n and all ω in Ω; and there is an a.s. finite-valued r.v. $L(\omega)$*

such that for all t in $[a, b]$, all ω in Ω and all x_1 and x_2 in R^n

$$| g_\rho{}^i(t, x_1, \omega) - g_\rho{}^i(t, x_2, \omega) | \leqq L(\omega) | x_1 - x_2 |,$$

$$| h^i_{\rho,\sigma}(t, x_1, \omega) - h^i_{\rho,\sigma}(t, x_2, \omega) | \leqq L(\omega) | x_1 - x_2 |,$$

$$| g_\rho{}^i(t, 0, \ldots, 0, \omega) | \leqq L(\omega),$$

$$| h^i_{\rho,\sigma}(t, 0, \ldots, 0, \omega) | \leqq L(\omega).$$

(v) *For each n-vector-valued process $X(\mathbf{t}, \boldsymbol{\omega})$ on $[a, b]$ that has a.s. continuous sample functions, the processes $g_\rho{}^i(\mathbf{t}, X(\mathbf{t}, \boldsymbol{\omega}), \boldsymbol{\omega})$ and $h^i_{\rho,\sigma}(\mathbf{t}, X(\mathbf{t}, \boldsymbol{\omega}), \boldsymbol{\omega})$ are separable and are continuous in probability a.e. in $[a, b]$.*

(vi) *For each n-vector-valued process $X(\mathbf{t}, \boldsymbol{\omega})$ adapted to $\mathfrak{F}_{\boldsymbol{\tau}}$, the processes $g^{\rho i}(\mathbf{t}, X(\mathbf{t}, \boldsymbol{\omega}), \boldsymbol{\omega})$ and $h^i_{\rho,\sigma}(\mathbf{t}, X(\mathbf{t}, \boldsymbol{\omega}), \boldsymbol{\omega})$ are adapted to $\mathfrak{F}_{\boldsymbol{\tau}}$.*

In the important special case in which the $g_\rho{}^i(\mathbf{t}, \mathbf{x}, \boldsymbol{\omega})$ and $h^i_{\rho,\sigma}(\mathbf{t}, \mathbf{x}, \boldsymbol{\omega})$ are independent of ω and are continuous functions of t and x alone, hypotheses (1-4,v) and (1-4,vi) are automatically satisfied. However, when the $g_\rho{}^i$, etc. depend directly on ω, as for example in the case of linear differential equations with random coefficients, hypotheses (1-4,v,vi) do not necessarily hold, and we have to accept them as assumptions.

It is convenient to use the enlarged σ-algebras $\bar{\mathfrak{F}}_\tau$ defined in Definition I(3-12); $\bar{\mathfrak{F}}_\tau$ consists of all sets B such that for some A in $\mathfrak{F}_\tau$, both $B \backslash A$ and $A \backslash B$ are negligible sets in the algebra $\mathfrak{A}$. If (1-4,vi) holds as stated, it also holds with $\bar{\mathfrak{F}}_{\boldsymbol{\tau}}$ in place of $\mathfrak{F}_{\boldsymbol{\tau}}$; and likewise for (1-4,iii).

From this point on we shall use an abbreviation. An equation or inequality involving affixes or variables will be understood to hold over the entire range of those indices or variables, unless mention is made to the contrary. Thus the statement $h^i_{\rho,\sigma}(t, x, \omega) \leqq F^i(t, x)$ is understood to mean

$$h^i_{\rho,\sigma}(t, x, \omega) \leqq F^i(t, x)$$

$$(i = 1, \ldots, n; \quad \rho, \sigma = 1, \ldots, r; \quad a \leqq t \leqq b; \quad x \in R^n; \quad \omega \in \Omega).$$

Likewise $\sum_\rho$ and $\sum_{\rho,\sigma}$ are understood to indicate summation over the whole range $\{1, \ldots, r\}$ of ρ and σ.

(1-5) *THEOREM* *Let the standing hypotheses (1-4) be satisfied. Let $x^i(a, \boldsymbol{\omega})$ be $\mathfrak{F}_a$-measurable r.v.'s. Then there exists a process $x(\mathbf{t}, \boldsymbol{\omega})$ adapted to $\bar{\mathfrak{F}}_{\boldsymbol{\tau}}$, with continuous sample functions, such that (1-3) is satisfied, and for each ω for which the $z^\rho(\mathbf{t}, \omega)$ are Lipschitzian and $L(\omega) < \infty$, Eq. (1-3) holds with the first-order integrals interpreted as Riemann–Stieltjes integrals, the second-order integrals being 0. Any two solutions of (1-3), both adapted to $\bar{\mathfrak{F}}_{\boldsymbol{\tau}}$, are equivalent.*

We use Picard's method. We first show that it is possible to define inductively a sequence of processes $x_0, x_1, x_2, \ldots$, each of which has a.s. continuous sample functions and is adapted to $\bar{\mathfrak{F}}_{\tau}$, such that for $a \leqq t \leqq b$ and $k \geqq 0$

$$
\text{(1-6)} \qquad x_0{}^i(t, \omega) = x^i(a, \omega),
$$

$$
x^i_{k+1}(t, \omega) = x^i(a, \omega) + \sum_{\rho} \int_a^t g_\rho{}^i(s, x_k(s, \omega), \omega)\, dz^\rho(s, \omega)
$$

$$
+ \sum_{\rho,\sigma} \int_a^t h^i_{\rho,\sigma}(s, x_k(s, \omega), \omega)\, dz^\rho(s, \omega)\, dz^\sigma(s, \omega).
$$

Since x_0 satisfies the requirements, we assume $x_0, \ldots, x_k$ defined and show that x_{k+1} can be chosen to satisfy the conditions. If f is any one of the $g_\rho{}^i$ or $h^i_{\rho,\sigma}$, by (1-4, iv) we have

$$
\text{(1-7)} \qquad | f(t, x, \omega) | \leqq L(\omega)[1 + | x |].
$$

So the integrands in the right member of (1-6) are a.s. bounded. They are adapted to $\bar{\mathfrak{F}}_{\tau}$ by (1-4,vi). By (1-4,v), all the integrands in (1-6) are continuous in probability at almost all points of $[a, b]$. By Theorems IV(1-7), IV(1-21) and IV(1-34) each of the indefinite integrals has a strict version with continuous sample functions.

For each t in $[a, b]$ and each belated partition Π of $[a, t]$, the Riemann sums

$$
\text{(1-8)} \qquad S(\Pi; g_\rho{}^i; z^\rho), \qquad S(\Pi; h^i_{\rho,\sigma}; z^\rho, z^\sigma)
$$

are $\bar{\mathfrak{F}}_t$-measurable r.v.'s, and they converge in probability to the corresponding integrals over $[a, t]$, so these too are $\bar{\mathfrak{F}}_t$-measurable r.v.'s. Hence x_{k+1} has all the required properties, and (1-6) defines a sequence of processes as described.

We first consider a special case:

(1-9) $\| x(a) \|$ is finite, and (1-4, iv) holds with a (finite) constant L_0 in place of $L(\omega)$.

For all processes $y(\mathbf{t}, \omega)$ on $[a, b]$, we define

$$
\text{(1-10)} \qquad N(t, y) = \sup\{\| y(s) \| : a \leqq s \leqq t\}.
$$

We shall now prove that with the C of III(2-2), and with $k = 1, 2, \ldots$,

$$
\text{(1-11)} \quad N(t, x_k - x_{k-1})^2 \leqq (1 + \| x(a) \|)^2 \{nC^2 L_0{}^2 (r + r^2)^2\}^k (t - a)^k / k!.
$$

In this proof and henceforth, whenever it is convenient we shall write $g_\rho{}^i(t, x)$ for $g_\rho{}^i(t, x(t, \omega), \omega)$ and $h^i_{\rho,\sigma}(t, x)$ for $h^i_{\rho,\sigma}(t, x(t, \omega), \omega)$.

Consider first $k = 1$. By (1-4, iv) and (1-6), if f satisfies the Lipschitz condition (1-9) we have

$$\text{(1-12)} \qquad \| f(t, x_0(t)) \| \leqq L_0(1 + \| x(a) \|).$$

So by (1-6), Lemma IV(4-13) and Corollary III(2-11),

$$\| x_1{}^i(t) - x_0{}^i(t) \|^2$$

$$= E\left(\left\{\sum_\rho \int_a^t g_\rho{}^i(s, x_0)\, dz^\rho(s) + \sum_{\rho,\sigma} \int_a^t h^i_{\rho,\sigma}(s, x_0)\, dz^\rho\, dz^\sigma\right\}^2\right)$$

$$\leqq (r + r^2)\left\{\sum_\rho \left\| \int_a^t g_\rho{}^i(s, x_0)\, dz^\rho \right\|^2 + \sum_{\rho,\sigma} \left\| \int_a^t h^i_{\rho,\sigma}(s, x_0)\, dz^\rho\, dz^\sigma \right\|^2\right\}$$

$$\leqq (r + r^2)^2 C^2 L_0{}^2 (1 + \| x(a) \|)^2 (t - a).$$

Adding these inequalities member by member for $i = 1, \ldots, n$ yields (1-11) for $k = 1$.

Assume now that (1-11) holds for a positive integer k. By Lemma IV(4-13) and Corollary III(2-11), with (1-6),

(1-13)

$$\| x^i_{k+1}(t) - x_k{}^i(t) \|^2$$

$$= E\left(\left\{\sum_\rho \int_a^t [g_\rho{}^i(s, x_k) - g_\rho{}^i(s, x_{k-1})]\, dz^\rho\right.\right.$$

$$\left.\left. + \sum_{\rho,\sigma} \int_a^t [h^i_{\rho,\sigma}(s, x_k) - h^i_{\rho,\sigma}(s, x_{k-1})]\, dz^\rho\, dz^\sigma\right\}^2\right)$$

$$\leqq (r^2 + r)\left\{\sum_\rho \left\| \int_a^t [g_\rho{}^i(s, x_k) - g_\rho{}^i(s, x_{k-1})]\, dz^\rho \right\|^2\right.$$

$$\left. + \sum_{\rho,\sigma} \left\| \int_a^t [h^i_{\rho,\sigma}(s, x_k) - h^i_{\rho,\sigma}(s, x_{k-1})]\, dz^\rho\, dz^\sigma \right\|^2\right\}$$

$$\leqq (r^2 + r)^2 C^2 L_0{}^2 \int_a^t N(s, x_k - x_{k-1})^2\, ds$$

$$\leqq (r^2 + r)^2 C^2 L_0{}^2 (1 + \| x(a) \|)^2 \{n C^2 L_0{}^2 (r + r^2)^2\}^k (t - a)^{k+1}/(k+1)!.$$

Summing over $i = 1, \ldots, n$ shows that (1-11) holds with $k + 1$ in place of k, so (1-11) holds for all positive integers k.

From (1-11) it follows at once that

$$\sum_{k=1}^{\infty} N(b, x_k - x_{k-1}) < \infty. \tag{1-14}$$

That is, the sums

$$x(a, \omega) + \sum_{k=1}^{p} (x_k(t, \omega) - x_{k-1}(t, \omega)) \tag{1-15}$$

converge uniformly in L_2-norm on $[a, b]$. Let $x(t, \omega)$ be their limit; then

$$\lim_{p\to\infty} x_p(t, \omega) = x(t, \omega) \tag{1-16}$$

in L_2-distance, uniformly on $[a, b]$. So x is adapted to $\bar{\mathfrak{F}}_{\mathbf{t}}$ and is L_2-continuous. By Lemma IV(4-13) and Corollary III(2-11), with (1-6), for each fixed t, $x(t, \omega)$ satisfies (1-3) a.s. For each indefinite integral in the right member, we choose a strict version with continuous sample functions, as is possible by Theorems IV(1-7), IV(1-21) and IV(1-34). The right member thus obtained is a process $y^i(\mathbf{t}, \omega)$ which for each t is equivalent to $x^i(t, \omega)$. So if we substitute y for x in each integrand in (1-3) the integrals are unchanged, as we noted in Section 4 of Chapter II; y^i is still a version of the right member after the substitution. That is, we can choose a version of x (the process just called y) that satisfies (1-3) and has continuous sample functions.

If x and $\bar{x}$ are any two processes adapted to $\bar{\mathfrak{F}}_{\mathbf{t}}$ and satisfying (1-3), the first equation in (1-13) remains valid if we replace x_{k+1}, x_k in the left member by x, $\bar{x}$ respectively and replace x_k, x_{k-1} in the right member by x, $\bar{x}$ respectively. The remaining estimates in (1-13) continue to hold with this substitution, and in place of (1-11) we obtain

$$N(t, x - \bar{x})^2 \leqq (1 + \| x(a) \|)^2\{nC^2L_0^2(r + r^2)^2\}^k(t - a)^k/k!, \tag{1-17}$$

valid for all k. This implies that the left member is 0, so x and $\bar{x}$ are equivalent. The theorem is proved under the added hypothesis (1-9).

Returning to the hypotheses of the theorem, for each positive integer M there is a set Ω_M in $\mathfrak{A}$ such that

$$\text{if } \omega \in \Omega_M,\ | x(a, \omega) | \leqq M \text{ and } L(\omega) \leqq M, \tag{1-18}$$

and we can choose these sets so that

$$\emptyset = \Omega_0 \subset \Omega_1 \subset \Omega_2 \subset \cdots, \qquad \text{and} \qquad P(\Omega_M) \to 1. \tag{1-19}$$

Let D be a countable dense subset of R^n. For each x_1 and x_2 ($\neq x_1$) in D, the processes $g_\rho{}^i(\mathbf{t}, x_j, \omega)$ and $h^i_{\rho,\sigma}(\mathbf{t}, x_j, \omega)$ ($j = 1, 2$) are adapted to $\bar{\mathfrak{F}}_{\mathbf{t}}$ by

(1-4,vi) and separable by (1-4,v). Then for x_1 and x_2 ($\neq x_1$) in D, the processes

$$| g_\rho{}^i(\mathbf{t}, x_1, \boldsymbol{\omega}) - g_\rho{}^i(\mathbf{t}, x_2, \boldsymbol{\omega}) | / | x_1 - x_2 |$$

and

$$| h^i_{\rho,\sigma}(\mathbf{t}, x_1, \boldsymbol{\omega}) - h^i_{\rho,\sigma}(\mathbf{t}, x_2, \boldsymbol{\omega}) | / | x_1 - x_2 |$$

are adapted to $\bar{\mathfrak{F}}_{\boldsymbol{\tau}}$ and are separable. If for each t in $[a, b]$ and ω in Ω we define $L_1(t, \omega)$ to be the supremum of the numbers

$$| f(s, x_1, \omega) - f(s, x_2, \omega) | / | x_1 - x_2 |, \qquad | f(s, 0, \ldots, 0, \omega) |$$

for all f in the set $\{g_\rho{}^i, h^i_{\rho,\sigma}\}$, all s in $[a, t]$ and all x_1 and x_2 ($\neq x_1$) in D, this is an $\bar{\mathfrak{F}}_t$-measurable r.v. Then, since each f is continuous in x for fixed t and ω in $\bigcup \Omega_M$,

(1-20) $$| f(t, x_1, \omega) - f(t, x_2, \omega) | \leqq L_1(t, \omega) \, | x_1 - x_2 |,$$

$$| f(t, 0, \ldots, \omega) | \leqq L_1(t, \omega)$$

for all t, all ω in $\bigcup \Omega_M$ and all x_1, x_2 in R^n. Moreover

(1-21) if $a \leqq s \leqq t \leqq b$ and $\omega \in \Omega$, then

$$L_1(s, \omega) \leqq L_1(t, \omega) \leqq L(\omega).$$

We now define for $M = 1, 2, \ldots$

(1-22) $$\phi_M(t, \omega) = 1 \qquad \text{if} \quad L_1(t, \omega) \leqq M$$

$$= 0 \qquad \text{if} \quad L_1(t, \omega) > M.$$

Then ϕ_M is adapted to $\bar{\mathfrak{F}}_{\boldsymbol{\tau}}$ and is nonincreasing in t for each ω, and

(1-23) $$\phi_M(b, \omega) = 1 \qquad \text{for all } \omega \text{ in } \Omega_M.$$

So $E(\phi_M(t))$ is a nonincreasing function such that

(1-24) $$1 \geqq E(\phi_M(t)) \geqq P(\Omega_M) \qquad (a \leqq t \leqq b).$$

It follows that $\phi_M(t)$ is continuous in probability at all but denumerably many points of $[a, b]$.

By (1-20) and (1-22), the functions

(1-25) $$\phi_M g_\rho{}^i, \qquad \phi_M h^i_{\rho,\sigma}$$

satisfy (1-4,iv) with M in place of $L(\omega)$. They clearly satisfy (1-4,vi); and by (1-24) and the sentence following it, they satisfy (1-4,v).

We define $x_M{}^i(a, \omega) = \operatorname{mid}\{x^i(a, \omega), M, -M\}$. Then by the part of the proof already completed, the equation

$$x^i(t, \omega) = x_M{}^i(a, \omega) + \sum_{\rho=1}^{r} \int_a^t \phi_M(s, \omega) g_\rho{}^i(s, x(s, \omega), \omega)\, dz^\rho(s, \omega) \tag{1-26}$$

$$+ \sum_{\rho,\sigma=1}^{r} \int_a^t \phi_M(s, \omega) h^i_{\rho,\sigma}(s, x(s, \omega), \omega)\, dz^\rho(s, \omega)\, dz^\sigma(s, \omega)$$

has a solution on $[a, b]$ with continuous sample functions. We call this solution x_M, and we define processes $x^i(\mathbf{t}, \omega)$ $(i = 1, \ldots, n)$ by setting

$$\begin{aligned} x^i(t, \omega) &= x_M{}^i(t, \omega) \qquad (\omega \in \Omega_M \setminus \Omega_{M-1},\ M = 1, 2, 3, \ldots) \\ &= 0 \qquad (\omega \in \Omega \setminus \bigcup_M \Omega_M). \end{aligned} \tag{1-27}$$

If $\epsilon > 0$, we can and do choose an M' such that

$$P(\Omega_{M'}) > 1 - \epsilon/2. \tag{1-28}$$

By (1-26), there is a positive δ such that for every M in $\{1, \ldots, M'\}$, every t in $[a, b]$ and every belated partition Π of $[a, t]$ with mesh $\Pi < \delta$

$$P\{\omega \in \Omega_M - \Omega_{M-1} : | x_M{}^i(t) - x_M{}^i(a) - \sum_\rho S(\Pi; \phi_M g_\rho{}^i(t, x_M); z^\rho) - \sum_{\rho,\sigma} S(\Pi; \phi_M h^i_{\rho,\sigma}(t, x_M); z^\rho, z^\sigma) \mid > \epsilon/2\} < \epsilon/2M'. \tag{1-29}$$

But if $1 \leqq M \leqq M'$ and $\omega \in \Omega_M$, by (1-23)

$$\phi_M(t, \omega) = 1 \quad (a \leqq t \leqq b) \qquad \text{and} \qquad x_M{}^i(a) = x^i(a).$$

So the factor ϕ_M can be replaced by 1 and $x_M(a)$ by $x(a)$ in (1-29). Then by summing over M from 1 to M' we obtain

$$P\{\omega \in \Omega_{M'} : | x^i(t) - x^i(a) - \sum_{\rho=1}^{r} S(\Pi; g_\rho{}^i(t, x); z^\rho) - \sum_{\rho,\sigma=1}^{r} S(\Pi; h^i_{\rho,\sigma}(t, x); z^\rho, z^\sigma) \mid > \epsilon/2\} < \epsilon/2. \tag{1-30}$$

From (1-30) and (1-28) it follows that the sum

$$\sum_{\rho=1}^{r} S(\Pi; g_\rho{}^i(t, x); z^\rho) + \sum_{\rho,\sigma=1}^{r} S(\Pi; h^i_{\rho,\sigma}(t, x); z^\rho, z^\sigma)$$

converges in probability to $x^i(t) - x^i(a)$, so that (1-3) is satisfied. Since each $x_M{}^i$ had continuous sample functions, the same is true of x^i.

If there were two solutions x_1, x_2 adapted to $\bar{\mathfrak{F}}_\tau$ such that for some t in

$[a, b]$ the set $\{x_1(t) \neq x_2(t)\}$ had positive measure, there would be some M for which the set

$$\Omega_M' = \Omega_M \cap \{x_1(t) \neq x_2(t)\}$$

has positive P-measure. On this set $\phi_M = 1$, so both $x_1(t, \omega)$ and $x_2(t, \omega)$ coincide at all points of $\Omega_M' \backslash N$ (N negligible) with the unique solution of (1-26). This contradicts the definition of Ω_M', so x_1 and x_2 are equivalent processes.

The proof of the next statement is left as an exercise.

(1-31) *In the deterministic case, if $f^i(\mathbf{t}, \mathbf{x})$ is continuous on $[a, b] \times R^n$ and is Lipschitzian in x ($i = 1, \ldots, n$), for every bounded measurable $(\phi^1(\mathbf{t}), \ldots, \phi^n(\mathbf{t}))$ the system*

$$x^i(t) = x^i(a) + \int_a^t f^i(s, x(s))\, ds + \phi^i(t)$$

can be proved solvable by Picard's method. Hence if $g_\rho^1(\mathbf{t}, \omega), \ldots, g_\rho^n(\mathbf{t}, \omega)$ are of bounded variation on $[a, b]$ for each ω, then for each ω such that $f^i(\mathbf{t}, \mathbf{x}, \omega)$ is Lipschitzian in x and the $z^\rho(\mathbf{t}, \omega)$ are continuous, the system

$$x^i(t, \omega) = x^i(a, \omega) + \int_a^t f^i(s, x(s, \omega), \omega)\, ds + \sum_{\rho=1}^{r} \int_a^t g_\rho^i(s, \omega)\, dz^\rho(s, \omega) \tag{1-32}$$

has a solution. (In this case the concept of stochastic integral is irrelevant; the sample curves are treated individually.)

2 Linear differential equations and their adjoints

Consider the system of stochastic differential equations

$$(2\text{-}1)\quad x^i(t) = x^i(a) + \sum_{\rho,h} \int_a^t [A^i_{\rho,h}(s) x^h(s) + C_\rho{}^i(s)]\, dz^\rho(s) + \sum_{\rho,\sigma,h} \int_a^t [B^i_{\rho,\sigma,h}(s) x^h(s) + D^i_{\rho,\sigma}(s)]\, dz^\rho(s)\, dz^\sigma(s),$$

wherein we assume

(2-2) the $x^i(a)$ are $\mathfrak{F}_a$-measurable r.v.'s, and

(2-3) the processes $A^i_{\rho,h}$, $B^i_{\rho,\sigma,h}$, $C_\rho{}^i$, $D^i_{\rho,\sigma}$ all are separable, have a.s. bounded sample functions and are adapted to $\mathfrak{F}_{\boldsymbol{\tau}}$ on $[a, b]$, and are continuous in probability at almost all points of $[a, b]$.

We verify readily that all hypotheses of Theorem (1-5) are satisfied, so Eqs. (2-1) have solutions with all the properties specified in that theorem.

In the special case in which all the $z^\rho(\mathbf{t}, \omega)$ are Lipschitzian and Ω consists of only one point, by means of a well-known and useful theorem on ordinary linear systems we can deduce that if y is a solution of the adjoint system

(2-4) $$dy^i = -\sum_{\rho,h} A^h_{\rho,i}(s)y^h(s)\,dz^\rho(s),$$

the equation

(2-5) $$\sum_i x^i(t)y^i(t) = \sum_i x^i(a)y^i(a) + \sum_{i,\rho}\int_a^t C_\rho{}^i(s)y^i(s)\,dz^\rho(s)$$

holds on $[a, b]$; of course the second-order integrals in (2-1) are all 0 in this case. In particular, a set of solutions of (2-4) that is linearly independent at a remains so, and can be used to express the solutions of (2-4) as indefinite integrals.

Warfield [13] has shown that (2-1) always possesses such an adjoint system, but it involves second-order integrals even when (2-1) does not. This system has the form

(2-6)

$$y^i(t) = y^i(a) - \sum_{h,\rho}\int_a^t A^h_{\rho,i}(s)y^h(s)\,dz^\rho(s)$$

$$+\sum_{h,\rho,\sigma}\int_a^t \{\sum_k A^k_{\rho,i}(s)A^h_{\sigma,k}(s) - B^h_{\rho,\sigma,i}(s)\}y^h(s)\,dz^\rho(s)\,dz^\sigma(s).$$

(2-7) *THEOREM* *Let* $x(a, \omega)$, $y(a, \omega)$ *be* $\mathfrak{F}_a$*-measurable. Let* $x(\mathbf{t}, \omega)$, $y(\mathbf{t}, \omega)$ *be solutions of* (2-1), (2-6), *respectively, both having a.s. continuous sample functions. Then the stochastic differential equation*

(2-8) $$\sum_i x^i(t,\omega)y^i(t,\omega)$$

$$= \sum_i x^i(a,\omega)y^i(a,\omega) + \sum_{i,\rho}\int_a^t C_\rho{}^i(s,\omega)y^i(s,\omega)\,dz^\rho(s,\omega)$$

$$+ \sum_{i,\rho,\sigma}\int_a^t \Big[D^i_{\rho,\sigma}(s,\omega) - \sum_k C_\rho{}^k(s,\omega)A^i_{\sigma,k}(s,\omega)\Big]$$

$$\times\, y^i(s,\omega)\,dz^\rho(s,\omega)\,dz^\sigma(s,\omega)$$

is satisfied. If for the integrals in (2-8) we choose versions with a.s. continuous sample functions, there is a set $\Omega_0 \subset \Omega$ with $P\Omega_0 = 1$ such that for each ω in Ω_0, (2-8) holds for all t in $[a, b]$. In particular, if

(2-9) $$C_\rho{}^i = D^i_{\rho,\sigma} = 0,$$

for all ω in a set Ω_0 with $P\Omega_0 = 1$ we have

(2-10) $$\sum_i x^i(t,\omega)y^i(t,\omega) = \sum_i x^i(a,\omega)y^i(a,\omega) \qquad (a \leqq t \leqq b).$$

We define a function on R^{2n} by

(2-11) $$F(X^1, \ldots, X^{2n}) = \sum_{i=1}^n X^i X^{n+i}.$$

If we give $y^1, \ldots, y^n$ the alternative names $x^{n+1}, \ldots, x^{2n}$, by Itô's differentiation formula [Theorem IV(2-4)], Eq. (2-8) is satisfied. Suppose that for the integrals in (2-8) we choose versions such that for all ω except those in a set N_1 with $PN_1 = 0$ the sample functions of the integrals are continuous on $[a, b]$. Since (2-8) is satisfied as a stochastic differential equation, for each t in $[a, b]$ there is a set $N(t)$ with $PN(t) = 0$ such that the right and left members of (2-8) have the same value on $\Omega\backslash N(t)$. Now let N_2 be the union of the $N(t)$ for all rational t in $[a, b]$. Then $P(N_1 \cup N_2) = 0$, and for each ω in $\Omega_0 = \Omega\backslash N_1 \cup N_2$ the two members of (2-8) are continuous functions on $[a, b]$ that coincide at all rational t in $[a, b]$, and therefore are identically equal on $[a, b]$.

Equation (2-10) is an immediate consequence of this when (2-9) is satisfied.

(2-12) *COROLLARY Let $x_v(\mathbf{t},\omega) = (x_v{}^1(\mathbf{t},\omega), \ldots, x_v{}^n(\mathbf{t},\omega))$ $(v = 1, \ldots, n)$ be processes with continuous sample functions such* *that each* $x_v{}^i(a, \omega)$ *is* $\mathfrak{F}(a)$*-measurable and each* x_v *is a solution of the homogeneous system*

$$x^i(t) = x^i(a) + \sum_{\rho,h} \int_a^t A^i_{\rho,h}(s) x^h(s)\, dz^\rho(s) \tag{2-13}$$

$$+ \sum_{\rho,\sigma,h} \int_a^t B^i_{\rho,\sigma,h}(s) x^h(s)\, dz^\rho(s)\, dz^\sigma(s),$$

the coefficients satisfying (2-3). *Let* Ω_1 *be the set of* ω *such that the* $x_v(a, \omega)$ *are linearly independent. Then there is a negligible set* N *such that for all* ω *in* $\Omega_1 \backslash N$, *the vectors* $x_1(t, \omega), \ldots, x_n(t, \omega)$ *are linearly independent for all* t *in* $[a, b]$.

For each ω in Ω_1 there is a matrix $(Y_w{}^j(\omega) : j, w = 1, \ldots, n)$ such that

$$\sum_{j=1}^n Y_w{}^j(\omega) x_v{}^j(a, \omega) = \delta_{w,v},$$

where as usual $\delta_{w,v} = 1$ if $w = v$ and $= 0$ if $w \neq v$. The set Ω_1 belongs to $\mathfrak{F}(a)$, and $Y_w{}^j$ is $\mathfrak{F}(a)$-measurable on it. Let $y_w(\mathbf{t}, \boldsymbol{\omega})$ be a process with continuous sample functions that satisfies (2-6) with initial values

$$\begin{aligned} y_w{}^j(a, \omega) &= Y_w{}^j(\omega) \qquad &&\text{if} \quad \omega \in \Omega_1, \\ &= 0 &&\text{if} \quad \omega \in \Omega \backslash \Omega_1. \end{aligned}$$

By Theorem (2-7), Eq. (2-10) holds on $\Omega \backslash N$, where N is negligible. So for all ω in $\Omega_1 \backslash N$, we have

$$\sum_{i=1}^n x_v{}^i(t, \omega) y_w{}^i(t, \omega) = \delta_{w,v} \qquad (a \leqq t \leqq b).$$

This implies that the $x_v(t, \omega)$ are linearly independent.

In Corollary (2-12) it is not necessary that there be as many as n processes $x_v(\mathbf{t}, \boldsymbol{\omega})$. If there are n_1 $(<n)$ of them, and Ω_1 is as in Corollary (2-12), we can adjoin $n - n_1$ solutions of the homogeneous equation (2-13) in such a way that all n have initial values that are linearly independent for all ω in Ω_1. Then on $\Omega_1 \backslash N$ all n processes are linearly independent for $a \leqq t \leqq b$; in particular, the original n_1 processes have this property. As a special case, if x_1 is a solution of the equation with continuous sample functions, and $| x_1(a, \omega) | \neq 0$ for ω in Ω_1, then $| x_1(t, \omega) | \neq 0$ $(a \leqq t \leqq b)$

for all ω in $\Omega_1 \backslash N$. This generalizes what we have previously established for solutions of Eq. IV(3-11).

We shall say that a system of n n-vector-valued processes $x_v(\mathbf{t}, \boldsymbol{\omega}) = (x_v^1(\mathbf{t}, \boldsymbol{\omega}), \ldots, x_v^n(\mathbf{t}, \boldsymbol{\omega}))$ is a **basis** of solutions of the homogeneous equation (2-13) if they have continuous sample functions, are strict solutions of (2-13), and are linearly independent for all t in $[a, b]$ and all ω in Ω. By Corollary (2-12), if we choose linearly independent $x_v(a, \omega)$ [for example, $x_v^i(a, \omega) = \delta_{iv}$] there are processes $x_v(t, \omega)$ with these initial values that have continuous sample functions, satisfy (2-13) strictly and are linearly independent for all t if $\omega \in \Omega \backslash N$, where N is negligible. This N can be chosen to have no points in common with the set of ω for which the $z^\rho(\mathbf{t}, \omega)$ are Lipschitzian. So if on N we redefine the $x_v(\mathbf{t}, \omega)$, setting say $x_v^i(t, \omega) \equiv x_v^i(a, \omega)$ for $a \leqq t \leqq b$, we obtain a basis of solutions. Therefore a basis of solutions always exists.

Let $X_1, \ldots, X_n$ be a basis of solutions of (2-13). The matrix with columns $X_1(t, \omega), \ldots, X_n(t, \omega)$ is nonsingular for all t and ω. Let $Y_1(t, \omega), \ldots, Y_n(t, \omega)$ be the columns of the transpose of the inverse of the X-matrix; then

$$\text{(2-14)} \qquad \sum_i X_v^i(t, \omega) Y_w^i(t, \omega) = \delta_{vw}.$$

If for each w in $\{1, \ldots, n\}$ the process $y_w(\mathbf{t}, \boldsymbol{\omega})$ is a strict solution of (2-6) with initial values $Y_w(a, \boldsymbol{\omega})$, having continuous sample functions, by Theorem (2-7)

$$\sum_i X_v^i(t, \omega) y_w^i(t, \omega) = \delta_{vw} \qquad (a \leqq t \leqq b)$$

for all ω in a set Ω_0 with $P(\Omega_0) = 1$. In particular this holds for all ω such that the $z^\rho(\mathbf{t}, \omega)$ are Lipschitzian. Since X is nonsingular, the last two equations imply that $Y_w^i(t, \omega) = y_w^i(t, \omega)$ if $\omega \in \Omega_0$, so that Y_w is also a strict solution of (2-6).

Let $x^i(a, \boldsymbol{\omega})$ be $\mathfrak{F}(a)$-measurable r.v.'s, and let $x(\mathbf{t}, \boldsymbol{\omega})$ be a process with continuous sample functions that is a strict solution of (2-1). By Theorem (2-7), for all ω in a set Ω_0 with $P(\Omega_0) = 1$ and for all t in $[a, b]$, Eq. (2-8) holds with Y_w^i in place of y^i. Since the Y-matrix is the transpose of the inverse of the X-matrix,

$$\sum_{w=1}^{n} Y_w^i(t, \omega) X_w^j(t, \omega) = \delta_{ij}.$$

So from (2-8), with Y_w^i in place of y^i, we obtain by multiplying by $X_w^j(t, \omega)$

and summing on w

(2-15)

$$x^j(t, \omega) = \sum_{i,w} x^i(a, \omega) Y_w{}^i(a, \omega) X_w{}^j(t, \omega)$$

$$+ \sum_{i,w,\rho} X_w{}^j(t, \omega) \int_a^t C_\rho{}^i(s, \omega) Y_w{}^i(s, \omega)\, dz^\rho(s, \omega)$$

$$+ \sum_{i,w,\rho,\sigma} X_w{}^j(t, \omega) \int_a^t [D^i_{\rho,\sigma}(s, \omega) - \sum_k C_\rho{}^k(s, \omega) A^i_{\sigma,k}(s, \omega)]$$

$$\times Y_w{}^i(s, \omega)\, dz^\rho(s, \omega)\, dz^\sigma(s, \omega).$$

Equation (2-15), with (2-14), enables us to solve (2-1) by quadratures when we know a basis of solutions of the homogeneous equations (2-13). When $z^1(\mathbf{t}, \omega) = \mathbf{t}$ and there are no other z^ρ, it specializes to a well-known formula in the theory of ordinary differential equations.

3 An approximation lemma

There are several well-known procedures for approximating the solutions of differential equations (1-1) by computation of the approximation at the successive division points of a partition Π. One of the most venerable is that of Cauchy (or of Euler). In this method, we choose a partition of $[a, b]$ with division points $a = t_1, t_2, \ldots, t_{m+1} = b$. We choose $\bar{x}^i(a)$ to be the given initial value of x^i, and having defined $\bar{x}^i$ at $t_1, \ldots, t_k$ we define it at t_{k+1} by extending $\bar{x}^i$ linearly on $[t_k, t_{k+1}]$ with initial value $\bar{x}^i(t_k)$ and slope $f^i(t_k, \bar{x}(t_k))$. Maruyama [8] has extended this method to Eqs. (1-3) in which the $h^i_{\rho,\sigma}$ are 0 and the integrals are Itô integrals. In the next section we shall show that Maruyama's approximation procedure is applicable to (1-3). However, for this, as well as for some proofs in the next chapter, it is convenient first to prove two lemmas.

We shall use the notation of the preceding section. Given a Cauchy partition

(3-1) $$\Pi = (t_1, \ldots, t_{m+1}; t_1, \ldots, t_m)$$

of $[a, b]$, for any function F on $[a, b]$, we define

$$\Delta_j F = F(t_{j+1}) - F(t_j).$$

Given any t in $[a, b]$, we define

(3-2) $\tau(t, \Pi)$ = *greatest number in the set* $\{t_1, \ldots, t_m\}$ *that is* $\leqq t$.

We are primarily interested in the values of the approximation $\bar{x}^i$ at

$t_1, \ldots, t_{m+1}$; the values at other points are obtained from these by some sort of interpolation. For the purposes of this section, it is convenient to choose $\bar{x}$ to be constant on each of the intervals $[t_1, t_2), [t_2, t_3), \ldots, [t_m, t_{m+1}]$, so that

$$\bar{x}^i(t) = \bar{x}^i(\tau(t, \Pi)) \qquad (a \leqq t \leqq b). \tag{3-3}$$

We can then state the following lemma.

(3-4) *LEMMA Assume that to each Cauchy partition* $\Pi = (t_1, \ldots, t_{m+1}; t_1, \ldots, t_m)$ *of* $[a, b]$ *there corresponds a process* $x_\Pi(\mathbf{t}, \boldsymbol{\omega})$ *on* $[a, b]$ *with the following properties.*

(i) *For each* ω, $x_\Pi{}^i(\mathbf{t}, \omega)$ *is constant on each of the intervals* $[t_1, t_2), [t_2, t_3), \ldots, [t_m, t_{m+1}]$ *of* Π.
(ii) $x_\Pi{}^i(a, \boldsymbol{\omega}) = x^i(a, \boldsymbol{\omega})$.
(iii) $x_\Pi{}^i$ *is adapted to* $\mathfrak{F}(\boldsymbol{\tau})$.
(iv) *If we define*

$$\epsilon_\Pi{}^i(t_j) = x_\Pi{}^i(t_{j+1}) - x_\Pi{}^i(t_j) - \sum_\rho g_\rho{}^i(t_j, x_\Pi(t_j))\, \Delta_j z - \sum_{\rho,\sigma} h^i_{\rho,\sigma}(t_j, x_\Pi(t_j))\, \Delta_j z^\rho\, \Delta_j z^\sigma,$$

there is a positive-valued function ϕ *on* $(0, \infty)$ *such that* $\phi(r) \to 0$ *as* $r \to 0$ *and, for* $k = 1, \ldots, m$,

$$\left\| \sum_{j=1}^k | \epsilon_\Pi(t_j) | \right\| \leqq \phi(\text{mesh } \Pi)[1 + \sup\{\| x_\Pi(s) \| : a \leqq s \leqq t_{k+1}\}].$$

Assume further that the standing hypotheses (1-4) *and hypothesis* (1-9) *are satisfied, and that* x *is a solution of* (1-3) *with a.s. continuous sample functions. Then as* mesh $\Pi \to 0$, $\| x_\Pi(t) - x(t) \|$ *converges to* 0 *uniformly on* $[a, b]$.

Before beginning the proof we remark that the conclusion of this lemma remains valid, with the same proof, if hypothesis (iv) is weakened by replacing the left member of the inequality by

$$\left\| \sum_{j=1}^k \epsilon_\Pi(t_j) \right\|$$

In the process of proving Theorem (1-5) we showed that when (1-9) is satisfied, the solution of (1-3) is L_2-bounded on $[a, b]$. Let $M - 1$ be an upper bound for $\| x(t) \|$ on $[a, b]$, and let ϵ be positive. Define

$$N(t) = \sup\{\| x_\Pi(s) - x(s) \| : a \leqq s \leqq t\}, \tag{3-5}$$

$$(3\text{-}6) \qquad X^i(t_k) = x^i(a) + \sum_{j=1}^{k-1} \sum_{\rho} g_\rho{}^i(t_j, x(t_j))\, \Delta_j z^\rho + \sum_{j=1}^{k-1} \sum_{\rho,\sigma} h^i_{\rho,\sigma}(t_j, x(t_j))\, \Delta_j z^\rho\, \Delta_j z^\sigma.$$

By Theorems III(2-3) and II(4-3), there is a positive δ_1 such that

(3-7) *if mesh* $\Pi < \delta_1$, *then* $\| X(t_k) - x(t_k) \| < \epsilon/n$ $(k = 1, \ldots, m+1)$.

Since $\phi(r)$ tends to 0 with r, there is a positive δ_2 such that

(3-8) if $0 < r < \delta_2$, then $\phi(r) < \min\{\epsilon/Mn, 1/2n\}$.

Since $x(t)$ is continuous in L_2-norm by Corollary III(2-11) and Lemma IV(4-13), there is a positive δ_3 such that

(3-9) if s and t are in $[a, b]$ and $| s - t | < \delta_3$, then $\| x(s) - x(t) \| < \epsilon/n$.

We now suppose

$$(3\text{-}10) \qquad \text{mesh}\ \Pi < \min\{\delta_1, \delta_2, \delta_3\}.$$

For each t in $[a, b]$, $\tau(t, \Pi)$ is a certain t_k in $\{t_1, \ldots, t_m\}$, and $x_\Pi(t) = x_\Pi(t_k)$. Therefore

$$\begin{aligned}(3\text{-}11) \qquad x_\Pi{}^i(t) - x^i(t) &= X^i(t_k) - x^i(t_k) + x^i(t_k) - x^i(t) \\ &\quad + \sum_{j=1}^{k-1} \{x_\Pi{}^i(t_{j+1}) - x_\Pi{}^i(t_j) - \sum_\rho g_\rho{}^i(t_j, x_\Pi(t_j))\, \Delta_j z^\rho \\ &\quad - \sum_{\rho,\sigma} h^i_{\rho,\sigma}(t_j, x_\Pi(t_j))\, \Delta_j z^\rho\, \Delta_j z^\sigma\} \\ &\quad + \sum_{j=1}^{k-1} \{\sum_\rho [g_\rho{}^i(t_j, x_\Pi(t_j)) - g_\rho{}^i(t_j, x(t_j))]\, \Delta_j z^\rho \\ &\quad + \sum_{\rho,\sigma} [h^i_{\rho,\sigma}(t_j, x_\Pi(t_j)) - h^i_{\rho,\sigma}(t_j, x(t_j))]\, \Delta_j z^\rho\, \Delta_j z^\sigma\}.\end{aligned}$$

By (1-4,iv) and (1-9), with (3-5), we have

$$(3\text{-}12) \qquad \| g_\rho{}^i(t_j, x_\Pi(t_j)) - g_\rho{}^i(t_j, x(t_j)) \| \leqq L_0 N(t_j),$$

so by Lemma III(2-2)

$$\text{(3-13)}\qquad \left|\left|\sum_{j=1}^{k-1}\sum_{\rho}[g_\rho{}^i(t_j, x_\Pi(t_j)) - g_\rho{}^i(t_j, x(t_j))]\,\Delta_j z^\rho\right|\right|$$

$$\leqq rC\left\{\sum_{j=1}^{k-1} L_0{}^2 N(t_j)^2\,\Delta_j t\right\}^{1/2}$$

$$\leqq rCL_0\left\{\int_a^{t_k} N(s)^2\,ds\right\}^{1/2}.$$

In the same way, by Lemma IV(4-13),

$$\text{(3-14)}\qquad \left|\left|\sum_{j=1}^{k-1}\sum_{\rho,\sigma}[h^i_{\rho,\sigma}(t_j, x_\Pi(t_j)) - h^i_{\rho,\sigma}(t_j, x(t_j))]\,\Delta_j z^\rho\,\Delta_j z^\sigma\right|\right|$$

$$\leqq r^2CL_0\left\{\int_a^{t_k} N(s)^2\,ds\right\}^{1/2}.$$

Since $M - 1$ is an upper bound for $||\,x(t)\,||$, we have

$$\text{(3-15)}\qquad 1 + \sup\{||\,x_\Pi(s)\,|| : a \leqq s \leqq t_k\} \leqq M + N(t_k) \leqq M + N(t).$$

So by hypothesis (iv) and (3-8)

$$\text{(3-16)}\qquad \left|\left|\sum_{j=1}^{k-1}\epsilon_\Pi(t_j)\right|\right| \leqq \phi(\text{mesh }\Pi)(M + N(t)) \leqq \epsilon/n + N(t)/2n.$$

If we insert estimates (3-7), (3-9), (3-13), (3-14), and (3-16) in (3-11), we obtain

$$\text{(3-17)}\qquad ||\,x(t) - x_\Pi(t)\,|| \leqq n\left[3\epsilon/n + N(t)/2n + (r + r^2)CL_0\left\{\int_a^t N(s)^2\,ds\right\}^{1/2}\right].$$

The right member of (3-17) is a nondecreasing function of t, so (3-17) remains valid if we replace t in the right member by any larger number in $[a, b]$. Equivalently, (3-17) remains valid if we replace t in the left member by any smaller number in $[a, b]$. Hence it remains valid if we replace the left member by $N(t)$. Therefore, if we define

$$B = 2n(r + r^2)CL_0,$$

we obtain

$$N(t) \leqq 6\epsilon + B\left\{\int_a^t N(s)^2\,ds\right\}^{1/2}. \tag{3-18}$$

This implies

$$N(t)^2 \leqq 72\epsilon^2 + 2B^2 \int_a^t N(s)^2\,ds. \tag{3-19}$$

In order to utilize (3-19), we first observe

(3-20) *if* $\alpha > 0$, *and* $y(t)$ *is right-continuous on* $[a, b]$ *and satisfies*

$$y(t) \leqq 2B^2 \int_a^t y(s)\,ds + \alpha,$$

then

$$y(t) \leqq \alpha \exp 2B^2(t - a) \qquad (a \leqq t \leqq b).$$

For if this is false, there is a positive γ such that for some t in $[a, b]$,

$$y(t) \geqq (\alpha + \gamma) \exp 2B^2(t - a). \tag{3-21}$$

By the right-continuity of y, there is a least number c for which this holds. Then (3-21) is false for $a \leqq t \leqq c$, so by hypothesis

$$\begin{aligned} y(c) &\leqq 2B^2 \int_a^c (\alpha + \gamma) \exp 2B^2(s - a)\,ds + \alpha \\ &= (\alpha + \gamma)[\exp 2B^2(c - a) - 1] + \alpha \\ &< (\alpha + \gamma) \exp 2B^2(c - a). \end{aligned}$$

So (3-21) fails at c, contrary to the definition of c. This establishes (3-20).

Since x is continuous and x_Π right-continuous in L_2-norm, $N(t)^2$ is right-continuous, and by (3-19) and (3-20)

$$N(t)^2 \leqq 72\epsilon^2 \exp 2B^2(t - a). \tag{3-22}$$

In particular,

$$N(b) \leqq [72^{1/2} \exp B^2(b - a)]\epsilon, \tag{3-23}$$

and so x_Π converges to x uniformly in L_2-norm as mesh $\Pi \to 0$. Lemma (3-4) is proved.

Lemma (3-4) implies that under its hypotheses, for each t in $[a, b]$ the difference $|x_\Pi(t) - x(t)|$ tends to 0 in probability as mesh $\Pi \to 0$. A much stronger statement is that the supremum of $|x_\Pi(t) - x(t)|$ on

$[a, b]$ tends in probability to 0 as mesh $\Pi \to 0$. We shall now prove that this stronger statement is in fact correct.

(3-24) *LEMMA Let the hypotheses of Lemma (3-4) be satisfied. Then the r.v. defined by*

$$\sup\{|x_\Pi(t, \omega) - x(t, \omega)| : a \leqq t \leqq b\}$$

converges in probability to 0 as mesh $\Pi \to 0$, *and converges to 0 at each ω such that the $z^\rho(\mathbf{t}, \omega)$ are Lipschitzian on $[a, b]$.*

Assume first that $g_\rho{}^i$ and $h^i_{\rho,\sigma}$ are bounded. Define

$$Z^{\rho,\sigma}(t) = \int_a^t 1\, dz^\rho(s)\, dz^\sigma(s) \qquad (a \leqq t \leqq b). \tag{3-25}$$

By Theorem IV(1-21), we can and do choose versions of these integrals that are strict and have a.s. continuous sample functions; in particular, the sample functions are continuous on $[a, b]$ for all ω such that the $z^\rho(\mathbf{t}, \omega)$ are Lipschitzian. Now by (3-25) and (3-6) we can rewrite (3-11) in the form

$$x_\Pi{}^i(t) - x^i(t) = A^i(t) + B^i(t) + C^i(t) + (x^i(t_k) - x^i(t)) + \sum_{j=1}^{k-1} \epsilon_\Pi{}^i(t_j) + D^i(t) + E^i(t), \tag{3-26}$$

where

$$A^i(t) = \sum_\rho \sum_{j=1}^{k-1} \left[g_\rho{}^i(t_j, x(t_j))\, \Delta_j z^\rho - \int_{t_j}^{t_{j+1}} g_\rho{}^i(s, x(s))\, dz^\rho(s) \right], \tag{3-27}$$

$$B^i(t) = \sum_{\rho,\sigma} \sum_{j=1}^{k-1} h^i_{\rho,\sigma}(t_j, x(t_j))[\Delta_j z^\rho\, \Delta_j z^\sigma - \Delta_j Z^{\rho,\sigma}],$$

$$C^i(t) = \sum_{\rho,\sigma} \sum_{j=1}^{k-1} \int_{t_j+1}^{t_j} [h^i_{\rho,\sigma}(t_j, x(t_j)) - h^i_{\rho,\sigma}(s, x(s))]\, dz^\rho(s)\, dz^\sigma(s),$$

$$D^i(t) = \sum_\rho \sum_{j=1}^{k-1} [g_\rho{}^i(t_j, x_\Pi(t_j)) - g_\rho{}^i(t_j, x(t_j))]\, \Delta_j z^\rho,$$

$$E^i(t) = \sum_{\rho,\sigma} \sum_{j=1}^{k-1} [h^i_{\rho,\sigma}(t_j, x_\Pi(t_j)) - h^i_{\rho,\sigma}(t_j, x(t_j))]\, \Delta_j z^\rho\, \Delta_j z^\sigma.$$

Since

$$\text{(3-28)} \qquad \sup\left\{\left|\sum_{j=1}^{k-1} \epsilon_\Pi{}^i(t_j, \omega)\right| : k = 1, \ldots, m\right\} \leqq \sum_{j=1}^{m} |\epsilon_\Pi{}^i(t_j, \omega)|,$$

by hypothesis (iv) of Lemma (3-4) the left member of (3-28) converges in probability to 0 as mesh $\Pi \to 0$.

Since t_k was defined to be $\tau(t, \Pi)$ [cf. (3-2)], $|t - t_k| \leqq$ mesh Π; and since x^i is continuous on $[a, b]$ with probability 1, we have a.s.

$$\text{(3-29)} \qquad \sup |x^i(t_k) - x^i(t)| \to 0 \quad \text{as} \quad \text{mesh } \Pi \to 0.$$

We now prove an auxiliary statement.

(3-30) *If $y_1, \ldots, y_m$ are r.v.'s such that each y_j is $\mathfrak{F}(t_j)$-measurable and $E(y_j{}^2) < \infty$, and $\rho \in \{1, \ldots, r\}$, and*

$$Y_k = \sum_{j=1}^{k} (y_j \, \Delta_j z^\rho + K |y_j| \Delta_j t),$$

then $(Y_1, \ldots, Y_m)$ is a submartingale with respect to $(\mathfrak{F}(t_1), \ldots, \mathfrak{F}(t_m))$. This remains valid if $\Delta_j z^\rho$ is replaced by $\Delta_j z^\rho \, \Delta_j z^\sigma$ [ρ, σ in $\{1, \ldots, r\}$].

We need only prove that a.s.

$$E(y_j \, \Delta_j z^\rho + K |y_j| \Delta_j t \mid \mathfrak{F}(t_j)) \geqq 0.$$

But since y_j is $\mathfrak{F}(t_j)$-measurable, this follows readily from (1-4, iii). A similar proof holds with $\Delta_j z^\rho \, \Delta_j z^\sigma$ in place of $\Delta_j z^\rho$.

If for fixed i and ρ we define

$$y_j = g_\rho{}^i(t_j, x_\Pi(t_j)) - g_\rho{}^i(t_j, x(t_j)),$$

by Lemma (3-4), (1-4, iv), and (1-9) we have [with notation (3-1) for Π]

$$\sup\{\| y_j \| : j = 1, \ldots, m\} \to 0 \quad \text{as} \quad \text{mesh } \Pi \to 0.$$

So by Lemma III(2-2), as mesh $\Pi \to 0$

$$\left\| \sum_{j=1}^{m} y_j \, \Delta_j z^\rho \right\| \to 0 \quad \text{and} \quad \left\| \sum_{j=1}^{m} |y_j| \Delta_j t \right\| \to 0.$$

Therefore $\| Y_m \| \to 0$. So if we define

$$\text{(3-31)} \qquad c_\Pi = \{2E(Y_m{}^+)\}^{1/2}$$

we also have $c_\Pi \to 0$. By Theorem I(3-11),

$$P(\sup\{Y_k : k = 1, \ldots, m\} > c_\Pi) \leqq c_\Pi/2,$$

and therefore, since $\sum_1^k y_j \Delta_j z^\rho \leqq Y_k$,

(3-32)
$$P\left(\sup\left\{\sum_{j=1}^{k} y_j\, \Delta_j z^\rho : k = 1, \ldots, m\right\} > c_\Pi\right) \leqq c_\Pi/2.$$

This holds, with analogous c_Π', if y_j is replaced by $-y_j$. Therefore

$$P\left(\sup\left\{\left|\sum_{j=1}^{k} y_j\, \Delta_j z\right| : k = 1, \ldots, m\right\} > \max\{c_\Pi, c_\Pi'\}\right) \leqq \max\{c_\Pi, c_\Pi'\}.$$

Hence as mesh $\Pi \to 0$

(3-33)
$$\sup\{|\, D^i(t)\,| : a \leqq t \leqq b\} \to 0 \quad \text{in probability.}$$

By a similar proof, as mesh $\Pi \to 0$

(3-34)
$$\sup\{|\, E^i(t)\,| : a \leqq t \leqq b\} \to 0 \quad \text{in probability.}$$

(3-35) *If $y(t)$ is a process adapted to $\mathfrak{F}(\tau)$, and is bounded and almost everywhere continuous in L_2-norm on $[a, b]$, then the process $Y'(t)$ defined by*

$$Y'(t) = \int_a^t y(s)\, dz^\rho(s) + K \int_a^t |\, y(s)\,|\, ds$$

is a submartingale on $[a, b]$. So is the process defined by

$$Y''(t) = \int_a^t y(s)\, dz^\rho(s)\, dz^\sigma(s) + K \int_a^t |\, y(s)\,|\, ds.$$

Let Π [with notation (3-1)] be a Cauchy partition of $[a, b]$. By (3-30), a.s.

$$E\left(\sum_{j=1}^{k-1} [y(t_j)\, \Delta_j z^\rho + K\,|\, y(t_j)\,|\, \Delta_j t] \,|\, \mathfrak{F}(a)\right) \geqq 0.$$

Letting mesh $\Pi \to 0$ yields

$$E\left(\int_a^b y(s)\, dz^\rho(s) + K \int_a^b |\, y(s)\,|\, ds \,|\, \mathfrak{F}(a)\right) \geqq 0$$

a.s. This applies equally well if we replace a by u and b by v, where $a \leqq u < v \leqq b$. Therefore Y' is a submartingale. A similar proof applies to Y''.

The sum over $\rho = 1, \ldots, r$ of the processes defined by

(3-36)
$$\int_a^t [g_\rho{}^i(\tau(s, \Pi), x(\tau(s, \Pi))) - g_\rho{}^i(s, x(s))]\, dz^\rho(s)$$

coincides with $A^i(t)$ at the points $t_1, \ldots, t_m$. For every t in $[a, b]$, $A^i(t)$ is the value of A^i at a certain one $t_k = \tau(t, \Pi)$ of these points, so to prove that $\sup\{|A^i(t)|: a \leqq t \leqq b\}$ tends to 0 in probability it suffices to prove that each of the processes (3-36) has a supremum on $[a, b]$ that tends to 0 in probability as mesh $\Pi \to 0$. If we define

$$y(t) = g_\rho{}^i(\tau(t, \Pi), x(\tau(t, \Pi))) - g_\rho{}^i(t, x(t)),$$

by assumption (p. 170) $y(t)$ is bounded, and by hypothesis (1-4,v) $\| y(t) \|$ tends to 0 at almost all points of $[a, b]$ as mesh $\Pi \to 0$. So by the dominated convergence theorem

$$\int_a^b E(y(s)^2)\, ds \to 0$$

as mesh $\Pi \to 0$. By Corollary III(2-11),

$$\left\| \int_a^b y(s)\, dz^\rho(s) \right\|_2 \to 0 \qquad \text{and} \qquad \int_a^b E(|y(s)|)\, ds \to 0$$

as mesh $\Pi \to 0$. The r.v.'s

$$Y_k = \int_a^{t_{k+1}} y(s)\, dz^\rho(s) + K \int_a^{t_{k+1}} |y(s)|\, ds$$

form a submartingale by (3-35), and we see by the preceding inequalities that $\| Y_m \|_1 \to 0$ as mesh $\Pi \to 0$. So, again defining c_Π by (3-31), we see that $c_\Pi \to 0$ as mesh $\Pi \to 0$. From this we deduce, just as we deduced (3-33), that as mesh $\Pi \to 0$

(3-37) $\qquad \sup\{|A^i(t)|: a \leqq t \leqq b\} \to 0$ in probability.

In a like manner, as mesh $\Pi \to 0$

(3-38) $\qquad \sup\{|C^i(t)|: a \leqq t \leqq b\} \to 0$ in probability.

To discuss the remaining terms, the $B^i(t)$, we first prove an auxiliary statement.

(3-39) *Let $y_1, \ldots, y_m$ be r.v.'s such that $E(y_j{}^2) < \infty$ and y_j is $\mathfrak{F}(t_j)$-measurable ($j = 1, \ldots, m$). Define*

$$Q = 2K^{3/2}$$

and

$$Y_k = \sum_{j=1}^{k} [y_j\, \Delta_j Z^{\rho,\sigma} - y_j\, \Delta_j z^\rho \Delta_j z^\sigma + Q(\text{mesh } \Pi)^{1/2} |y_j|\, \Delta_j t]$$

($k = 1, \ldots, m$). Then $(Y_1, \ldots, Y_m)$ is a submartingale relative to $(\mathfrak{F}(t_1), \ldots, \mathfrak{F}(t_m))$.

It is sufficient to prove for $j = 1, \ldots, m$

$$E(y_j \, \Delta_j Z^{\rho,\sigma} - y_j \, \Delta_j z^\rho \, \Delta_j z^\sigma + Q(\text{mesh } \Pi)^{1/2} \, | \, y_j \, | \, \Delta_j t \, | \, \mathfrak{F}(t_j)) \geqq 0 \tag{3-40}$$

a.s. Let $\Pi' = (u_1, u_2, \ldots, u_{q+1}; u_1, \ldots, u_q)$ be a Cauchy partition of $[t_j, t_{j+1}]$. We define

$$\Pi_p' = (u_1, u_p, u_{p+1}, \ldots, u_{q+1}; u_1, u_p, u_{p+1}, \ldots, u_q). \tag{3-41}$$

Then

$$\Pi_2' = \Pi' \qquad \text{and} \qquad \Pi'_{q+1} = (t_j, t_{j+1}; t_j).$$

Since a.s.

$$E(| z^\rho(u_p) - z^\rho(u_1) | \, \big| \, \mathfrak{F}(u_1)) \leqq \{E([z^\rho(u_p) - z^\rho(u_1)]^2 \, \big| \, \mathfrak{F}(u_1))\}^{1/2}$$

$$\leqq \{K(u_p - u_1)\}^{1/2},$$

we deduce

$$\left| E([z^\rho(u_p) - z^\rho(u_1)][z^\sigma(u_{p+1}) - z^\sigma(u_p)] \, | \, \mathfrak{F}(u_1)) \right|$$

$$= \left| E([z^\rho(u_p) - z^\rho(u_1)]E(z^\sigma(u_{p+1}) - z^\sigma(u_p) \, | \, \mathfrak{F}(u_p)) \, | \, \mathfrak{F}(u_1)) \right|$$

$$\leqq K(u_{p+1} - u_p) E(| z^\rho(u_p) - z^\rho(u_1) | \, \big| \, \mathfrak{F}(u_1))$$

$$\leqq K^{3/2}(u_p - u_1)^{1/2}(u_{p+1} - u_p)$$

$$\leqq (Q/2)(\text{mesh } \Pi)^{1/2}(u_{p+1} - u_p).$$

A similar inequality holds with ρ and σ interchanged. Hence

$$\left| E(S(\Pi_p'; 1; z^\rho, z^\sigma) - S(\Pi'_{p+1}; 1; z^\rho, z^\sigma) \, | \, \mathfrak{F}(t_j)) \right|$$

$$\leq Q(\text{mesh } \Pi)^{1/2}(u_{p+1} - u_p).$$

Summing over $p = 2, \ldots, q$ yields

$$| E(S(\Pi_2'; 1; z^\rho, z^\sigma) - S(\Pi'_{q+1}; 1; z^\rho, z^\sigma) \, | \, \mathfrak{F}(t_j)) | \tag{3-42}$$

$$\leqq Q(\text{mesh } \Pi)^{1/2}(t_{j+1} - t_j).$$

But

$$S(\Pi'_{q+1}; 1; z^\rho, z^\sigma) = \Delta_j z^\rho \, \Delta_j z^\sigma$$

and

$$S(\Pi_2'; 1; z^\rho, z^\sigma) = S(\Pi'; 1; z^\rho, z^\sigma),$$

which converges in L_1-norm to $\Delta_j Z^{\rho,\sigma}$ as mesh $\Pi' \to 0$. So if we let mesh $\Pi' \to 0$, from (3-42) we obtain

$$E(\Delta_j Z^{\rho,\sigma} - \Delta_j z^\rho \, \Delta_j z^\sigma + Q(\text{mesh } \Pi)^{1/2} \Delta_j t \mid \mathfrak{F}(t_j)) \geqq 0$$

a.s. Since y_j is $\mathfrak{F}(t_j)$-measurable, we readily deduce that (3-40) holds, and so (3-39) is established.

With notation (3-1) and (3-2) we define

$$y(t) = h^i_{\rho,\sigma}(t, x(t)), \qquad y_j = y(t_j). \tag{3-43}$$

Then $y(\mathbf{t})$ is L_2-continuous, and by Theorem III(3-2)

$$S(\Pi; y; z^\rho, z^\sigma) \to \int_a^b y(s)\, dz^\rho(s)\, dz^\sigma(s) \tag{3-44}$$

in L_1-norm as mesh $\Pi \to 0$. By Lemma IV(4-13) the sum

$$S(\Pi; y; Z^{\rho,\sigma}) = \int_a^b y(\tau(s, \Pi))\, dz^\rho(s)\, dz^\sigma(s)$$

tends in L_2-norm to the right member of (3-44). So

$$\left\| \sum_{j=1}^m [y_j \Delta_j Z^{\rho,\sigma} - y_j \Delta_j z^\rho \Delta_j z^\sigma \right\|_1 \to 0.$$

It follows readily that $\| Y_m \|_1 \to 0$, whence $E(Y_m{}^+) \to 0$. Just as we deduced (3-33) from (3-30), we prove

$$\sup\{| B^i(t) | : a \leqq t \leqq b\} \to 0 \quad \text{in probability.}$$

This completes the proof when $g_\rho{}^i$ and $h^i_{\rho,\sigma}$ are bounded.

For the general case, when $0 < \epsilon < 1$, there is an M such that the set $\Omega_M = \{\sup_t | x(t) | \geqq M - 1\}$ has $P\Omega_M < \epsilon/2$. If $\varphi_M(\mathbf{x})$ is infinitely differentiable, is 1 for $| x | < M$, and is 0 for $| x | > M + 1$, we denote by (1-3_M) equations (1-3) with their right members multiplied by $\varphi_M(x)$, by x_M their solution, and by $x_{M\Pi}$ the corresponding approximation. As already proved, the set $\Omega_\Pi = \{\sup_t | x_{M\Pi}(t) - x_M(t) | < \epsilon/2\}$ has $P\Omega_\Pi > 1 - \epsilon/2$ if mesh Π is small. By Corollary II(2-22), $x_M(\mathbf{t}, \omega) = x(\mathbf{t}, \omega)$ and $x_{M\Pi}(\mathbf{t}, \omega) = x_\Pi(\mathbf{t}, \omega)$ on $\Omega_\Pi \backslash (\Omega_M \cup N)$, where $PN = 0$. So

$$P\{\sup_t | x_\Pi(t) - x(t) | < \epsilon/2\} > 1 - \epsilon.$$

The proof of Lemma (3-24) is complete.

By means of (3-20) and the end of Section 1, we can prove:

(3-45) *Assume that in Eq. (1-32), for each ω in Ω, the $f^i(\mathbf{t}, \mathbf{x}, \omega)$ are continuous in (t, x) and Lipschitzian in x, and the $z^\rho(\mathbf{t}, \omega)$ are continuous, and the $g_\rho{}^i(\mathbf{t}, \mathbf{x}, \omega)$ are independent of x and of bounded variation in t. Then for each ω and each positive ϵ, there is a positive*

δ such that if $\bar{z}^1(\mathbf{t}), \ldots, \bar{z}^r(\mathbf{t})$ are continuous functions such that

$$| \bar{z}^\rho(t) - z^\rho(t, \omega) | < \delta \qquad (a \leqq t \leqq b; \quad \rho = 1, \ldots, r),$$

and $x^1(\mathbf{t}), \ldots, x^n(\mathbf{t})$ are the solutions of (1-32) and $\bar{x}^1(\mathbf{t}), \ldots, \bar{x}^n(\mathbf{t})$ are the solutions of the equation with $\bar{z}^\rho$ in place of z^ρ, it is true that

$$| x(t) - \bar{x}(t) | < \epsilon \qquad (a \leqq t \leqq b).$$

We leave the proof as an exercise.

Statement (3-45) shows that for equations of this especially simple type the solutions are stable, in the sense discussed in II(1-4). Also, in this case Eqs. (1-32) are already in canonical form. In Section 5 of Chapter VI we shall show that some systems with $g_\rho{}^i$ depending on x have the same kind of stability.

4 The Cauchy–Maruyama approximation

The extension to Eqs. (1-3) of the approximation method of Maruyama, which is itself an extension of the classical method of Cauchy (or Euler), can be described thus. Let Π be a Cauchy partition of $[a, b]$, with notation (3-1). First we define

(4-1) $$X_\Pi{}^i(a) = x^i(a).$$

Next, $X_\Pi(t)$ having been defined on $[a, t_j]$, we define it on $(t_j, t_{j+1}]$ by

(4-2) $$X_\Pi{}^i(t) = X_\Pi{}^i(t_j) + \sum_\rho g_\rho{}^i(t_j, X_\Pi(t_j))[z^\rho(t) - z^\rho(t_j)] + \sum_{\rho,\sigma} h^i_{\rho,\sigma}(t_j, X_\Pi(i_j))[z^\rho(t) - z^\rho(t_j)][z^\sigma(t) - z^\sigma(t_j)].$$

(4-3) THEOREM *Assume that the standing hypotheses (1-4) are satisfied. Let $x(t)$ be a strict-sense solution of Eqs. (1-3) with a.s. continuous sample functions. Then as* mesh $\Pi \to 0$,

(i) *$X_\Pi(t)$ converges to $x(t)$ uniformly in near-L_2-norm on $[a, b]$, and $X_\Pi(t, \omega_0)$ converges uniformly to $x(t, \omega_0)$ for each ω_0 such that the $z^\rho(\mathbf{t}, \omega_0)$ are all Lipschitzian and $L(\omega_0)$ is finite. Moreover,*

(ii) *if condition (1-9) holds, $X_\Pi(t)$ converges to $x(t)$ uniformly in L_2-norm;*

(iii) *the r.v.*

$$\sup\{| X_\Pi(t, \omega) - x(t, \omega) | : a \leqq t \leqq b\}$$

converges to 0 in probability.

We first assume that (1-9) is satisfied. With $\tau(t, \Pi)$ defined as in (3-2), the process

$$x_\Pi(\mathbf{t}) = X_\Pi(\tau(\mathbf{t}, \Pi)) \qquad (a \leqq t \leqq b) \tag{4-4}$$

satisfies the hypotheses of Lemma (3-4) with $\epsilon_\Pi = 0$. By that lemma, $x_\Pi(t)$ converges to $x(t)$ uniformly in L_2-norm as mesh $\Pi \to 0$. Since $\| x(t) \|$ is bounded, by restricting ourselves to partitions of sufficiently small mesh we bring it about that

$$\| x_\Pi(t) \| \leqq M \tag{4-5}$$

for a finite M. Then by (1-4) and (1-9)

$$\| g_\rho{}^i(t, x_\Pi(t)) \| \leqq L_0(1 + M), \tag{4-6}$$
$$\| h^i_{\rho,\sigma}(t, x_\Pi(t)) \| \leqq L_0(1 + M).$$

If $t_j \leqq t \leqq t_{j+1}$,

(4-7)

$$X_\Pi{}^i(t) - x_\Pi{}^i(t) = \sum_\rho g_\rho{}^i(t_j, x_\Pi(t_j))[z^\rho(t) - z^\rho(t_j)]$$
$$+ \sum_{\rho,\sigma} h^i_{\rho,\sigma}(t_j, x_\Pi(t_j))[z^\rho(t) - z^\rho(t_j)][z^\sigma(t) - z^\sigma(t_j)].$$

But

$$\| g_\rho{}^i(t_j, x_\Pi(t_j))[z^\rho(t) - z^\rho(t_j)] \|^2$$
$$= E(g_\rho{}^i(t_j, x_\Pi(t_j))^2 E([z^\rho(t) - z^\rho(t_j)]^2 \mid \mathfrak{F}(t_j)))$$
$$\leqq K(t - t_j) L_0{}^2(1 + M)^2,$$

and since $t - t_j \leqq$ mesh Π, this tends to 0 uniformly on $[a, b]$ as mesh $\Pi \to 0$. The other terms in (4-7) satisfy a similar inequality, so $\| X_\Pi(t) - x_\Pi(t) \|$ converges uniformly to 0 as mesh $\Pi \to 0$. This establishes conclusion (ii).

If hypotheses (1-4) are satisfied, but (1-9) is not, we define $x_M{}^i(a, \omega) = \text{mid}\{x^i(a, \omega), M, -M\}$, and we define the sets Ω_M as in (1-18) and the functions ϕ_M as in (1-22). We again denote by $x_M{}^i$ the solution of (1-26). Since the functions

$$x_M{}^i(a, \omega), \qquad \phi_M g^i, \qquad \phi_M h^i_{\rho,\sigma} \tag{4-8}$$

satisfy all the hypotheses including (1-9), by the part of the proof already completed, as mesh $\Pi \to 0$ the Cauchy–Maruyama approximations $X_{\Pi,M}$ to the solution of (1-26) converge in L_2-norm to x_M, uniformly on $[a, b]$.

But except on a set of ω with arbitrarily small P-measure we have

$$x_M(a, \omega) = x(a, \omega), \qquad \phi_M(\mathbf{t}, \omega) = 1.$$

So except on that small set we have

$$X_\Pi(\mathbf{t}, \omega) = X_{\Pi,M}(\mathbf{t}, \omega), \qquad x(\mathbf{t}, \omega) = x_M(\mathbf{t}, \omega);$$

and therefore X_Π converges to x uniformly in near-L_2-norm.

If ω_0 is a point of Ω for which all $z^\rho(\mathbf{t}, \omega_0)$ are Lipschitzian, $x(\mathbf{t}, \omega_0)$ continues to be a solution of (1-3) with respect to the probability measure that assigns measure 1 to the single point ω_0 and 0 to $\Omega\backslash\{\omega_0\}$. Uniform convergence in L_2-norm, or in near-L_2-norm, is then merely uniform convergence of $X_\Pi(t, \omega_0)$ to $x(t, \omega_0)$. By the preceding proof, this is the case for all such ω_0.

If the hypotheses of (3-24) are satisfied, they are also satisfied by the functions (4-8). Then by Lemma (3-24), the r.v.'s

$$\sup\{| x_{\Pi,M}(t, \omega) - x_M(t, \omega) | : a \leqq t \leqq b\}$$

converge in probability to 0 as mesh $\Pi \to 0$. But except on a set whose measure is arbitrarily close to 0 for large M, we have

$$x_{\Pi,M}(t, \omega) = x_\Pi(t, \omega), \qquad x_M(t, \omega) = x(t, \omega),$$

so

$$\sup\{| x_\Pi(t, \omega) - x(t, \omega) | : a \leqq t \leqq b\} \tag{4-9}$$

also tends in probability to 0. If $\epsilon > 0$, except on a set with P-measure less than ϵ we have $| x(t, \omega) | \leqq M - 1$ $(a \leqq t \leqq b)$ for some M, and except on a set of P-measure less than ϵ we have $| x_\Pi(t, \omega) - x(t, \omega) | < 1$. So except on a set of P-measure less than 2ϵ we have $| x_\Pi(t, \omega) | < M$ $(a \leqq t \leqq b)$, whence

$$| g_\rho{}^i(t, x_\Pi(t, \omega), \omega) | \leqq L(\omega)[M + 1] \qquad (a \leqq t \leqq b),$$

and likewise for $h^i_{\rho,\sigma}$. The factors

$$z^\rho(t) - z^\rho(t_j), [z^\rho(t) - z^\rho(t_j)][z^\sigma(t) - z^\sigma(t_j)]$$

tend uniformly to 0 with mesh Π for almost all ω in Ω, so by (4-7) the supremum of the difference

$$| X_\Pi{}^i(t, \omega) - x_\Pi{}^i(t, \omega) |$$

tends to 0 in probability. Since (4-9) does also, conclusion (iii) is established.

Although when (1-9) holds, $x_\Pi(t)$ converges uniformly in L_2-norm to $x(t)$ as mesh $\Pi \to 0$, the convergence is disappointingly slow. To see this we consider the system

$$dx^1 = dz, \qquad dx^2 = 2x^1\,dz. \tag{4-10}$$

with initial values $x^1(0) = x^2(0) = 0$, and with z a standard Brownian-motion process having $z(0) = 0$. Let Π be a Cauchy partition of $[0, 1]$ into subintervals of equal length. Then

$$x_\Pi{}^1(t) = x^1(t) = z(t),$$

$$x_\Pi{}^2(1) = \sum_{j=1}^{m} 2z(t_j)[z(t_{j+1}) - z(t_j)]$$

$$= z(1)^2 - \sum_{j=1}^{m} [z(t_{j+1}) - z(t_j)]^2.$$

The last term tends to -1 in L_2-norm as mesh $\Pi \to 0$. The L_2-distance e between the last term and its limit satisfies

$$e^2 = E\left\{\sum_{j=1}^{m} [z(t_{j+1}) - z(t_j)]^2 - 1\right\}^2.$$

By elementary properties of the Brownian motion the last expression is computed to be

$$2\sum_{1}^{m} (t_{j+1} - t_j)^2 = 2 \text{ mesh } \Pi.$$

So

$$\| e \| = (2 \text{ mesh } \Pi)^{1/2}. \tag{4-11}$$

VI Equations in Canonical Form

1 Invariance under change of coordinates

In Chapter II we asserted that if a system, when subjected to noises $z^1, \ldots, z^r$ with Lipschitzian sample functions, evolves in time according to Eqs. II(1-5), then in order to study its evolution in the presence of more general types of noises we should select equations, the canonical extension of Eqs. II(1-5), defined in Definition II(3-2), as the mathematical model of the system. With any reasonably well-behaved functions $g^i_{\rho\sigma}$, Eqs. II(3-1) have solutions independent of the $g^i_{\rho\sigma}$ [and hence satisfying II(1-5)] when the z^ρ have Lipschitzian sample functions. However, until now we have only minor evidence that the canonical choice II(3-3) is advantageous. In Section 2 of Chapter II we saw that in one example, the solution of the stochastic differential equations in canonical form has a stability under modification of the z^ρ that is not possessed by the equations with other choices for the $g^i_{\rho\sigma}$ (including the traditional choice 0). In Section 2 of Chapter V we saw that with Lipschitzian z^ρ any two solutions of the respective equations

$$dx^i = \sum_{\rho,h} A^i_{\rho,h}(t)x^h(t)\, dz^\rho(t), \tag{1-1}$$

$$dy^i = -\sum_{\rho,h} A^h_{\rho,i}(t)y^h(t)\, dz^\rho(t)$$

with a.s. Lipschitzian sample functions have the property that $\sum x^i(t)y^i(t)$ is a.s. constant. By Theorem V(2-7), the canonical extensions of the Eqs. (1-1) retain this property not merely for a.s. sample-continuous processes, but also for all processes $z^1, \ldots, z^r$ that satisfy the standing hypotheses V(1-4).

In this chapter we shall exhibit several distinctive properties of equations in canonical form. These virtues range all the way from simple computational convenience to an especial suitability for retaining adequate agreement with experiment when the noises are idealized to "white noise."

In this section we show that canonical equations possess an essential property that is not in general possessed by other equations: the solutions of the equations do not depend on the coordinate system in which we choose to express them.

Suppose that a system is described by a set of state variables $x^2, \ldots, x^n$ (we have chosen to number these beginning with 2 rather than 1 merely for future notational convenience), and that when the system is subjected to random disturbances $z^2(\mathbf{t}, \boldsymbol{\omega}), \ldots, z^r(\mathbf{t}, \boldsymbol{\omega})$ with Lipschitzian sample functions it evolves in accordance with a system of equations

$$dx^i = \sum_{\rho=1}^{r} g_\rho{}^i(t, x, \boldsymbol{\omega})\, dz^\rho(t, \boldsymbol{\omega}), \tag{1-2}$$

wherein as usual we write $z^1(t) = t$. If we make a change of coordinates

$$y^i = \Psi^i(x^2, \ldots, x^n) \qquad (i = 2, \ldots, n), \tag{1-3}$$

whose inverse is

$$x^i = \Phi^i(y^2, \ldots, y^n) \qquad (i = 2, \ldots, n), \tag{1-4}$$

and if the Φ^i and Ψ^i are twice continuously differentiable, every solution $x(\mathbf{t}, \boldsymbol{\omega})$ of (1-2) transforms into a function $y(\mathbf{t}, \boldsymbol{\omega})$,

$$y^i(t, \omega) = \Psi^i(x^2(t, \omega), \ldots, x^n(t, \omega)), \tag{1-5}$$

that satisfies

$$dy^i = \sum_{\rho=1}^{r} G_\rho{}^i(t, y, \boldsymbol{\omega})\, dz^\rho(t, \boldsymbol{\omega}), \tag{1-6}$$

where

$$G_\rho^{\,i}(t, y, \omega) = \sum_{h=2}^{n} \Psi^i_{x^h}(\Phi(y)) g_\rho^{\,h}(t, \Phi(y), \omega). \tag{1-7}$$

But if we now idealize the model by replacing the z^ρ ($\rho \geqq 2$) by independent standard Wiener processes, and $(x^2(\mathbf{t}, \boldsymbol{\omega}), \ldots, x^n(\mathbf{t}, \boldsymbol{\omega}))$ is a corresponding solution of (1-2), its transform $y(\mathbf{t}, \boldsymbol{\omega})$ given by (1-5) will not in general satisfy (1-6) [with (1-7)]. Instead, by the Itô differentiation lemma [IV(2-4)], together with III(7-1), we find that it satisfies (suppressing ω)

$$\begin{aligned} dy^i = {} & \sum_{\rho=2}^{r} G_\rho^{\,i}(t, y)\, dz^\rho \\ & + \frac{1}{2} \sum_{h,k=2}^{n} \sum_{\rho=2}^{r} \Psi^i_{x^h x^k}(\Phi(y)) g_\rho^{\,h}(t, \Phi(y)) g_\rho^{\,k}(t, \Phi(y))\, dt. \end{aligned} \tag{1-8}$$

Even for as simple a change as from rectangular to polar coordinates in the plane, with all $g_\rho^{\,i} = 1$, the last term does not vanish.

This is a well-known difficulty, and has been called a "paradox." It is in fact something different. In order for a method of constructing mathematical models to be satisfactory, it would seem to be fundamental that its predictions should be unambiguous, and not subject to change at the user's whim. The simple rule: "write the equations that hold for disturbances with smooth sample functions, and for more general disturbances interpret all first-order integrals as Itô integrals" fails to pass this test if the equations are written in the customary form (1-2). We shall now show that if the equations (1-2) are replaced by their canonical extensions the difficulty disappears.

The canonical extension of (1-2) is

$$dx^i = \sum_\rho g_\rho^{\,i}(t, x, \boldsymbol{\omega})\, dz^\rho + \frac{1}{2} \sum_{\rho,\sigma} g^i_{\rho\sigma}(t, x, \boldsymbol{\omega})\, dz^\rho\, dz^\sigma, \tag{1-9}$$

where

$$g^i_{\rho\sigma}(t, x, \omega) = \sum_{h=2}^{n} g^i_{\rho x^h}(t, x, \omega) g_\sigma^{\,h}(t, x, \omega). \tag{1-10}$$

When the z^ρ have Lipschitzian sample functions, this is equivalent to (1-2). We shall show that the property of being a solution of this equation is invariant, not merely under the change of coordinates (1-3) and (1-4), but under the more general change

$$t = t, \qquad y^i = \Psi^i(t, x^2, \ldots, x^n) \qquad (i = 2, \ldots, n), \tag{1-11}$$

with inverse

$$t = t, \qquad x^i = \Phi^i(t, y^2, \ldots, y^n) \qquad (i = 2, \ldots, n), \tag{1-12}$$

the Ψ^i and Φ^i being adequately smooth.

However, the details of this proof, and of all other proofs in this chapter, tend to be lengthy. To shorten the formulas somewhat we introduce a new "state variable" x^1. This x^1 is required to satisfy Eq. (1-2) with $i = 1$, the $g_\rho{}^i$ being defined by

$$g_1{}^1 = 1, \qquad g_\rho{}^1 = 0 \qquad (\rho = 2, \ldots, r), \tag{1-13}$$

and the initial value $x^1(a)$ being a. Then identically

$$x^1(t) = t. \tag{1-14}$$

Equations (1-2) are then equivalent to

$$dx^i = \sum_\rho g_\rho{}^i(x)\, dz^\rho \qquad (i = 1, \ldots, n), \tag{1-15}$$

where $x = (x^1, \ldots, x^n)$. The canonical extension of (1-15) is

$$dx^i = \sum_\rho g_\rho{}^i(x)\, dz^\rho + \frac{1}{2} \sum_{\rho,\sigma} g^i_{\rho\sigma}(x)\, dz^\rho\, dz^\sigma \qquad (i = 1, \ldots, n), \tag{1-16}$$

where now

$$g^i_{\rho\sigma}(x) = \sum_{h=1}^{n} g^i{}_{\rho x^h}(x) g_\sigma{}^h(x). \tag{1-17}$$

These equations are not identical in form with (1-9). There is an extra equation in (1-16), namely that with $i = 1$. But by (1-13) all $g^1_{\rho\sigma}$ are 0, so this equation has the unique solution (1-14). For $i > 1$, Eq. (1-16) differs from (1-9) in that the sums defining $g^i_{\rho\sigma}$ in (1-17) contain terms

$$g^i{}_{\rho x^1}(x) g_\sigma{}^1(x) \tag{1-18}$$

absent from (1-10). But if $\sigma > 1$ these terms are 0 by (1-13); and when (1-16) is written in integrated form, the term corresponding to (1-18) with $\sigma = 1$ is

$$\int_a^t g^i{}_{\rho x^1}(x(s)\, dz^\rho(s)\, dz^1(s).$$

Since $z^1(s) = s$, this integral vanishes, and (1-16) [with (1-17)] has the same solutions as (1-9) [with (1-10)].

If we define

(1-19) $$\Psi^1(x^1, \ldots, x^n) = x^1 \qquad (x \in R^n)$$

and regard x^1 as being another name for t, Eqs. (1-11) can be written as

(1-20) $$y^i = \Psi^i(x) \qquad (i = 1, \ldots, n),$$

with $y^1 = x^1$. The inverse (1-12) can be written as

(1-21) $$x^i = \Phi^i(y) \qquad (i = 1, \ldots, n).$$

If the $z^\rho(\mathbf{t}, \boldsymbol{\omega})$ are all Lipschitzian on $[a, b]$, and $x(\mathbf{t}, \boldsymbol{\omega})$ is a solution of (1-15) for each ω, we define $y(\mathbf{t}, \boldsymbol{\omega})$ by

(1-22) $$y^i(t, \omega) = \Psi^i(x(t, \omega)) \qquad (i = 1, \ldots, n; a \leqq t \leqq b).$$

We shall assume henceforth that the Ψ^i and Φ^i are twice continuously differentiable. Then for each ω the y^i satisfy

(1-23) $$dy^i = \sum_\rho G_\rho{}^i(y)\, dz^\rho,$$

where [as in (1-7)]

(1-24) $$G_\rho{}^i(y) = \sum_{h=1}^{n} \Psi^i{}_{x^h}(\Phi(y))g_\rho{}^h(\Phi(y)).$$

The canonical extension of (1-23) is

(1-25) $$dy^i = \sum_\rho G_\rho{}^i(y)\, dz^\rho + \frac{1}{2} \sum_{\rho,\sigma} G^i_{\rho\sigma}(y)\, dz^\rho\, dz^\sigma,$$

where

(1-26) $$G^i_{\rho\sigma}(y) = \sum_h G^i{}_{\rho y^h}(y) G_\sigma{}^h(y).$$

Let us assume that the hypotheses of Theorem V(1-5) are satisfied, and that the random n-vector $x(a, \boldsymbol{\omega})$ is $\mathfrak{F}(a)$-measurable and has finite second moment. Then Eq. (1-16) has a solution $x(\mathbf{t}, \boldsymbol{\omega})$ with initial value $x(a, \boldsymbol{\omega})$, the solution being unique up to equivalence. Let us again define $y(\mathbf{t}, \boldsymbol{\omega})$ by (1-22). Then, by the Itô differentiation lemma [IV(2-4)], $y^i(\mathbf{t}, \boldsymbol{\omega})$ satisfies

(1-27) $$\begin{aligned} dy^i = {} & \sum_\rho G_\rho{}^i(y)\, dz^\rho \\ & + \frac{1}{2} \sum_{\rho,\sigma} \Big\{ \sum_{h,k=1} \Psi^i{}_{x^h x^k}(\Phi(y))g_\rho{}^h(\Phi(y))g_\sigma{}^k(\Phi(y)) \\ & + \sum_{h=1}^{n} \Psi^i{}_{x^h}(\Phi(y))g^h_{\rho\sigma}(\Phi(y)) \Big\}\, dz^\rho\, dz^\sigma. \end{aligned}$$

From (1-20) and (1-21),

$$\Phi^i(\Psi(x)) = x^i; \tag{1-28}$$

so by differentiation

$$\sum_{h=1}^{n} \Phi^i{}_{y^h}(\Psi(x))\Psi^h{}_{x^j} = \delta_j{}^i \qquad (i, j = 1, \ldots, n), \tag{1-29}$$

where as usual $\delta_j{}^i = 1$ if $i = j$ and $= 0$ if $i \neq j$. If we substitute (1-24) in (1-26) and perform the indicated differentiation, by (1-29) we obtain

$$\begin{aligned} G^i_{\rho\sigma}(y) = &\sum_{h,k} \Psi^i{}_{x^h x^k}(\Phi(y))g_\rho{}^h(\Phi(y))g_\sigma{}^k(\Phi(y)) \\ &+ \sum_h \Psi^i{}_{x^h}(\Phi(y))g^h_{\rho\sigma}(\Phi(y)). \end{aligned} \tag{1-30}$$

So Eqs. (1-27) and (1-25) are the same, and the transform $y(\mathbf{t}, \omega)$ of the solution $x(\mathbf{t}, \omega)$ of the canonical equation (1-16) form is the solution of the transformed equation (1-15).

Any acceptable extension of Eqs. (1-2) must have the same solutions as does (1-2) when the z^ρ are Lipschitzian and must have solutions invariant under change of coordinates. We have seen that the canonical extension satisfies these demands. It is natural to ask whether it is the only set of equations that satisfies them. We now show that it is not, but still any acceptable extension must be closely related to the canonical extension.

Trivially, if we replace each $g^i_{\rho\sigma}$ by another function $h^i_{\rho\sigma}$ such that

$$h^i_{\rho\sigma} + h^i_{\sigma\rho} = g^i_{\rho\sigma} + g^i_{\sigma\rho},$$

the right member of (1-16) is unchanged, and likewise for (1-25). The "new" equations are still the canonical equations in a slightly different notation.

Suppose now that for each coordinate system obtainable from the x-system by (1-11) and (1-12) there is a set of functions $K^i_{\rho,\sigma}(\mathbf{t}, \mathbf{x}, \omega)$ satisfying the standing hypotheses V(1-4,iv,v,vi); the set corresponding to the x-system we denote by $k^i_{\rho,\sigma}$. Then when all z^ρ are Lipschitzian, the equations

$$dx^i = \sum_\rho g_\rho{}^i(x)\, dz^\rho + \frac{1}{2} \sum_{\rho,\sigma} \{g^i_{\rho\sigma} + k^i_{\rho\sigma}\}\, dz^\rho\, dz^\sigma \tag{1-31}$$

have the same solutions as (1-16). If for each ρ and σ the n-vector with components $K^i_{\rho\sigma} + K^i_{\sigma\rho}$ is a contravariant vector, that is if

$$K^i_{\rho\sigma}(y) + K^i_{\sigma\rho}(y) = \sum_h \Psi^i{}_{x^h}(\Phi(y))[k^h_{\rho\sigma} + k^h_{\sigma\rho}], \tag{1-32}$$

we can repeat the computations in (1-27)–(1-30) to show that the solutions of (1-31), when expressed in y coordinates, satisfy

$$dy^i = \sum_{\rho} G_{\rho}{}^i(y)\, dz^{\rho} + \frac{1}{2} \sum_{\rho,\sigma} \{G^i_{\rho\sigma}(y) + K^i_{\rho\sigma}(y)\}\, dz^{\rho}\, dz^{\sigma}. \tag{1-33}$$

If we restrict our attention to Eqs. (1-31), (1-32) in which the functions $g_{\rho}{}^i$, etc. do not depend directly on ω, it is possible to show by a somewhat tedious computation that in order for the transforms of solutions of (1-31) to satisfy (1-33) when the $z^2, \ldots, z^r$ are Wiener processes, (1-32) must be satisfied. We omit the computational details. The interesting point is that the vital requirement of invariance of solutions under change of coordinates is by itself almost enough to force us to use equations in canonical form; the only arbitrariness left is the possibility of adding second-order terms of the restricted type, (1-32). In later sections we shall find other reasons for preferring the canonical extension.

2 *Runge–Kutta approximations*

To compute the solution of an ordinary differential equation, we could use the Cauchy (or Euler) method, discussed in the preceding chapter; but its convergence is slow. Runge and Kutta have devised a family of methods, all preferable to the Cauchy method. We shall shortly see that the simplest of these methods can easily be extended to stochastic differential equations. Like the Cauchy method, for each partition Π it gives an estimate by proceeding from division point to division point of Π. When applied to a stochastic differential equation such as (1-2), without second-order terms, the method gives estimates that converge as mesh Π tends to 0. But the surprising feature is that the estimates converge, not to a solution of (1-2) itself, but to a solution of its canonical extension. This establishes another distinctive property of the canonical form, and also enables us to compute solutions of canonical extensions without ever having to compute the functions $g^i_{\rho\sigma}$.

If we wish to approximate the solution of equations

$$dx^i = g^i(t, x)\, dt \qquad (a \leqq t \leqq b) \tag{2-1}$$

with initial value $x^i(a)$, given a Cauchy partition

$$\Pi = (t_1, t_2, \ldots, t_{m+1}; t_1, \ldots, t_m)$$

of $[a, b]$ we can first compute

$$\bar{x}^i(t_2) = x^i(a) + g^i(a, x(a)) \cdot (t_2 - t_1).$$

In the Cauchy method we call this $x_\Pi{}^i(t_2)$ and continue on to t_3. But since $g^i\,(a, x(a))$ and $g^i(t_2, \bar{x}(t_2))$ are good estimates for the derivative of x^i at the beginning and end of the interval $[a, t_2]$, it is plausible that their arithmetic mean would be better than either of them as an estimate for the mean of the derivative of x^i on $[a, t_2]$. So as an improved estimate for the value of the solution at t_2 we choose

$$x_\Pi{}^i(t_2) = x^i(a) + [g^i(a, x(a)) + g^i(t_2, \bar{x}(t_2))](t_2 - a)/2.$$

Continuing step by step we obtain $x_\Pi(t_j)$ for each division point of Π. It is well-known that for smooth g^i, as mesh Π tends to 0 this, the simplest type of Runge–Kutta estimate, converges to the solution of (2-1) much more rapidly than does the Cauchy approximation.

We shall now adapt this method to stochastic differential equations, but shall confine ourselves to equations of the form (1-2). As before, $z^1(t) = t$, and for $i = 2, \ldots, n$

$$g_\rho{}^i = (g_\rho{}^i(t, x, \omega) : a \leqq t \leqq b, x \in R^{n-1}, \omega \in \Omega).$$

Given any Cauchy partition Π [with notation as just after (2-1)] of $[a, b]$, we define a corresponding Runge–Kutta process x_Π at the successive t_j by first setting

$$x_\Pi{}^i(t_1) = x^i(a), \tag{2-2}$$

where the right member is the assigned initial value of x, and then successively determining each $x_\Pi(t_{j+1})$ from $x_\Pi(t_j)$ by defining

$$\bar{x}_\Pi{}^i(t_{j+1}) = x_\Pi{}^i(t_j) + \sum_\rho g_\rho{}^i(t_j, x_\Pi(t_j))\, \Delta_j z^\rho, \tag{2-3}$$

$$x_\Pi{}^i(t_{j+1}) = x_\Pi{}^i(t_j) + \sum_\rho [g_\rho{}^i(t_j, x_\Pi(t_j)) + g_\rho{}^i(t_{j+1}, \bar{x}_\Pi(t_{j+1}))]\, \Delta_j z^\rho/2.$$

On the interior of each interval $[t_j, t_{j+1}]$ we define

$$x_\Pi{}^i(t) = x_\Pi{}^i(t_j). \tag{2-4}$$

We shall now prove that under moderate hypotheses, these x_Π do in fact converge to a limit process x as mesh $\Pi \to 0$. As we have already remarked, the striking peculiarity of the convergence is that the limit x is not in general a solution of (1-2). Instead, it is a solution of the canonical extension of (1-2).

Again for typographical convenience we use the notation introduced in (1-13)–(1-17).

The hypothesis that the $g_\rho{}^i$ are Lipschitzian [at least in $(x^2, \ldots, x^n)$] is inescapable; even in the deterministic case it is needed to ensure uniqueness of solutions. Since (2-3) involves two different values of t, we replace V(1-4,iv) by the stronger assumption that the $g_\rho{}^i$ are Lipschitzian, as well as differentiable, in $(x^1, \ldots, x^n)$. Then the $|g^i_{\rho\sigma}|$ have an upper bound linear in $|x|$. It is not much more than this to assume them Lipschitzian in x. So V(1-4,iv) will be replaced by a stronger assumption:

(2-5) For each ω in Ω the $g_\rho{}^i(\mathbf{x}, \omega)$ and their partial derivatives of order $\leqq 3$ with respect to the x^i are defined and continuous for all x, and there is a number L (independent of ω) such that $|g_\rho{}^i(0, \omega)| \leqq L$, $|g^i{}_{\rho x^h}(x, \omega)| \leqq L$ and $|g^i{}_{\rho\sigma x^h}(x, \omega)| \leqq L$ for i, h in $\{1, \ldots, n\}$, ρ, σ in $\{1, \ldots, r\}$, ω in Ω, x^1 in $[a, b]$ and $(x^2, \ldots, x^n)$ in R^{n-1}.

[In Theorem (2-9) no use is made of the third-order partial derivatives of the $g_\rho{}^i$.]

From (2-5) we obtain

$$|g_\rho{}^i(x, \omega)| \leqq L(1 + \sum_i |x^i|). \tag{2-6}$$

Let $y^2(\mathbf{t}, \boldsymbol{\omega}), \ldots, y^n(\mathbf{t}, \boldsymbol{\omega})$ be L_2-continuous processes on $[a, b]$. We write $y^1(t, \omega)$ for t; then if f is any of the functions $g_\rho{}^i$, $g^i_{\rho\sigma}$, for $a \leqq s \leqq t$ and ω in Ω we have

$$|f(y(t, \omega), \omega) - f(y(s, \omega), \omega)| \leqq Ln^{1/2} |y(t, \omega) - y(s, \omega)|. \tag{2-7}$$

Hence

$$\|f(y(t)) - f(y(s))\| \leqq Ln^{1/2} \|y(t) - y(s)\|, \tag{2-8}$$

and $f(y(\mathbf{t}, \boldsymbol{\omega}), \boldsymbol{\omega})$ is continuous in L_2-norm on $[a, b]$.

(2-9) *THEOREM Let conditions* V(1-4) *and* (2-5) *be satisfied. Assume that there exists a nondecreasing function* $(\psi(r): r > 0)$ *tending to* 0 *with* r *such that if* $a \leqq s < t \leqq b$ *and* $\rho = 1, \ldots, r$, *then a.s.*

$$E([z^\rho(t) - z^\rho(s)]^6 \mid \mathfrak{F}(s)) \leqq (t - s)\psi(t - s). \tag{2-10}$$

For each Cauchy partition Π *of* $[a, b]$, *let* $x_\Pi(t)$ *be the process defined by* (2-2), (2-3), *and* (2-4), *wherein* $x(a, \omega)$ *is* $\mathfrak{F}_a$*-measurable and* $\|x(a)\| < \infty$; *and let* $x(t)$ *be a version of the solution of the canonical extension* (1-9) *of Eq.* (1-2).

Then as mesh $\Pi \to 0$, $x_\Pi(t)$ *converges to* $x(t)$ *uniformly in* L_2*-norm on* $[a, b]$. *Moreover, if* $x(t)$ *is chosen to have a.s. continuous sample functions (as is always possible), the r.v.*

$$\sup\{|x_\Pi(t, \boldsymbol{\omega}) - x(t, \boldsymbol{\omega})| : a \leqq t \leqq b\}$$

converges to 0 *in probability and converges to* 0 *at each* ω *such that all* $z^\rho(\mathbf{t}, \omega)$ *are Lipschitzian.*

Let Π be a Cauchy partition of $[a, b]$. To simplify typography we define

$$y_j^i(\omega) = x_\Pi^i(t_j, \omega) \tag{2-11}$$

and

$$G_j^i = \sum_\sigma g_\sigma^i(y_j)\, \Delta_j z^\sigma. \tag{2-12}$$

Then (2-3) can be rewritten in the form

$$y_{j+1}^i = y_j^i + \frac{1}{2} \sum_\rho \{g_\rho^i(y_j) + g_\rho^i(y_j + G_j)\}\, \Delta_j z^\rho. \tag{2-13}$$

Another notational simplification that is useful to us is to use θ in an equation as a symbol to denote a function of all the independent variables involved in the equation, such that all values of θ lie in $(0, 1)$. The symbol θ may be used in different parts of the same proof, or even in several terms in the same equation, to mean different functions, but all have values in $(0, 1)$. Likewise Θ will (ambiguously) denote a function whose values lie in $[-1, 1]$. Thus, for example, the inequality

$$\Big| \sum_\sigma g_\sigma^i(y_j)\, \Delta_j z^\sigma \Big| \leqq L(1 + n^{1/2} | y_j |) \sum_\sigma | \Delta_j z^\sigma |$$

[which is a consequence of (2-6)] can be written as an equation

$$\sum_\sigma g_\sigma^i(y_j)\, \Delta_j z^\sigma = \Theta L(1 + n^{1/2} | y_j |) \sum_\sigma | \Delta_j z^\sigma |. \tag{2-14}$$

By the theorem of mean value, from (2-13) we obtain

$$y_{j+1}^i = y_j^i + G_j^i + \frac{1}{2} \sum_{\rho, h} g^i{}_{\rho x^h}(y_j + \theta G_j) G_j^h\, \Delta_j z^\rho. \tag{2-15}$$

To estimate the last term we write

$$\begin{aligned} &\sum_h g^i{}_{\rho x^h}(y_j + \theta G_j) g_\sigma^h(y_j) \\ &\quad = g_{\rho\sigma}^i(y_j + \theta G_j) - \sum_h g^i{}_{\rho x^h}(y_j + \theta G_j)[g_\sigma^h(y_j + \theta G_j) - g_\sigma^h(y_j)]. \end{aligned} \tag{2-16}$$

This, (2-5), (2-6), and (2-14) imply

$$\begin{aligned} &\sum_h g^i{}_{\rho x^h}(y_j + \theta G_j) g_\sigma^h(y_j) \\ &\quad = g_{\rho\sigma}^i(y_j) + \Theta L \sum_h | G_j^h | + \Theta L^2 n \sum_h | G_j^h | \\ &\quad = g_{\rho\sigma}^i(y_j) + \Theta(L^2 n + L^3 n^2)(1 + n^{1/2} | y_j |) \sum_\tau | \Delta_j z^\tau |. \end{aligned} \tag{2-17}$$

So (2-15) takes the form

$$\text{(2-18)} \quad y_{j+1}^i = y_j{}^i + \sum_\rho g_\rho{}^i(y_j)\ \Delta_j z^\rho + \frac{1}{2}\sum_{\rho,\sigma} g_{\rho\sigma}^i(y_j)\ \Delta_j z^\rho\ \Delta_j z^\sigma + \epsilon_j{}^i,$$

where

$$\text{(2-19)} \quad \epsilon_j{}^i = \Theta \sum_{\rho,\sigma,\tau} (L^2 n + L^3 n^2)(1 + n^{1/2}\,|\,y_j\,|)\,|\,\Delta_j z^\rho\ \Delta_j z^\sigma\ \Delta_j z^\tau\,|.$$

For brevity we write Δ_j for $|\,\Delta_j z^\rho\ \Delta_j z^\sigma\ \Delta_j z^\tau\,|$. Then

$$\Delta_j = [(\Delta_j z^\rho)^2]^{1/4}[(\Delta_j z^\sigma)^2]^{1/4}[(\Delta_j z^\tau)^2]^{1/4}[(\Delta_j z^\rho)^6]^{1/12}[(\Delta_j z^\sigma)^6]^{1/12}[(\Delta_j z^\tau)^6]^{1/12}.$$

The sum of the exponents on the brackets is 1, so by Hölder's inequality [I(2-12)] we have a.s.

$$\begin{aligned}\text{(2-20)} \quad E(\Delta_j\,|\,\mathfrak{F}(t_j)) &\leqq [E((\Delta_j z^\rho)^2\,|\,\mathfrak{F}(t_j))]^{1/4}\cdots[E((\Delta_j z^\tau)^6\,|\,\mathfrak{F}(t_j))]^{1/12}\\ &\leqq [K\Delta_j(t)]^{3/4}[\Delta_j t\psi(\Delta_j t)]^{1/4}\\ &= K^{3/4}\psi(\Delta_j t)^{1/4}\,\Delta_j t.\end{aligned}$$

Likewise

$$\Delta_j^2 = [(\Delta_j z^\rho)^6]^{1/3}[(\Delta_j z^\sigma)^6]^{1/3}[(\Delta_j z^\tau)^6]^{1/3},$$

so by Hölder's inequality we have a.s.

$$\begin{aligned}\text{(2-21)} \quad E(\Delta_j^2\,|\,\mathfrak{F}(t_j)) &\leqq [E((\Delta_j z^\rho)^6\,|\,\mathfrak{F}(t_j))]^{1/3}\\ &\qquad\times [E((\Delta_j z^\sigma)^6\,|\,\mathfrak{F}(t_j))]^{1/3}[E((\Delta_j z^\tau)^6\,|\,\mathfrak{F}(t_j))]^{1/3}\\ &\leqq \psi(\Delta_j t)\ \Delta_j t.\end{aligned}$$

So by Lemma III(1-1)

$$\begin{aligned}\left\|\sum_{j=1}^{k-1}|\,\epsilon_j\,|\right\| &\leqq r^3(L^2 n + L^3 n^2)(L + n^{1/2}\sup\{\|\,y_1\,\|,\ldots,\|\,y_{k-1}\,\|\})\\ &\qquad\times\left[K^{3/4}\psi(\text{mesh }\Pi)^{1/4}\sum_{j=1}^{k-1}\Delta_j t + \left\{\psi(\text{mesh }\Pi)\sum_{j=1}^{k-1}\Delta_j t\right\}^{1/2}\right].\end{aligned}$$

By Lemmas V(3-4) and V(3-24) and (2-8), the conclusions of Theorem (2-9) are established.

In deterministic problems with no random noises, the Runge–Kutta approximations converge to the solution more rapidly than the Cauchy approximations. This is not always the case when there are random noises. Consider for example the equations

$$\text{(2-22)} \qquad dx^1 = dz^1, \qquad dx^2 = 2x^1\,dz^2 \qquad [x^1(0) = x^2(0) = 0],$$

where z^1 and z^2 are independent Wiener processes. We subdivide [0, 1] into

n equal subintervals by points $t_1 = 0, t_2, \ldots, t_{n+1} = 1$. Equations (2-3) yield

$$x_\Pi^2(t_{j+1}) = x_\Pi^2(t_j) + [z^1(t_j) + z^1(t_{j+1})]\,\Delta_j z^2. \tag{2-23}$$

The solution of (2-22) is obviously

$$x^1(t) = z^1(t), \qquad x^2(t) = 2\int_0^t z^1(s)\,dz^2(s),$$

assuming as we may that $z^1(0) = 0$. The error in the estimate (2-23) for $x^2(1)$ is

$$\begin{aligned} e &= \sum_{j=1}^{n} \left\{ [z^1(t_j) + z^1(t_{j+1})]\,\Delta_j z^2 - 2\int_{t_j}^{t_{j+1}} z^1(s)\,dz^2(s) \right\} \\ &= \sum_{j=1}^{n} \int_{t_j}^{t_{j+1}} [z^1(t_j) + z^1(t_{j+1}) - 2z^1(s)]\,dz^2(s). \end{aligned}$$

The integral from t_j to t_{j+1} can be computed by means of a Cauchy partition of $[t_j, t_{j+1}]$ whose mesh we allow to tend to 0. We find easily that

$$E\left(\left\{\int_{t_j}^{t_{j+1}} [z^1(t_{j+1}) + z^1(t_j) - 2z^1(s)]\,dz^2(s)\right\}^2\right) = (t_{j+1} - t_j)^2.$$

Since the integrals are independent r.v.'s,

$$\begin{aligned} \| e \|^2 &= \sum (t_{j+1} - t_j)^2 \\ &= \text{mesh } \Pi. \end{aligned}$$

So the convergence of the Runge–Kutta approximations has the same slow rate $O\,([\text{mesh } \Pi]^{1/2})$ as that of the Cauchy approximations.

In Section 4 we shall show that for certain problems a modified Cauchy procedure closely related to the Runge–Kutta procedure produces faster convergence; but it gives no improvement in example (2-22).

3 *Comparison of ordinary and stochastic differential equations*

If there were a theorem asserting the unconditioned stability of a system such as (1-9), it would say that for all z^ρ in some reasonably large class and for all z_1^ρ in that class whose sample functions are uniformly close to those of the correspondingly numbered z^ρ, the solutions of (1-9) as written and of (1-9) with z_1^ρ in place of z^ρ would be near each other. No such theorem is possible. Consider the system

$$(3\text{-}1)\qquad dx^2 = dz^2, \qquad dx^3 = dz^3, \qquad dx^4 = x^3\, dz^2 \qquad (0 \leqq t \leqq 2\pi),$$

with initial values $x^2(0) = x^3(0) = x^4(0) = 0$. If the z^ρ vanish identically, so do the $x^i(t)$. But if a is any real number and n a positive integer, and the $z_n{}^\rho$ are defined by

$$(3\text{-}2)\qquad z_n{}^2(t, \omega) = an^{-1} \sin n^2 t, \qquad z_n{}^3(t, \omega) = n^{-1} \cos n^2 t,$$

then as n increases $z_n{}^\rho(t, \omega)$ converges uniformly to 0 ($\rho = 2, 3$) independently of ω, whereas $x^4(2\pi) = \pi a$ for all positive integers n. No adjusting with higher order integrals can help, since the $z_n{}^\rho$ are infinitely differentiable.

However, it is possible to show that under the hypotheses of Theorem (2-9), we can at least assert that *there exist* processes with Lipschitzian (even infinitely differentiable) sample functions, with each sample function uniformly close to the sample function $z^\rho(\mathbf{t}, \omega)$ of the given process, and with solutions of (1-9) uniformly close in near-L_2-norm to the solution of (1-9) with the given z^ρ. In fact, the Lipschitzian approximations to the sample functions of the z^ρ are defined by interpolation in the latter. This is a partial generalization of a theorem of Wong and Zakai [15].

As an aid in the proof we shall prove a theorem about "stopped" processes a theorem that has some interest in its own right. Let $z(\mathbf{t})$ be a process on $[a, b]$ adapted to a family $\mathfrak{F}_\tau$ of σ-subalgebras of $\mathfrak{A}$. A "stopping time," or "Markov time," for $z(\mathbf{t})$ is a r.v. $M(\boldsymbol{\omega})$ such that the values of $M(\boldsymbol{\omega})$ all lie in $[a, b]$, and for each t in $[a, b]$ the set $\{M \leqq t\}$ belongs to $\mathfrak{F}(t)$. A process $z_M(\mathbf{t})$ is obtained by "stopping z at time M" if for each ω in Ω,

$$(3\text{-}3)\qquad \begin{aligned} z_M(t, \omega) &= z(t, \omega) && (a \leqq t \leqq M(\omega)) \\ &= z(M(\omega), \omega) && (M(\omega) < t \leqq b). \end{aligned}$$

We shall now show that if z satisfies the standing hypotheses V(1-4), so does z_M, although perhaps with a larger constant K.

(3-4) *LEMMA* *If* $z(\mathbf{t}, \boldsymbol{\omega})$ *is a process on* $[a, b]$ *that satisfies* V(1-4), *and* Ω_C *is the subset of* Ω *such that* $z(\mathbf{t}, \omega)$ *is continuous when* $\omega \in \Omega_C$, *and* $M(\boldsymbol{\omega})$ *is a stopping time for* z, *there is a process* $Z(\mathbf{t}, \boldsymbol{\omega})$ *on* $[a, b]$ *such that* $Z(\mathbf{t}, \omega) = z_M(\mathbf{t}, \omega)$ *for all* ω *in* Ω_C *and* Z *satisfies* V(1-4), *except that* K *is replaced by* $16K$.

For each positive integer n and each v in $[a, b]$, we define $S_n(v)$ to consist of v and all those numbers

$$(3\text{-}5)\qquad t_j{}^n = a + (j - 1)(b - a)/2^n \qquad (j = 1, \ldots, 2^n + 1)$$

that are less than v. For each such n and v and each ω in Ω, we define $t^*(v, \omega; n)$ to be the least member of $S_n(v)$ that is not less than $\min\{v, M(\omega)\}$; in symbols,

$$t^*(v, \omega; n) = \min\{S_n(v) \cap [\min\{v, M(\omega)\}, b]\}. \tag{3-6}$$

If $v > a$, and t_k^n is the second-largest member of $S_n(v)$, $t^*(v, \omega; n)$ takes on the values $t_1^n, \ldots, t_k^n, v$ on the respective sets

$$\{M = t_1^n\}, \quad \{t_1^n < M \leqq t_2^n\}, \ldots, \{t_{k-1}^n < M \leqq t_k^n\}, \quad \{M > t_k^n\}. \tag{3-7}$$

These sets all belong to $\mathfrak{F}(v)$, so if $v > a$

(3-8) the r.v. $z(t^*(v, \boldsymbol{\omega}; n), \boldsymbol{\omega})$ is $\mathfrak{F}(v)$-measurable.

If $v = a$, $t^*(v, \boldsymbol{\omega}; n)$ is constantly a, so (3-8) is valid for all v in $[a, b]$.

Next we define

$$t^n(s) = \min\{t \in S_n(b) : t \geqq s\}. \tag{3-9}$$

For each v in $[a, b]$ and each ω in Ω, if $v < t^n(M(\omega))$, there is no member of $S(v)$ except v itself that is as great as $\min\{v, M(\omega)\}$, so

$$\text{if} \quad v < t^n(M(\omega)), \qquad t^*(v, \omega; n) = v.$$

But if $v \geqq t^n(M(\omega))$, then $\min\{v, M(\omega)\} = M(\omega)$, and $t^n(M(\omega))$ is a member of $S_n(v) \cap [\min\{v, M(\omega)\}, b]$. Any member of $S_n(v)$ that is less than $t^n(M(\omega))$ is less than $M(\omega)$, by definition of $t^n(s)$, so

$$\text{if} \quad v \geqq t^n(M(\omega)), \qquad t^*(v, \omega; n) = t^n(M(\omega)).$$

Combining the last two statements yields

$$z(t^*(v, \omega; n), \omega) = z(\min\{v, t^n(M(\omega))\}, \omega). \tag{3-10}$$

For each ω in Ω_C, the right member of (3-10) is a continuous function of v on $[a, b]$. Moreover,

$$M(\omega) \leqq t^n(M(\omega)) \leqq M(\omega) + (b - a)/2^n,$$

so

$$0 \leqq \min\{v, t^n(M(\omega))\} - \min\{v, M(\omega)\} \leqq (b - a)/2^n. \tag{3-11}$$

Statements (3-10) and (3-11) imply that

(3-12) for each ω in Ω_C, $z(t^*(v, \omega; n), \omega)$ converges uniformly on $[a, b]$ to $z_M(v, \omega)$ as $n \to \infty$.

We now define, for each v in $[a, b]$ and ω in Ω,

$$Z(v, \omega) = \limsup_{n\to\infty} z(t^*(v, \omega; n), \omega). \tag{3-13}$$

Then $Z(t, \omega)$ is a process adapted to $\mathfrak{F}_\tau$ by (3-8), and by (3-12)

(3-14) for each ω in Ω_C, $Z(t, \omega) = z_M(t, \omega)$.

The numbers in $S_n(b)$ define sets

(3-15) $$J_0^n = \{a\}, \qquad J_i^n = (t_i^n, t_{i+1}^n] \qquad (i = 1, \ldots, 2^n).$$

Then for $i = 0, \ldots, 2^n$ the set

(3-16) $$\Omega(n, i) = \{\omega \in \Omega : M(\omega) \in J_i^n\}$$

belongs to $\mathfrak{F}(t_{i+1}^n)$; and if $I_i^n(\omega) =$ indicator function of $\Omega(n, i)$, then

(3-17) I_i^n is $\mathfrak{F}(t_{i+1}^n)$-measurable, and the sum of all the functions I_i^n is 1.

Let u, v be numbers such that $a \leqq u \leqq v \leqq b$. Then for some h and k,

(3-18) $$u \in J_h^n, \qquad v \in J_k^n.$$

By definition of $t^*(v, \omega; n)$, or by (3-10),

(3-19) if $M(\omega) \in J_i^n$ and $i < h$, then

$$z(t^*(v, \omega; n), \omega) - z(t^*(u, \omega; n), \omega) = 0;$$

if $M(\omega) \in J_i^n$ and $h \leqq i < k$, then

$$z(t^*(v, \omega; n), \omega) - z(t^*(u, \omega; n), \omega) = z(t_{i+1}^n, \omega) - z(u, \omega);$$

if $M(\omega) \in J_i^n$ and $i \geqq k$, then

$$z(t^*(v, \omega; n), \omega) - z(t^*(u, \omega; n), \omega) = z(v, \omega) - z(u, \omega).$$

Statements (3-19) can be combined into

(3-20) $$\begin{aligned} &z(t^*(v, \omega; n), \omega) - z(t^*(u, \omega; n), \omega) \\ &\quad = [1 - \sum_{j<h} I_j^n(\omega)][z(v, \omega) - z(u, \omega)] \\ &\quad - \sum_{j=h}^{k-1} I_j^n(\omega)[z(v, \omega) - z(t_{j+1}^n, \omega)]. \end{aligned}$$

Let c be 2 or 4. Since $I_j^n = 0$ or 1 and $I_i^n I_j^n = 0$ if $i \neq j$, by (3-20) and the elementary inequality

(3-21) $$(x + y)^c \leqq 2^{c-1}(x^c + y^c)$$

we obtain, a.s.

$$
\begin{aligned}
\text{(3-22)}\quad & E([z(t^*(v,\omega;n),\omega) - z(t^*(u,\omega;n),\omega)]^c \mid \mathfrak{F}(u)) \\
& \leqq 2^{c-1}E([1 - \sum_{j<h} I_j{}^n(\omega)][z(v,\omega) - z(u,\omega)]^c \mid \mathfrak{F}(u)) \\
& + 2^{c-1}\sum_{j=h}^{k-1} E(E(I_j{}^n[z(v,\omega) - z(t_{j+1}^n,\omega)]^c \mid \mathfrak{F}(t_{j+1}^n)) \mid \mathfrak{F}(u)).
\end{aligned}
$$

By (3-17) and V(1-4), a.s.

$$
\begin{aligned}
\text{(3-23)}\quad & E([z(t^*(v,\omega;n),\omega) - z(t^*(u,\omega;n),\omega)]^c \mid \mathfrak{F}(u)) \\
& \leqq 2^{c-1}K(v-u)E(1 - \sum_{j<h} I_j{}^n \mid \mathfrak{F}(u)) \\
& \quad + 2^{c-1}K\sum_{j=h}^{k-1}(v - t_{j+1}^n)E(I_j{}^n \mid \mathfrak{F}(u)) \\
& \leqq 2^cK(v-u).
\end{aligned}
$$

If A is any set that belongs to $\mathfrak{F}(u)$, we conclude from (3-23) that

$$
\text{(3-24)}\quad \int_A [z(t^*(v,\omega;n),\omega) - z(t^*(u,\omega;n),\omega)]^c\, P(d\omega) \leqq 2^cK(v-u)P(A).
$$

By Fatou's lemma and (3-12), since $P(\Omega_c) = 1$,

$$
\text{(3-25)}\quad \int_A [z_M(v,\omega) - z_M(u,\omega)]^c\, P(d\omega) \leqq 2^cK(v-u)P(A).
$$

By this and Definition I(2-7) of conditional expectation

$$
\text{(3-26)}\quad \int_A E([z_M(v,\omega) - z_M(u,\omega)]^c \mid \mathfrak{F}(u))\, P(d\omega) \leqq 2^cK(v-u)P(A).
$$

Since (3-26) holds for all A in $\mathfrak{F}(u)$ and the integrand is $\mathfrak{F}(u)$-measurable, a.s.

$$
\text{(3-27)}\quad E([z_M(v,\omega) - z_M(u,\omega)]^c \mid \mathfrak{F}(u)) \leqq 2^cK(v-u).
$$

So the second and third inequalities in V(1-4,iii) hold with $16K$ in place of K.

For the first inequality in V(1-4, iii) we have by (3-20)

(3-28)

$$
\begin{aligned}
&| E(z(t^*(v, \omega; n), \omega) - z(t^*(u, \omega; n), \omega) \mid \mathfrak{F}(u)) | \\
&\quad \leqq | E([1 - \sum_{j<h} I_j(\omega)][z(v, \omega) - z(u, \omega)] \mid \mathfrak{F}(u)) | \\
&\qquad + \sum_{j=h}^{k-1} | E(E(I_j{}^n(\omega)[z(v, \omega) - z(t^n_{j+1}, \omega)] \mid \mathfrak{F}(t^n_{j+1})) \mid \mathfrak{F}(u)) | \\
&\quad \leqq K(u - v) E(1 - \sum_{j<h} I_j(\omega) \mid \mathfrak{F}(u)) \\
&\qquad + \sum_{j=h}^{k-1} K(v - t^n_{j+1}) E(I_j{}^n(\omega) \mid \mathfrak{F}(u)) \\
&\quad \leqq 2K(v - u).
\end{aligned}
$$

As we deduced (3-27) from (3-23), from (3-28) we deduce that a.s.

$$| E(z_M(v, \omega) - z_M(u, \omega) \mid \mathfrak{F}(u)) | \leqq 2K(v - u);$$

so the first inequality in V(1-4,iii) holds with $2K$ in place of K. The proof of Lemma (3-4) is complete.

Let $z^1(t), \ldots, z^r(t)$ be a set of processes on $[a, b]$; we shall assume

$$z^1(t) = t \qquad (a \leqq t \leqq b). \tag{3-29}$$

Let $u^1(\mathbf{y}), \ldots, u^r(\mathbf{y})$ be a collection of continuously differentiable non-decreasing functions on $[0, 1]$ such that $u^\rho(0) = 0$ and $u^\rho(1) = 1$; we shall further assume

$$u^1(y) = y \qquad (0 \leqq y \leqq 1). \tag{3-30}$$

Then for each Cauchy partition Π of $[a, b]$, with division points $t_1, \ldots, t_{k+1}$, we define

(3-31)

$$
\begin{aligned}
z_\Pi{}^\rho(t, \omega) &= z^\rho(t_j, \omega) + u^\rho([t - t_j]/[t_{j+1} - t_j])[z^\rho(t_{j+1}, \omega) - z^\rho(t_j, \omega)] \\
&\qquad (\rho = 1, \ldots, r; \quad j = 1, \ldots, k).
\end{aligned}
$$

Then

$$z_\Pi{}^1(t, \omega) = t \qquad (a \leqq t \leqq b). \tag{3-32}$$

If we have chosen each u^ρ $(\rho = 2, \ldots, r)$ to be the restriction to $[0, 1]$ of an infinitely differentiable function with $u(y) = 0$ $(y < 0)$ and $u(y) = 1$ $(y > 1)$, the $z_\Pi{}^\rho$ will be infinitely differentiable on $[a, b]$.

If $x(a, \omega)$ is an $\mathfrak{F}(a)$-measurable r.v., it is possible to solve for each ω the ordinary differential equation

$$(3\text{-}33) \qquad dx^i/dt = \sum_{\rho=1} g_\rho{}^i(t, x(t, \omega), \omega)[\partial z_\Pi{}^\rho(t, \omega)/\partial t],$$

with the initial values $x^i(a, \omega)$. Let us denote the solution of (3-33) by $X_\Pi(\mathbf{t}, \omega)$.

We shall first show that as mesh $\Pi \to 0$, the solutions $X_\Pi(\mathbf{t}, \omega)$ of (3-33) converge in near-L_2-norm to a limit, but as in the case of the Runge–Kutta approximations this limit is not a solution of the stochastic differential equation (1-2) that resembles (3-33). The limit can in fact depend essentially on the choice of the u^ρ. However, we shall see that in what may be considered the usual case, this limit is in fact the canonical extension of (1-2).

For $\rho, \sigma = 1, 2, \ldots, r$, let us define

$$(3\text{-}34) \qquad J_{\rho\sigma} = \int_0^1 [1 - u^\rho(y)]u^{\sigma\prime}(y)\, dy.$$

We can now state a lemma that contains the central part of our convergence theorem. It is convenient as in Section 1 to define $x^1(t) = t$, and to define $g_\rho{}^1$ by (1-13); then, since $X_\Pi{}^i$ satisfies (3-33),

$$(3\text{-}35) \quad X_\Pi{}^i(i, \omega) - X_\Pi{}^i(a, \omega) = \sum_\rho \int_a^t g_\rho{}^i(X_\Pi(s, \omega))\dot{z}_\Pi{}^\rho(s, \omega)\, ds,$$

where $\dot{z}_\Pi{}^\rho$ means $\partial z_\Pi{}^\rho/\partial t$.

(3-36) *LEMMA Let the standing hypotheses* V(1-4) *be satisfied. Assume also that there exists a nondecreasing function* ψ *on* $[0, \infty)$, *with* $\psi(r) \to 0$ *as* $r \to 0$, *such that if* $a \leqq s \leqq t \leqq b$ *and* $\rho = 1, 2, \ldots, r$, *then a.s.*

$$(*) \quad E([z^\rho(t) - z^\rho(s)]^6 \mid \mathfrak{F}(s)) \leqq (t - s)\psi(t - s).$$

Let $u^1, \ldots, u^r$ *be continuously differentiable and nondecreasing on* $[0, 1]$, *with* $u^\rho(0) = 0$, $u^\rho(1) = 1$, *and with* $u^1(y) = y$. *For each Cauchy partiiion* Π *of* $[a, b]$, *let* $z_\Pi{}^\rho$ *be defined by* (3-31), *and let* $X_\Pi(\mathbf{t}, \omega)$ *be the solution of* (3-33) *with initial value* $x(a, \omega)$. *Let*

$x(\mathbf{t}, \boldsymbol{\omega})$ be the solution of the equation

$$x^i(t) = x^i(a) + \sum_\rho \int_a^t g_\rho{}^i(s, x)\, dz^\rho(s) + \sum_{\rho,\sigma} J_{\rho\sigma} \int_a^t g^i_{\rho\sigma}(s, x)\, dz^\rho(s)\, dz^\sigma(s). \tag{3-37}$$

Then as mesh $\Pi \to 0$, $X_\Pi(t)$ *converges uniformly in near-L_2-norm to $x(t)$; and if $x(t)$ is chosen to have a.s. continuous sample functions, the r.v.*

$$\sup\{|X_\Pi(t, \boldsymbol{\omega}) - x(t, \boldsymbol{\omega})| : a \leqq t \leqq b\} \tag{3-38}$$

converges in probability to 0.

We prove this first under two supplementary hypotheses:

(3-39) there exists an N_1 such that

$$|z^\rho(t, \omega)| \leqq N_1 \qquad (\rho = 1, \ldots, r; \quad a \leqq t \leqq b; \quad \omega \in \Omega),$$

(3-40) the $x^i(a, \boldsymbol{\omega})$ have finite second moments.

Let Π be a Cauchy partition of $[a, b]$, with division points $t_1, \ldots, t_{l+1}$. As usual, for every function F on $[a, b]$ we define $\Delta_j F$ to mean $F(t_{j+1}) - F(t_j)$. Also, merely for typographical convenience, we use the notation

$$Y(t, \omega) = X_\Pi(t, \omega). \tag{3-41}$$

For $j = 1, \ldots, l$ we compute $Y(t) - Y(t_j)$ on $[t_j, t_{j+1}]$ by (3-33), and make the change of variables

$$s = t_j + y\, \Delta_j t \qquad (0 \leqq y \leqq 1). \tag{3-42}$$

Then for each ω (which we suppress as usual)

$$\begin{aligned} Y^i(t_j + y\, \Delta_j t) - Y^i(t_j) &= \sum_\rho \int_0^y [g_\rho{}^i(Y(t_j + v\, \Delta_j t)) - g_\rho{}^i(Y(t_j))][du^\rho(v)/dv]\, \Delta_j z^\rho\, dv \\ &\quad + \sum_\rho g_\rho{}^i(Y(t_j))\, \Delta_j z^\rho u^\rho(y). \end{aligned} \tag{3-43}$$

By hypothesis,

$$\begin{aligned} |g_\rho{}^i(x, \omega)| &\leqq L(1 + |x|), \\ |g_\rho{}^i(x_1, \omega) - g_\rho{}^i(x_2, \omega)| &\leqq L|x_1 - x_2|. \end{aligned} \tag{3-44}$$

Also, there is a number M' such that

$$|du^\rho(y)/dy| \leqq M' \qquad (\rho = 1, \ldots, r; \quad 0 \leqq y \leqq 1). \tag{3-45}$$

So by (3-43),

$$\begin{aligned} |Y(t_j + y\,\Delta_j t) - Y(t_j)| &\leqq nLM'\Big(\sum_\rho |\Delta_j z^\rho|\Big) \int_0^y |Y(t_j + v\,\Delta_j t) - Y(t_j)|\,dv \\ &\quad + L(1 + |Y(t_j)|) \sum_\rho |\Delta_j z^\rho|. \end{aligned} \tag{3-46}$$

Let us define

$$N(y) = \sup\{|Y(t_j + v\,\Delta_j t) - Y(t_j)| : 0 \leqq v \leqq y\}. \tag{3-47}$$

Then (3-46) remains valid if we replace the integrand by $N(v)$, and also if we replace the upper limit of the integral by any larger number in [0, 1], or equivalently replace y in the left member by any smaller number in [0, 1]. Hence

$$N(y) \leqq nLM'\Big(\sum_\rho |\Delta_j z^\rho|\Big) \int_0^y N(v)\,dv + L(1 + |Y(t_j)|) \sum_\rho |\Delta_j z^\rho|. \tag{3-48}$$

By V(3-20), this and (3-39) imply

$$N(y) \leqq L(1 + |Y(t_j)|) \sum_\rho |\Delta_j z^\rho| \exp 2rnLM'N_1. \tag{3-49}$$

Next, we define, as in Lemma V(3-4),

$$\epsilon_j{}^i = \Delta_j Y^i - \sum_\rho g_\rho{}^i(Y(t_j))\,\Delta_j z^\rho - \sum_{\rho,\sigma} J_{\rho\sigma} g^i_{\rho\sigma}(Y(t_j))\,\Delta_j z^\rho\,\Delta_j z^\sigma. \tag{3-50}$$

By integration by parts in (3-43), with $y = 1$, we obtain

$$\begin{aligned} \epsilon_j{}^i = \sum_{\rho,\sigma} \int_0^1 &[g^i_{\rho\sigma}(Y(t_j + v\,\Delta_j t)) - g^i_{\rho\sigma}(Y(t_j))] \\ &\times [1 - u^\rho(v)][du^\sigma(v)/dv]\,\Delta_j z^\rho\,\Delta_j z^\sigma\,dv. \end{aligned} \tag{3-51}$$

Since $g^i_{\rho\sigma}$ satisfies a Lipschitz condition of constant L, and (3-45) holds, if we define

$$C_2 = L^2 M' \exp 2rnLM'N_1, \tag{3-52}$$

it follows from (3-51) and (3-49) that

$$|\epsilon_j^i| \leqq \sum_{\rho,\sigma,\upsilon} C_2(1 + |Y(t_j)|)\,|\Delta_j z^\rho\, \Delta_j z^\sigma\, \Delta_j z^\upsilon|. \tag{3-53}$$

By Lemma III(1-1) and (2-20) and (2-21)

$$\left\|\left|\sum_{j=1}^{k-1} |\epsilon_j^i|\right\|\right| \leqq C_2 r^3 \sum_{j=1}^{k-1} (1 + \|Y(t_j)\|) K^{3/4} \psi(\Delta_j t)^{1/4}\, \Delta_j t + C_2 r^3 \left\{\sum_{j=1}^{k-1} (1 + \|Y(t_j)\|)^2 \psi(\Delta_j t)\, \Delta_j t\right\}^{1/2}. \tag{3-54}$$

If we restrict Π by requiring $\psi(\text{mesh }\Pi) \leqq 1$, this implies

$$\left\|\left|\sum_{j=1}^{k-1} |\epsilon_j^j|\right\|\right| \leqq C_2 r^3 \psi(\text{mesh }\Pi)^{1/4}[1 + \sup\{\|Y(t_j))\|: j = 1, \ldots, k-1\}] \times [K^{3/4}(b-a) + (b-a)^{1/2}]. \tag{3-55}$$

We have now verified that the hypotheses of Lemmas V(3-4) and V(3-24) are satisfied. By those lemmas, with (3-50), the process $y_\Pi(\mathbf{t})$ defined by

$$y_\Pi{}^i(t, \omega) = Y^i(t_j, \omega) \qquad (t_j \leqq t < t_{j+1}) \tag{3-56}$$

(with $<$ replaced by $\leqq$ if $t_{j+1} = b$) converges uniformly in L_2-norm to the solution $x(t)$ of the equation

$$x^i(t) = x^i(a) + \sum_\rho \int_a^t g_\rho{}^i(s, x)\, dz^\rho(s) + \sum_{\rho,\sigma} J_{\rho\sigma} \int_a^t g^i_{\rho\sigma}(s, x)\, dz^\rho(s)\, dz^\sigma(s), \tag{3-57}$$

and the r.v.

$$\sup\{|y_\Pi(t, \boldsymbol{\omega}) - x(t, \boldsymbol{\omega})|: a \leqq t \leqq b\} \tag{3-58}$$

converges in probability to 0, as mesh $\Pi \to 0$.

For each positive ϵ there is an M_2 such that $|x(t, \omega)| < M_2$ $(a \leqq t \leqq b)$ for all ω except those in a set Ω_1 with $P(\Omega_1) < \epsilon/3$. We define

$$\text{Osc}(\delta, \omega) = \sup\{\sum_\rho |z^\rho(t, \omega) - z^\rho(s, \omega)|: a \leqq s \leqq t \leqq b, t - s < \delta\}.$$

As in the proof of Theorem III(5-1), we see that this is a r.v. that tends to 0 a.s. as $\delta \to 0$. Therefore there is a positive δ_1 such that the set

$$\Omega_2 = \{\omega \in \Omega: \text{Osc}(\delta_1, \omega) \geqq \epsilon/L(2 + M_2) \exp 2rnLM'N_1\}$$

has $P(\Omega_2) < \epsilon/3$. Since the r.v. (3-58) converges in probability to 0, there is a positive δ_2 such that if mesh $\Pi < \delta_2$, the r.v. (3-58) is less than 1 except on a set $\Omega(\Pi)$ with $P(\Omega(\Pi)) < \epsilon/3$. If ω is not in $\Omega_1 \cup \Omega(\Pi)$, for all division points t_j of Π we have

$$| Y(t_j, \omega) | = | [y_\Pi(t_j, \omega) - x(t_j, \omega)] + x(t_j, \omega) | \leqq 1 + M_2.$$

By (3-49), if Π has a mesh less than $\min\{\delta_1, \delta_2\}$ and ω is not in the set $\Omega_1 \cup \Omega_2 \cup \Omega(\Pi)$, whose P-measure is less than ϵ, for all t in $[a, b]$

$$| Y(t, \omega) - y_\Pi(t, \omega) | \leqq L(1 + Y(t_j, \omega) |) \cdot \mathrm{Osc}(\delta_1, \omega) \exp 2rnLM'n_1 < \epsilon.$$

So $\sup\{| Y(t, \omega) - y_\Pi(t, \omega) | : a \leqq t \leqq b\}$ converges in probability to 0 with mesh Π. By this and the convergence of (3-58) to 0, as mesh $\Pi \to 0$

$$\text{(3-59)} \quad \sup\{| Y(t, \omega) - x(t, \omega) | : a \leqq t \leqq b\} \to 0 \quad \text{in probability.}$$

In a like manner, $\| Y(t) - y_\Pi(t) \|$ tends uniformly to 0, so as mesh $\Pi \to 0$

$$\text{(3-60)} \qquad \| Y(t) - x(t) \| \to 0 \quad \text{uniformly on } [a, b].$$

The lemma, with the strengthened conclusion (3-60), is established under the supplementary hypotheses (3-39) and (3-40).

Returning to the hypotheses of the lemma, we first observe that there is no loss of generality in assuming that the $z^\rho(\mathbf{t}, \omega)$ are continuous for all ω. For there is a set Ω_C with $P(\Omega_C) = 1$ on which they are all continuous. If we restrict z^ρ to Ω_C, and for each τ replace $\mathfrak{F}(\tau)$ by the σ-algebra of sets of the form $A \cap \Omega_C$, A in $\mathfrak{F}(\tau)$, the hypotheses of the lemma are satisfied by these processes also. There is no loss of generality in assuming also that $z^\rho(a, \omega) = 0$ for all ρ and ω. If $\epsilon > 0$, there is an N_1 such that except on a set $\Omega'(\epsilon)$ with $P(\Omega'(\epsilon)) < \epsilon/2$ we have

$$| z^\rho(t, \omega) | < N_1 \qquad (\rho = 1, \ldots, r; \quad a \leqq t \leqq b).$$

For each ω we define $M(\omega)$ to be the greatest number t in $[a, b]$ such that

$$| z^\rho(s, \omega) | < N_1 \qquad (a \leqq s < t; \quad \rho = 1, \ldots, r).$$

Then M is a stopping time. Denote by $z_M{}^\rho$ the process z^ρ stopped at $M(\omega)$. Then by Lemma (3-4), $z_M{}^\rho$ satisfies V(1-4), and by definition of M it satisfies (3-39). Moreover,

$$\text{(3-61)} \quad \text{if } \omega \in \Omega \backslash \Omega'(\epsilon), z_M(\mathbf{t}, \omega) = z(\mathbf{t}, \omega).$$

There is a number V such that the set Ω_V on which $| x(a, \omega) | > V$ has $P(\Omega_V) < \epsilon/4$. We define

$$x_V{}^i(a, \omega) = \mathrm{mid}\{x^i(a, \omega), V, -V\};$$

we define $x_M(\mathbf{t}, \boldsymbol{\omega})$ to be the solution of (3-37) with initial value $x_V{}^i(a, \boldsymbol{\omega})$ and with $z_M{}^\rho$ in place of z^ρ; and we define $X_{M\Pi}$ to be the solution of (3-33) with the same replacements. Then by the part of the proof already completed (3-60) holds for these processes, so

$$E(| X_{M\Pi}(t) - X_M(t) |^2) \to 0$$

uniformly on $[a, b]$ as mesh $\Pi \to 0$. But by Theorem II(2-22), except on $\Omega_V \cup \Omega'(\epsilon) \cup N$, where N is negligible, we have

$$X_{M\Pi}(\mathbf{t}, \omega) = X_\Pi(\mathbf{t}, \omega), \qquad x_M(\mathbf{t}, \omega) = x(\mathbf{t}, \omega).$$

So if I is the indicator function of the set $\Omega\backslash\Omega_V \cup \Omega'(\epsilon)$, whose P-measure is greater than $1 - \epsilon$, the quantity

$$E(I \mid X_\Pi(t) - x(t) |^2)$$

converges to 0 uniformly on $[a, b]$ as mesh Π tends to 0. Likewise, if mesh Π is small enough, except on a set $\Omega''(\Pi)$ with $P(\Omega''(\Pi)) < \epsilon/4$ we have

$$\sup\{| X_{M\Pi}(t, \omega) - x_M(t, \omega) | : a \leqq t \leqq b\} < \epsilon,$$

so by (3-61) the r.v. (3-38) is less than ϵ except on $\Omega_V \cup \Omega'(\epsilon) \cup \Omega''(\Pi)$, whose measure is less than ϵ. The proof is complete.

We have consistently attempted to hold our hypotheses concerning the noise processes z^ρ to a minimum. Hypotheses about the coefficients $g_\rho{}^i$ are capable of unambiguous verification or rejection, since we know the $g_\rho{}^i$ precisely. But in applications we usually have only scanty and imprecise information about the z^ρ, and conclusions based on probabilistic assumptions about them will be of uncertain applicability. Nevertheless, there is one type of assumption about the z^ρ that would seem plausible in many cases. If the various z^ρ are random disturbances coming from different sources that do not affect each other, then, even when conditioned by all information available at time s, the different increments $z^\rho(v) - z^\rho(u)$ $(s \leqq u < v \leqq b)$ should be independent of each other. In this case we can show that the conclusion of Lemma (3-36) remains valid if we replace all the $J_{\rho\sigma}$ by $\frac{1}{2}$. This is part of the conclusion of the next theorem

(3-62) *THEOREM Let all the hypotheses of Lemma* (3-36) *hold. Suppose further that for each ρ and σ in the set $\{2, \ldots, r\}$ for which neither of the two conditions*

(i) $g^i_{\rho\sigma}(t, \omega) = g^i_{\sigma\rho}(t, \omega)$ $(i = 1, \ldots, n; a \leqq t \leqq b; \omega \in \Omega)$;
(ii) *whenever $a \leqq s \leqq t \leqq b$, $z^\rho(t, \omega) - z^\rho(s, \omega)$ and $z^\sigma(t, \omega) - z^\sigma(s, \omega)$ are conditionally independent given $\mathfrak{F}(s)$*

is satisfied, we choose $u^\sigma = u^\rho$. *Then as* mesh $\Pi \to 0$, *the solution* X_Π *of* (3-33) *with initial values* $x^i(a, \omega)$ *converges in near-*L_2*-norm to the solution* $x(\mathbf{t}, \omega)$ *of*

$$x^i(t) = x^i(a) + \sum_\rho \int_a^t g_\rho{}^i(s, x(s))\, dz^\rho(s) + \frac{1}{2} \sum_{\rho,\sigma} \int_a^t g^i_{\rho\sigma}(s, x(s))\, dz^\rho(s)\, dz^\sigma(s); \tag{3-63}$$

and if $x(t)$ *is chosen to have a.s. continuous sample functions, the r.v.*

$$\sup\{|\, X_\Pi(t, \omega) - x(t, \omega)\, | : a \leqq t \leqq b\} \tag{3-64}$$

converges in probability to 0 *as* mesh $\Pi \to 0$.

This will follow at once from Lemma (3-36) when we show that under the hypotheses, Eqs. (3-63) and (3-37) have the same solutions. In both these equations the second-order integrals with $\rho = 1$ or $\sigma = 1$ vanish. By (3-34),

$$J_{\rho\sigma} + J_{\sigma\rho} = -\int_0^1 [1 - u^\rho][1 - u^\sigma]'\, dy - \int_0^1 [1 - u^\sigma][1 - u^\rho]'\, dy = 1,$$

so we can write

$$J_{\rho\sigma} = \tfrac{1}{2} + c, \qquad J_{\sigma\rho} = \tfrac{1}{2} - c.$$

If $g^i_{\rho\sigma} = g^i_{\sigma\rho}$, on substituting these expressions in (3-37) we find that the coefficient of c is 0, and the coefficients of $g^i_{\rho\sigma}$ and $g^i_{\sigma\rho}$ are $\frac{1}{2}$. If the increments of z^ρ and z^σ are conditionally independent, the second-order integral with integrand $g^i_{\rho\sigma}$ is 0, and replacing its coefficients $J_{\rho\sigma}$ by $\frac{1}{2}$ has no effect. If neither of these conditions holds, $u^\rho = u^\sigma$, so

$$J_{\rho\sigma} = \int_0^1 [1 - u^\rho(y)] u^{\rho\prime}(y)\, dy = \tfrac{1}{2}.$$

4 Rate of convergence of approximations to solutions

At the end of Chapter V we remarked that even for very simple systems of equations the difference between the solution of a stochastic differential equation and the Cauchy–Maruyama approximations can be disappointingly slow in its convergence to 0; and at the end of Section 2 we saw that this is true for Runge–Kutta approximations also. In this section we shall show that when the equations have the canonical form and some additional hypotheses are satisfied, a set of approximations closely related to the Cauchy–Maruyama approximations converges considerably faster. This has evident computational advantages; but besides this, it will be seen in Section 6 to furnish evidence that the canonical form of stochastic differentials is appropriate for constructing mathematical models of systems affected by random disturbances.

First, we shall strengthen V(1-4) by assuming

(4-1) With the K of V(1-4), there is a positive δ such that if $a \leqq s \leqq t \leqq b$ and $t - s < \delta$, then a.s.

$$E([z^\rho(t) - z^\rho(s)]^{12} \mid \mathfrak{F}(s)) \leqq K(t-s)^6 \qquad (\rho = 2, \ldots, r).$$

This is satisfied by the Wiener processes and by the processes with uniformly Lipschitzian sample functions, as well as by many others.

We have already pointed out, just before Theorem (3-63), that in many applications it is quite natural to assume that for each s in $[a, b]$, if $[u_1, v_1]$ and $[u_2, v_2]$ are subintervals of $[s, b]$, the increments $z^\rho(v_1) - z^\rho(u_1)$, $z^\sigma(v_2) - z^\sigma(u_2)$ $(\rho \neq \sigma)$ are conditionally independent, given $\mathfrak{F}(s)$. So, recalling Eq. I(2-18), when convenient we shall make the following assumption.

(4-2) If ρ, σ are two different members of $\{1, \ldots, r\}$ and $a \leqq s < b$, and $[u_1, v_1]$ and $[u_2, v_2]$ are subintervals of $[s, b]$, and p and q are positive integers with $p + q \leqq 12$, then a.s.

$$\begin{aligned} &E([z^\rho(v_1) - z^\rho(u_1)]^p[z^\sigma(v_2) - z^\sigma(u_2)]^q \mid \mathfrak{F}(s)) \\ &\quad = E([z^\rho(v_1) - z^\rho(u_1)]^p \mid \mathfrak{F}(s))E([z^\sigma(v_2) - z^\sigma(u_2)]^q \mid \mathfrak{F}(s)). \end{aligned}$$

A similar equation holds for products of increments of three different z^ρ.

Since we always assume that the $g^i_{\rho\sigma}$ are Lipschitzian, the functions $g^i_{\rho\sigma\tau}$ defined by

$$(4\text{-}3) \qquad g^i_{\rho\sigma\tau}(x) = \sum_h g^i{}_{\rho\sigma x^h}(x) g_\tau{}^h(x)$$

do not exceed a linear function of $|x|$; in fact, if (2-5) holds then by (2-6)

$$(4\text{-}4) \qquad |g^i_{\rho\sigma\tau}(x)| \leqq nL^2(1 + \sum_i |x^i|).$$

It is not a great deal more if we assume that they are Lipschitzian; this we shall do in Theorem (4-8). On the other hand, it is much more restrictive to require

$$g^i_{\rho\sigma}(x) = g^i_{\sigma\rho}(x) \qquad (i = 1, \ldots, n; \quad \rho, \sigma = 2, \ldots, r). \tag{4-5}$$

[Recall that we are defining $z^1(t)$ to be t, so that $z^2, \ldots, z^r$ are the random disturbances.] This condition is clearly satisfied in the important special case in which there is just one random noise ($r = 2$). It is trivially satisfied if the $g_\rho{}^i$ are independent of $x^2, \ldots, x^n$ (recall that x^1 is t). It is also satisfied if $r = n$, and $g_\rho{}^i(x) = 0$ unless $\rho = i$, and $g_i{}^i(x)$ depends only on x^i. There are other examples also. For instance, if (x^2, x^3, x^4) are the space coordinates of a point and (x^5, x^6, x^7) its velocity components, and the latter only are subject to random disturbances, while the $g_\rho{}^i$ ($i = 5, 6, 7$) depend only on (x^1, x^2, x^3, x^4), we find that all $g^i_{\rho\sigma}$ vanish.

Given an initial value $x(a, \omega)$ and a Cauchy partition

$$\Pi = (t_1, t_2, \ldots, t_{m+1}; t_1, \ldots, t_m) \tag{4-6}$$

of $[a, b]$, we shall define $x_\Pi{}^i$ successively on the intervals $[t_1, t_2]$, $[t_2, t_3]$, ... by setting

(4-7) if $t_j \leqq t \leqq t_{j+1}$,

$$x_\Pi{}^i(t) = x_\Pi{}^i(t_j) + \sum_\rho g_\rho{}^i(x_\Pi(t_j))[z^\rho(t) - z^\rho(t_j)]$$

$$+ \frac{1}{2} \sum_{\rho,\sigma} g^i_{\rho\sigma}(x_\Pi(t_j))[z^\rho(t) - z^\rho(t_j)][z^\sigma(t) - z^\sigma(t_j)]$$

$$+ \frac{1}{6} \sum_{\rho=2}^{r} g^i_{\rho\rho\rho}(x_\Pi(t_j))[z^\rho(t) - z^\rho(t_j)]^3.$$

This differs from the Cauchy–Maruyama approximation V(4-2) only in the last term, which presumably will be a small one in all cases of interest, but is essential in the proof of the next theorem.

(4-8) *THEOREM Let conditions* V(1-4) *and* (2-5) *be satisfied. Assume also that* (4-1), (4-2), *and* (4-5) *are satisfied, and that the absolute values of the first-order partial derivatives of the* $g^i_{\rho\sigma\tau}$ *are bounded by the same* L *as in* (2-5). *Let* $x(a, \omega)$ *be* $\mathfrak{F}(a)$*-measurable and have finite second moment, and let* $x(\mathbf{t}, \omega)$ *be the solution of the canonical equations* (1-9). *For each Cauchy partition* Π *of* $[a, b]$ *let* x_Π *be* *defined by* (4-7). *Then there exists a positive number* C *such that if mesh* $\Pi < \delta$ [δ *as in* (4-1)], *then*

$$\| x_\Pi(t) - x(t) \| \leqq C \text{ mesh } \Pi \qquad (a \leqq t \leqq b).$$

Moreover, the r.v.

$$(\sup\{| x_\Pi(t, \omega) - x(t, \omega) | : a \leqq t \leqq b\} \mid \omega \in \Omega)$$

converges in probability to 0 *as* mesh $\Pi \to 0$.

Since V(1-4) and (4-1) are satisfied, we can prove the following statement.

(4-9) If $a \leqq s \leqq t \leqq b$, and $2 \leqq C \leqq 12$, and Δ is the product of C factors each of the form $z^\rho(t) - z^\rho(s)$ for some ρ in $\{1, \ldots, r\}$, then a.s.

$$E(| \Delta | \,|\, \mathfrak{F}(s)) \leqq rK(t - s)^{C/2}.$$

Suppose first that

$$\Delta = [\Delta z^\rho]^C = [z^\rho(t) - z^\rho(s)]^C$$

for some ρ in $\{1, \ldots, r\}$; we omit the ρ. Define $c = (C - 2)/10$. Then by the Hölder inequality I(2-12)

$$\begin{aligned} E(| \Delta | \mid \mathfrak{F}(s)) &= E(([\Delta z]^2)^{1-c}([\Delta z]^{12})^c \mid \mathfrak{F}(s)) \\ &\leqq \{E([\Delta z]^2 \mid \mathfrak{F}(s))\}^{1-c}\{E([\Delta z]^{12} \mid \mathfrak{F}(s)\}^c \\ &\leqq \{K(t - s)\}^{1-c}\{K(t - s)^6\}^c \\ &= K(t - s)^{C/2}. \end{aligned}$$

If Δ is the product of C factors not necessarily equal, obviously

$$| \Delta | \leqq \sum_{\rho=1}^{r} | z^\rho(t) - z^\rho(s) |^C,$$

so by the part of the proof already completed

$$E(| \Delta | \mid \mathfrak{F}(s)) \leqq \sum_{\rho=1}^{r} E(| z^\rho(t) - z^\rho(s)|^C \mid \mathfrak{F}(s)) \leqq rK(t - s)^{C/2}.$$

If we define

$$\epsilon_\Pi{}^i(t_j) = \frac{1}{6} \sum_{\rho=1}^{r} g^i_{\rho\rho\rho}(x_\Pi(t_j))[z^\rho(t_{j+1}) - z^\rho(t_j)]^3,$$

we obtain from (4-4), (4-9), and Lemma III(1-1)

$$\left\|\sum_{j=1}^{k} | \epsilon_\Pi{}^i(t_j) |\right\|$$
$$\leq \tfrac{1}{6}n^2L^2 \sup\{1 + n \| x_\Pi(t_h) \| : 1 \leq h \leq k\}$$
$$\times \left\{2rK \sum_{j=1}^{k} (t_{j+1} - t_j)^{3/2} + (rK)^{1/2}\left(\sum_{j=1}^{k} (t_{j+1} - t_j)^3\right)^{1/2}\right\},$$

so that hypothesis (iv) of Lemma V(3-4) is satisfied. So by Lemma V(3-24) the function X_Π that coincides with x_Π at the division points of Π and is constant on the intervals of Π has the property that

$$\sup\{| X_\Pi(t, \omega) - x(t, \omega) | : a \leq t \leq b\}$$

converges in probability to 0 as mesh $\Pi \to 0$. The proof that this holds also with x_Π in place of X_Π is the same as in the proof of (3-59). This proves the last statement in the conclusion of Theorem (4-8); we now begin the more difficult proof of the first (and more important) conclusion.

As in Section 2 we shall use θ to denote a function depending on all the independent variables in the equation in which it occurs, and assuming values in (0, 1). Likewise Θ will denote a function with values in $[-1, 1]$. Also, we shall frequently need estimates which contain a factor that is the value of some polynomial in the data $n, r, K, L, b - a$ of the problem. Each such factor is easily computed, but they are numerous, and it becomes tedious to write out explicitly the forms of these factors. We shall reserve the letter C, with subscripts, for such factors. Thus, for example, from (2-5) and the theorem of the mean, for some $\bar{x}$ between 0 and x

$$| g^i_{\rho\sigma}(x, \omega) | \leq \sum_h | g^i{}_{\rho x^h}(0) g_\sigma{}^h(0) | + \sum_h | g^i{}_{\rho\sigma x^h}(\bar{x}) | | x^h |$$
$$\leq nL^2 + n^{1/2}L | x |,$$

which we write in the form

$$g^i_{\rho\sigma}(x, \omega) = \Theta(C_1 + C_2 | x |). \tag{4-10}$$

Similarly,

$$g_\rho{}^i(x, \omega) = \Theta(C_3 + C_4 | x |), \tag{4-11}$$
$$g^i_{\rho\sigma\tau}(x, \omega) = \Theta(C_5 + C_6 | x |).$$

For each t in $(a, b]$, define

$$N_1(t) = \sup\{\| x_\Pi(t_j) \| : t_j \text{ a division point of } \Pi,\ t_j \leq t\}. \tag{4-12}$$

Then

$$\| g_\rho{}^i(x_\Pi(t)) \| \leqq C_7(1 + N_1(t)), \tag{4-13}$$

and likewise for $g^i_{\rho\sigma}$ and $g^i_{\rho\rho\rho}$ (with perhaps different C_i). Since $N_1(t)$ is constant inside the intervals of Π,

$$\sum_{j=1}^{k-1} N_1(t_j)^2 \, \Delta_j t = \int_a^{t_k} N_1(s)^2 \, ds.$$

So by Lemma III(2-2)

$$\begin{aligned} \left\| \sum_{j=1}^{k-1} g_\rho{}^i(x_\Pi(t_j)) \, \Delta_j z^\rho \right\| &\leqq C \left\{ \sum_{j=1}^{k-1} C_7{}^2(1 + N_1(t_j))^2 \, \Delta_j t \right\}^{1/2} \\ &\leqq 2CC_7 \left\{ \sum_{j=1}^{k-1} (1 + N_1(t_j)^2) \, \Delta_j t \right\}^{1/2} \\ &\leqq C_8 + C_9 \left\{ \int_a^t N_1(s)^2 \, ds \right\}^{1/2}. \end{aligned} \tag{4-14}$$

A similar estimate applies to the other terms in (4-7), so by adding from $i = 1$ to $i = n$ we obtain

$$\| x_\Pi(t_j) \| \leqq \| x(a) \| + C_{10} + C_{11} \left\{ \int_a^t N_1(s)^2 \, ds \right\}^{1/2}. \tag{4-15}$$

This estimate remains valid if we replace t in the right member by any larger number, or equivalently if we replace j in the left member by any smaller positive integer. We may therefore replace the left member of (4-15) by $N_1(t)$. The resulting inequality has the same form as V(3-18). By the same argument as was used on V(3-18) to deduce V(3-22), we conclude

$$N_1(t)^2 \leqq C_{12}. \tag{4-16}$$

Let Π' be a Cauchy partition of $[a, b]$ containing all division points of Π and such that each interval $[t_j, t_{j+1}]$ of Π contains at most one division point of Π' in its interior. If there is such a point we name it s_j; if not, we define s_j to be t_{j+1}. We define

$$\begin{aligned} \Delta_j{}'t &= s_j - t_j, & \Delta_j''t &= t_{j+1} - s_j, \\ \Delta_j{}'z^\rho &= z^\rho(s_j) - z^\rho(t_j), & \Delta_j''z^\rho &= z^\rho(t_{j+1}) - z^\rho(s_j). \end{aligned} \tag{4-17}$$

Also, for typographical simplicity we introduce the symbols

$$x_j{}^i = x_\Pi{}^i(t_j), \qquad y_j{}^i = x^i_{\Pi'}(t_j), \qquad m_j{}^i = x^i_{\Pi'}(s_j). \tag{4-18}$$

By (4-7),

$$m_j{}^i = y_j{}^i + \sum_{\rho} g_\rho{}^i(y_j)\ \Delta_j{}'z^\rho + \frac{1}{2}\sum_{\rho,\sigma} g^i_{\rho\sigma}(y_j)\ \Delta_j{}'z^\rho\ \Delta_j{}'z^\sigma + \frac{1}{6}\sum_{\rho=2}^{r} g^i_{\rho\rho\rho}(y_j)\,(\Delta_j{}'z^\rho)^3, \tag{4-19}$$

and

$$y^i_{j+1} = m_j{}^i + \sum_{\rho} g_\rho{}^i(m_j)\Delta_j^{\prime\prime}z^\rho + \frac{1}{2}\sum_{\rho,\sigma} g^i_{\rho\sigma}(m_j)\ \Delta_j^{\prime\prime}z^\rho\ \Delta_j^{\prime\prime}{}^{\sigma} + \frac{1}{6}\sum_{\rho=2}^{r} g^i_{\rho\rho\rho}(m_j)\ (\Delta_j^{\prime\prime}z^\rho)^3, \tag{4-20}$$

while

$$x^i_{j+1} = x_j{}^i + \sum_{\rho} g_\rho{}^i(x_j)\ \Delta_j z^\rho + \frac{1}{2}\sum_{\rho,\sigma} g^i_{\rho\sigma}(x_j)\ \Delta_j z^\rho\ \Delta_j z^\sigma + \frac{1}{6}\sum_{\rho=2}^{r} g^i_{\rho\rho\rho}(x_j)\ (\Delta_j z^\rho)^3. \tag{4-21}$$

From (4-19) we deduce by the theorem of the mean

$$g_\rho{}^i(m_j) = g_\rho{}^i(y_j) + \sum_h g^i{}_{\rho x^h}(y_j + \theta[m_j - y_j]) \times\left\{\sum_\sigma g_\sigma{}^h(y_j)\ \Delta_j{}'z^\sigma + \frac{1}{2}\sum_{\sigma,\tau} g^h_{\sigma\tau}(y_j)\ \Delta_j{}'z^\sigma\ \Delta_j{}'z^\tau + \frac{1}{6}\sum_{\sigma=2}^{r} g^h_{\sigma\sigma\sigma}(y_j)[\Delta_j{}'z^\sigma]^3\right\}. \tag{4-22}$$

To estimate the right member we recall that the $g_\rho{}^i$ satisfy a Lipschitz condition in the x^h with constant L, so that their first-order partial derivatives are ΘL. By (2-16),

$$\begin{aligned}\sum_h g^i{}_{\rho x^h}(y_j + \theta[m_j - y_j])g_\sigma{}^h(y_j) &= g^i_{\rho\sigma}(y_j + \theta[m_j - y_j]) \\ &\quad - \sum_h g^i{}_{\rho x^h}(y_j + \theta[m_j - y_j])[g_\sigma{}^h(y_j + \theta[m_j - y_j]) - g_\sigma{}^h(y_j)] \\ &= g^i_{\rho\sigma}(y_j) + \Theta L \sum_h |\ m_j{}^h - y_j{}^h\ | + \Theta L^2 \sum |\ m_j{}^h - y_j{}^h\ |.\end{aligned}$$

Because of the bound on the partial derivatives,

$$g^i{}_{\rho x^h}(y_j + \theta[m_j - y_j])\{\tfrac{1}{2}g^h_{\sigma\tau}(y_j)\} = (\Theta L/2)\ |\ g^h_{\sigma\tau}(y_j)\ |,$$

$$g^i{}_{\rho x^h}(y_j + \theta[m_j - y_j])\{\tfrac{1}{6}g^h_{\sigma\sigma\sigma}(y_j)\} = (\Theta L/6)\ |\ g^h_{\sigma\sigma\sigma}(y_j)\ |.$$

By substituting these evaluations in (4-22) we obtain

$$(4\text{-}23)\quad g_\rho{}^i(m_j) = g_\rho{}^i(y_j) + \sum_\sigma \{g^i_{\rho\sigma}(y_j) + \sum_h [\Theta C_{13}(m_j{}^h - y_j{}^h)]\}\ \Delta_j{}'z^\sigma$$
$$+ \sum_{h,\sigma,\tau} [(\Theta L/2)g^h_{\sigma\tau}(y_j)]\ \Delta_j{}'z^\sigma\ \Delta_j{}'z^\tau$$
$$+ \sum_h \sum_{\sigma=2}^{r} [(\Theta L/6)g^h_{\sigma\sigma\sigma}(y_j)]\ (\Delta_j{}'z^\sigma)^3.$$

Similarly,

$$(4\text{-}24)\quad g^i_{\rho\sigma}(m_j) = g^i_{\rho\sigma}(y_j) + \sum_\tau \{g^i_{\rho\sigma\tau}(y_j) + \sum_h [\Theta C_{14}(m_j{}^h - y_j{}^h)]\}\ \Delta_j{}'z^\tau$$
$$+ \sum_{h,\tau,v} [(\Theta L/2)g^h_{\tau v}(y_j)]\ \Delta_j{}'z^\tau\ \Delta_j{}'z^v$$
$$+ \sum_h \sum_{\tau=2}^{r} [(\Theta L/6)g^h_{\tau\tau\tau}(y_j)]\ (\Delta_j{}'z^\tau)^3,$$

and

$$(4\text{-}25)\qquad g^i_{\rho\rho\rho}(m_j) = g^i_{\rho\rho\rho}(y_j) + \sum_h [\Theta C_{15}(m_j{}^h - y_j{}^h)].$$

If we substitute these and (4-19) in (4-20), and from the result subtract (4-21) member by member, and define

$$(4\text{-}26)\qquad \delta_{\rho\sigma\tau} = 0 \quad \text{if} \quad \rho = \sigma = \tau \geqq 2,$$
$$= 1 \quad \text{otherwise},$$

we obtain

$$(4\text{-}27)\quad (y^i_{j+1} - x^i_{j+1}) - (y_j{}^i - x_j{}^i) = \sum_\rho [g_\rho{}^i(y_j) - g_\rho{}^i(x_j)]\ \Delta_j z^\rho$$
$$+ \frac{1}{2}\sum_{\rho,\sigma} [g^i_{\rho\sigma}(y_j) - g^i_{\rho\sigma}(x_j)]\ \Delta_j z^\rho\ \Delta_j z^\sigma$$
$$+ \frac{1}{6}\sum_{\rho=2}^{r} [g^i_{\rho\rho\rho}(y_j) - g^i_{\rho\rho\rho}(x_j)]\ (\Delta_j z^\rho)^3$$
$$+ \sum_{v=1}^{6} R^i_{j,v},$$

where

$$R^i_{j,1} = \frac{1}{2} \sum_{\rho} (g^i_{1\rho}(y_j) - g^i_{\rho 1}(y_j))[\Delta_j' z^1 \, \Delta_j'' z^\rho - \Delta_j' z^\rho \, \Delta_j'' z^1], \tag{4-28}$$

$$R^i_{j,2} = \frac{1}{2} \sum_{\rho,\sigma,\tau} \delta_{\rho\sigma\tau} g^i_{\rho\sigma\tau}(y_j) \, \Delta_j' z^\tau \, \Delta_j'' z^\rho \, \Delta_j'' z^\sigma,$$

$$R^i_{j,3} = -\frac{1}{2} \sum_{\rho=2}^{r} g^i_{\rho\rho\rho}(y_j) \, (\Delta_j' z^\rho)^2 \, \Delta_j'' z^\rho,$$

$$R^i_{j,4} = \sum_{\rho,\sigma,\tau,h} C_{13} \Theta g_\tau{}^h(y_j) \, \Delta_j' z^\tau \, \Delta_j' z^\sigma \, \Delta_j'' z^\rho,$$

$$R^i_{j,5} = \sum_{\rho,\sigma,\tau,h} (\Theta L/2) g^h_{\sigma\tau}(y_j) \, \Delta' z^\sigma \, \Delta' z^\tau \, \Delta'' z^\rho$$

and $R^i_{j,6}$ is a collection of terms each containing a factor of the type C_α, a factor $g^i_{\rho\sigma}(y_j)$ or $g^i_{\rho\rho\rho}(y_j)$, a factor Θ, and either four, five, or six factors from the set $\{\Delta_j' z^1, \ldots, \Delta_j' z^r, \Delta_j'' z^1, \ldots, \Delta_j'' z^r\}$. Each factor Θ is $\mathfrak{F}(s_j)$-measurable; this is most easily seen by using the theorem of the mean with integral form of the remainder.

Since $z^1(t) = t$, $\Delta_j' z^1$ and $\Delta_j'' z^1$ are both at most mesh Π. By Lemma III(1-1), (4-10), and (4-16) applied to Π', we obtain

$$\left\| \sum_{j=1}^{k-1} R^i_{j,1} \right\| \leqq C_{16} \text{ mesh } \Pi. \tag{4-29}$$

If any two of ρ, σ, τ are different, or if all are 1, by (4-2) and (4-9) we obtain (assuming $K \geqq 1$)

$$| E(\Delta_j' z^\tau \, \Delta_j'' z^\sigma \, \Delta_j'' z^\rho \mid \mathfrak{F}(t_j)) | \leqq K^3 (\Delta_j t)^2, \tag{4-30}$$

$$E([\Delta_j' z^\tau \, \Delta_j'' z^\sigma \, \Delta_j'' z^\rho]^2 \mid \mathfrak{F}(t_j)) \leqq K^3 (\Delta_j t)^3.$$

If ρ, σ, τ are all equal and $\geqq 2$, the factor $\delta_{\rho\sigma\tau}$ is 0, so again by Lemma III(1-1), (4-11), and (4-16) applied to Π',

$$\left\| \sum_{j=1}^{k-1} R^i_{j,2} \right\| \leqq C_{17} \text{ mesh } \Pi. \tag{4-31}$$

Since

$$|E((\Delta_j' z^\rho)^2 (\Delta_j'' z^\rho) \mid \mathfrak{F}(t_j))| = \left| E((\Delta_j' z^\rho)^2 E(\Delta_j'' z^\rho \mid \mathfrak{F}(s_j)) \mid \mathfrak{F}(t_j)) \right| \tag{4-32}$$

$$\leqq K^2 (\Delta_j t)^2$$

and by (4-9)

$$E((\Delta_j'z^\rho)^4(\Delta_j''z^\rho)^2 \mid \mathfrak{F}(t_j)) \leqq K^3(\Delta_j t)^3, \tag{4-33}$$

we also obtain by Lemma III(1-1), (4-1), and (4-16)

$$\left|\left|\sum_{j=1}^{k-1} R_{j,3}^i\right|\right| \leqq C_{18} \text{ mesh } \Pi. \tag{4-34}$$

To estimate $\sum R_{j,4}^i$, we apply Lemma III(1-1) with (for each ρ)

$$u_j = \sum_{\sigma,\tau,h} C_{13}\Theta g_\tau{}^h(y_j)\,\Delta_j'z^\tau\,\Delta_j'z^\sigma, \qquad \Delta_j = \Delta_j''z^\rho.$$

Here u_j is $\mathfrak{F}(s_j)$-measurable, and by (4-16) and (4-9)

$$\begin{aligned} E(u_j^2) &= E(C_{13}^2 g_\tau{}^h(y_j)^2 E(\Theta[\Delta_j'z^\tau\,\Delta_j'z^\sigma]^2 \mid \mathfrak{F}(t_j)) \\ &\leqq C_{19}(\Delta_j't)^2, \end{aligned}$$

so by Lemma III(1-1)

$$\left|\left|\sum_{j=1}^{k-1} R_{j,4}^i\right|\right| \leqq C_{20} \text{ mesh } \Pi. \tag{4-35}$$

By a similar proof

$$\left|\left|\sum_{j=1}^{k-1} R_{j,5}^i\right|\right| \leqq C_{21} \text{ mesh } \Pi. \tag{4-36}$$

If Δ_j is the product of four, five, or six factors from the set $\{\Delta_j'z^1, \ldots, \Delta_j'z^r, \Delta_j''z^1, \ldots, \Delta_j''z^r\}$, then by (4-9) a.s.

$$E(|\,\Theta\Delta_j\,| \mid \mathfrak{F}(t_j)) \leqq K(\Delta_j t)^2,$$

$$E(|\,\Theta\Delta_j\,|^2 \mid \mathfrak{F}(t_j)) \leqq K(\Delta_j t)^4.$$

So by Lemma III(1-1) and (4-16)

$$\left|\left|\sum_{j=1}^{k-1} R_{j,6}^i\right|\right| \leqq C_{22} \text{ mesh } \Pi. \tag{4-37}$$

Now let

$$N(t) = \sup\{\|\, x_{\Pi'}(s) - x_\Pi(s)\,\| : a \leqq s \leqq t\}. \tag{4-38}$$

On each interval $[t_j, t_{j+1}]$ we have $N(t) \geqq N(t_j)$, so by Lemma III(2-2)

$$\left\| \sum_{j=1}^{k-1} [g_\rho{}^i(x_{\Pi'}(t_j)) - g_\rho{}^i(x_\Pi(t_j))] \Delta_j z^\rho \right\| \leqq \left\{ \int_a^{t_k} C_{23} N(t)^2 \, dt \right\}^{1/2}. \tag{4-39}$$

Similar estimates hold for the terms in the second and third sums in (4-27). So from (4-27) (summed over $j = 1, \ldots, k-1$), (4-29), (4-31), and (4-34)–(4-39) we obtain

$$\| x_{\Pi'}(t_k) - x_\Pi(t_k) \|^2 \leqq C_{24} \int_a^{t_k} N(t)^2 \, dt + C_{25}(\text{mesh } \Pi)^2. \tag{4-40}$$

The members of (4-40) are evaluated at a division point t_k of Π. But if t^* is any point of $(a, b]$ which is not a division point of Π, say $t_{k-1} < t^* < t_k$, we can insert it as a division point in Π and in Π' without affecting the reasoning. Then t^* takes the place of t_k in (4-40); that is, in (4-40) we may replace t_k by any t in $[a, b]$. The right member of (4-40) is nondecreasing, so (4-40) holds if we replace t_k in the right member by t and in the left member by any s such that $a \leqq s \leqq t$. Therefore

$$N(t)^2 \leqq C_{24} \int_a^t N(s)^2 \, ds + C_{25}(\text{mesh } \Pi)^2. \tag{4-41}$$

By V(3-20), from (4-41) it follows that

$$N(t)^2 \leqq C_{26}^2(\text{mesh } \Pi)^2. \tag{4-42}$$

Now we form a sequence $\Pi_1, \Pi_2, \ldots$ of Cauchy partitions, Π_1 being Π itself, and each Π_{h+1} being obtained from Π_h by using as division points all the division points of Π_h and the midpoints of all the intervals of Π_h. Then

$$\text{mesh } \Pi_h = 2^{1-h} \text{ mesh } \Pi. \tag{4-43}$$

If we write $x_h = x_{\Pi_h}$, by (4-42) and (4-43)

$$\sup\{\| x_{h+1}(s) - x_h(s) \| : a \leqq s \leqq b\} \leqq C_{26} 2^{1-h} \text{ mesh } \Pi.$$

Hence

$$\sup\{\| x_\Pi(s) - x_{h+1}(s) \| : a \leqq s \leqq b\} < 2C_{26} \text{ mesh } \Pi. \tag{4-44}$$

As h increases, x_{h+1} converges in L_2-norm to the solution x of (1-9), so if we define $C = 2C_{26}$ the remaining conclusion of Theorem (4-8) is established.

5 Continuous dependence of the solution on the disturbance

In this section we shall prove that the solutions of certain stochastic differential equations in selected form possess a strong type of the stability discussed in II(1-4). We also show that a.s., all disturbance functions $z^\rho(\mathbf{t}, \omega')$ that are uniformly close to a given $z^\rho(\mathbf{t}, \omega)$ will furnish solutions $x(\mathbf{t}, \omega')$ of those equations that are uniformly close to $x(t, \omega)$. Unfortunately, we can do this only for equations in which there is only a single random noise process z^ρ, and even then only with a supplementary hypothesis. The principal idea in the proof is an adaptation of a device used by Wong and Zakai [16] to establish essentially the same stability property in the case in which there is one state variable x^i, and the one disturbance z^ρ is a Wiener process. For the same x^i and z^ρ a similar mapping device was used by Lamperti [6] to define the solutions of stochastic differential equations with suitably restricted coefficient functions.

Since we are interested in uniform convergence on $[a, b]$, it is convenient to use the familiar "uniform norm" for (real- or vector-valued) continuous functions $X(t)$ on $[a, b]$,

$$\| X \|_\infty = \sup\{| X(t) | : a \leqq t \leqq b\}.$$

The main result of this section is not probabilistic, and the application will be to stochastic differential equations in which the $g_\rho{}^i$ do not depend directly on the ω. In this section it is convenient to abandon the condensed notation of the preceding sections, so that the equations under discussion have the form

$$\text{(5-1)} \quad x^i(t) = x^i(a) + \int_a^t f^i(s, x(s))\, ds + \sum_{\rho=1}^{r} \int_a^t g_\rho{}^i(s, x(s))\, dz^\rho(s)$$

$$(i = 1, \ldots, n).$$

We shall assume

(5-2) *G is a set in R^{n+1} which is the intersection of an open set with the strip $\{(t, x) : a \leqq t \leqq b\}$; and the functions f^i and $g_\rho{}^i$ $(i = 1, \ldots, n;$ $\rho = 1, \ldots, r)$ are defined and have continuous partial derivatives of first and second order on G.*

By a *Lipschitzian solution set* [for (5-1)] we shall mean an $(r + n)$-tuple $(z, x) = (z^1(t), \ldots, z^r(t),\ x^1(t), \ldots, x^n(t))$ of functions on $[a, b]$ such that the $z^\rho(t)$ are Lipschitzian, the graph of $x(t)$ is in G, and (5-1) is satisfied. By a *limit set* [for equations (5-1)] we shall mean an $(r + n)$-

tuple (z, x) of functions on $[a, b]$ such that the graph of $x(t)$ is in G, and there is a sequence of Lipschitzian solution sets (z_j, x_j) $(j = 1, 2, \ldots)$ such that as $j \to \infty$, both $\| z_j - z \|_\infty$ and $\| x_j - x \|_\infty$ tend to 0.

Limit sets do not in general satisfy Eqs. (5-1) in any sense that we have discussed. It is possible to use them, under suitable restrictions, to define a type of generalized solution of (5-1), but there are serious objections to doing so, especially when $r > 1$. Nevertheless, when there is just one random noise $(r = 1)$ they are useful to us, and we shall study this case. Since ρ has only the single value 1, we simplify notation by omitting it.

(5-3) *THEOREM Let hypothesis* (5-2) *be satisfied, with* $r = 1$. *Let* $(\tilde{z}, \tilde{x})$ *be a limit set for Eqs.* (5-1) *such that the* $g^i(t, \tilde{x}(t))$ $(i = 1, \ldots, n)$ *do not all vanish at any* t *in* $[a, b]$. *Then there exist positive numbers* δ *and* L *such that*

(i) *whenever* $\bar{x}$ *is an n-vector with* $| \bar{x} - \tilde{x}(a)) < \delta$ *and* $z(\mathbf{t})$ *is a continuous real-valued function with* $\| z - \tilde{z} \|_\infty < \delta$, *there is at most one limit set* (z, x) *with* $x(a) = \bar{x}$;

(ii) *whenever* $\bar{x}_1, \bar{x}_2, z_1, z_2$ *satisfy the conditions in* (i), *and* $x_1(\mathbf{t})$, $x_2(\mathbf{t})$ *are functions such that* (z_1, x_1) *and* (z_2, x_2) *are limit sets* with $x_1(a) = \bar{x}_1$ and $x_2(a) = \bar{x}_2$, then

$$\| x_1 - x_2 \|_\infty \leqq L\{| \bar{x}_1 - \bar{x}_2 | + \| z_1 - z_2 \|_\infty\}.$$

We first prove this under the supplementary hypotheses

(5-4) $G = \{(t, x) \in R^{n+1} : a \leqq t \leqq b\}$;

(5-5) the partial derivatives $g^i_{x^h}$ $(i, h = 1, \ldots, n)$ are bounded on G;

(5-6) there is a vector $(c_1, \ldots, c_n)$ and a positive number β such that

$$\sum_{i=1}^{n} c_i g^i(t, x) \geqq \beta \qquad (x \in G).$$

By dividing by $| c |$ and rotating axes we can bring (5-6) to the form

$$g^1(t, x) \geqq \beta > 0; \tag{5-7}$$

we shall suppose this done.

For each t in $[a, b]$ and each $(y^2, \ldots, y^n)$ in R^{n-1} there is a unique n-vector valued function $(\Phi(u, y^2, \ldots, y^n, t) : -\infty < u < \infty)$ that satisfies the equations

$$d\phi^i/du = g^i(t, \phi(u)) \tag{5-8}$$

and has initial values

$$\text{(5-9)} \qquad \Phi^1(0, y^2, \ldots, y^n, t) = 0,$$
$$\Phi^j(0, y^2, \ldots, y^n, t) = y^j \qquad (j = 2, \ldots, n).$$

These Φ^i are twice continuously differentiable in all variables. Given any x in R^n, there is a unique solution ϕ of equations (5-8) such that

$$\text{(5-10)} \qquad \phi^i(0) = x^i.$$

By (5-7), ϕ^1 is increasing, and tends to ∞ as $u \to \infty$ and to $-\infty$ as $u \to -\infty$. So there is a unique number y^1 such that

$$\phi^1(-y^1) = 0.$$

Define

$$\text{(5-11)} \qquad y^j = \phi^j(-y^1) \qquad (j = 2, \ldots, n).$$

Then the vector function $\phi(\mathbf{u} - y^1)$ satisfies (5-8) and has initial values

$$\text{(5-12)} \qquad \phi^1(0 - y^1) = 0,$$
$$\phi^j(0 - y^1) = y^j \qquad (j = 2, \ldots, n).$$

It therefore must coincide with $\Phi(\mathbf{u}, y^2, \ldots, y^n, t)$. That is, to each x in R^n there corresponds a unique y in R^n such that

$$\text{(5-13)} \qquad \Phi^i(y^1, \ldots, y^n, t) = x^i \qquad (x = 1, \ldots, n).$$

This y we name $\Psi(x, t)$; $\Psi(\mathbf{x}, t)$ is the inverse of $\Phi(\mathbf{y}, t)$.

By definition the Φ^i satisfy

$$\text{(5-14)} \qquad \partial\Phi^i(y, t)/\partial y^1 = g^i(t, \Phi(y, t)),$$

so by differentiating with respect to y^j we obtain

$$\text{(5-15)} \qquad \partial\Phi^i{}_{y^j}(y, t)/\partial y^1 = \sum_{k=1}^{n} g^i{}_{x^k}(t, \Phi(y, t))\Phi^k{}_{y^j}(y, t).$$

By (5-9)

$$\text{(5-16)} \qquad \Phi^i{}_{y^j}(0, y^2, \ldots, y^n, t) = \delta_j{}^i \qquad (i = 2, \ldots, n),$$

where as usual $\delta_j{}^i$ is 1 if $i = j$ and is 0 otherwise. So by (5-7) and (5-14)

$$\text{(5-17)} \qquad \det \Phi^i{}_{y^j}(0, y^2, \ldots, y^n, t) = \Phi^1{}_{y^1}(0, y^2, \ldots, y^n, t)$$
$$= g^1(t, \Phi(0, y^2, \ldots, y^n, t))$$
$$\geqq \beta.$$

Each n-tuple

$$(\Phi^1{}_{y^j}(y^1, y^2, \ldots, y^n, t), \ldots, \Phi^n{}_{y^j}(y^1, y^2, \ldots, y^n, t)) \tag{5-18}$$

is a solution of the system of linear equations (5-15), and their determinant is nonzero at $y^1 = 0$, so it remains nonzero for all y^1. By the implicit function theorem, the mapping $y = \Psi(x, t)$ inverse to $x = \Phi(y, t)$ is also twice continuously differentiable.

From (5-14) and the identity

$$\Psi^i(\Phi(y, t), t) = y^i \qquad (i = 1, \ldots, n)$$

we obtain

$$\sum_{h=1}^{n} \Psi^i{}_{x^h}(\Phi(y, t), t) g^h(t, y) = \delta_1{}^i. \tag{5-19}$$

Now let (z, x) be a Lipschitzian solution set, and define

$$y^i(t) = \Psi^i(x(t), t). \tag{5-20}$$

If we define

$$F^i(t, y) = \Psi^i{}_t(\Phi(y, t), t) + \sum_{h=1}^{n} \Psi^i{}_{x^h}(\Phi(y, t), t) f^h(t, \Phi(y, t)), \tag{5-21}$$

by (5-19) and (5-1) we obtain

$$dy^i/dt = F^i(t, y) + \delta_1{}^i \, dz/dt. \tag{5-22}$$

Now let $(\tilde{z}, \tilde{x})$ be a limit set for Eqs. (5-1), and define

$$\tilde{y}^i(t) = \Psi^i(\tilde{x}(t), t). \tag{5-23}$$

On the set

$$S = \{(t, y) : a \leqq t \leqq b, |y| \leqq \|\tilde{y}\|_\infty + 2\} \tag{5-24}$$

the functions F^i are continuously differentiable, so there exists a B_1 such that

$$|F(t, y_1) - F(t, y_2)| \leqq B_1 |y_1 - y_2| \qquad ((t, y_1), (t, y_2) \in S). \tag{5-25}$$

Moreover, there exists a $B_2 \geqq 1$ such that if $a \leqq t \leqq b$ and $|x_\alpha| \leqq |\tilde{x}(a)| + 2$ ($\alpha = 1, 2$), then

$$|\Psi(x_1, t) - \Psi(x_2, t)| \leqq B_2 |x_1 - x_2|. \tag{5-26}$$

We define

$$\delta = [\exp B_1(a - b)]/6B_2. \tag{5-27}$$

By definition of limit set and continuity of the Ψ^i, there exists a Lipschitzian solution set (z_1, x_1) such that

$$\| z_1 - \tilde{z} \|_\infty < \delta \qquad \text{and} \qquad \| x_1 - \tilde{x} \|_\infty < \delta \tag{5-28}$$

and, defining $y_1(t) = \Psi(x_1(t), t)$,

$$\| y_1 - \tilde{y} \|_\infty < 1. \tag{5-29}$$

Let $\bar{x}_2$ be a point in R^n and $z_2(\mathbf{t})$ a Lipschitzian function such that

$$| \bar{x}_2 - \tilde{x}(a) | < \delta \qquad \text{and} \qquad \| z_2 - \tilde{z} \|_\infty < \delta. \tag{5-30}$$

Since (5-5) holds, the equations

$$dx_2{}^i = f^i(t, x_2)\, dt + g^i(t, x_2)\, dz_2 \tag{5-31}$$

with initial conditions

$$x_2{}^i(a) = \bar{x}_2{}^i \tag{5-32}$$

have a solution $x_2(\mathbf{t})$ on $[a, b]$. We define

$$y_2{}^i(t) = \Psi^i(x_2(t), t); \tag{5-33}$$

these satisfy (5-22), as do the $y_1(t)$. Therefore

$$\begin{aligned} | y_2(t) - y_1(t) | \leqq{} & | \Psi(x_1(a), a) - \Psi(x_2(a), a) | \\ & + \int_a^t | F(s, y_1(s)) - F(s, y_2(s)) |\, ds \\ & + 2 \| z_1 - z_2 \|_\infty. \end{aligned} \tag{5-34}$$

Let c be the greatest number in $[a, b]$ such that

$$| y_2(t) | \leqq \| \tilde{y} \|_\infty + 2 \qquad (a \leqq t \leqq c). \tag{5-35}$$

By (5-34), (5-25), and (5-26), for all t in $[a, c]$

$$\begin{aligned} | y_2(t) - y_1(t) | \leqq{} & B_2 | \bar{x}_2 - x_1(a) | + B_1 \int_a^t | y_2(s) - y_1(s) |\, ds \\ & + 2 \| z_2 - z_1 \|_\infty. \end{aligned} \tag{5-36}$$

So by V(3-20), for $a \leqq t \leqq c$

$$\begin{aligned} | y_2(t) - y_1(t) | \leqq{} & [B_2 | \bar{x}_2 - x_1(a) | \\ & + 2 \| z_2 - z_1 \|_\infty] \exp B_1(t - a). \end{aligned} \tag{5-37}$$

In particular, by (5-30), (5-28), and (5-27)

$$|y_2(t) - y_1(t)| < 1 \qquad (a \leqq t \leqq c). \tag{5-38}$$

By (5-29), this is incompatible with the definition of c if $c < b$; so $c = b$, and (5-37) holds for $a \leqq t \leqq b$.

On the set

$$\{(t, y) \in R^{n+1}: a \leqq t \leqq b, |y| \leqq ||\tilde{y}||_\infty + 2\} \tag{5-39}$$

the vector function Φ is Lipschitzian; let B_3 be its Lipschitz constant. Since $y_1(t)$ and $y_2(t)$ are in the set (5-39), (5-37) holds, so

$$|x_2(t) - x_1(t)| \leqq [B_2 |x_2(a) - x_1(a)| + 2 ||z_2 - z_1||_\infty] B_3 \exp B_1(b - a). \tag{5-40}$$

If we define

$$L = 3B_2B_3 \exp B_1(b - a), \tag{5-41}$$

the inequality in conclusion (ii) of (5-3) is satisfied.

Since $||z_1 - \tilde{z}||_\infty$ and $||x_1 - \tilde{x}||_\infty$ can be chosen arbitrarily small, (5-40) [with (5-41)] implies

$$||x_2 - \tilde{x}||_\infty \leqq L(|x_2(a) - \tilde{x}(a)| + ||z_2 - \tilde{z}||_\infty). \tag{5-42}$$

If (z, x) and (z, x^*) are limit sets such that $x(a) = x^*(a)$ and $|x(a) - \tilde{x}(a)| < \delta$ and $||z - \tilde{z}||_\infty < \delta$, there are sequences $(z_{1,j}, x_{1,j})$ and $(z_{2,j}, x_{2,j})$ $(j = 1, 2, 3, \ldots)$ of Lipschitzian solution sets such that as $j \to \infty$

$$||z_{1,j} - z||_\infty \to 0, \qquad ||x_{1,j} - x||_\infty \to 0,$$
$$||z_{2,j} - z||_\infty \to 0, \qquad ||x_{2,j} - x^*||_\infty \to 0. \tag{5-43}$$

So for all large j the $z_{1,j}$, $z_{2,j}$, $x_{1,j}(a)$, and $x_{2,j}(a)$ satisfy the conditions imposed on z_1, z_2, $x_1(a)$, and $x_2(a)$ in the preceding paragraphs, and therefore by (5-40)

$$||x_{1,j} - x_{2,j}||_\infty \leqq L(|x_{1,j}(a) - x_{2,j}(a)| + ||z_{1,j} - z_{2,j}||_\infty). \tag{5-44}$$

The right member tends to 0 as $j \to \infty$, so by (5-43) x and x^* are identical. Conclusion (i) is established, under the supplementary hypotheses (5-4), (5-5), and (5-6).

If (z_1, x_1) and (z_2, x_2) are limit sets such that

$$|x_\alpha(a) - \tilde{x}(a)| < \delta \quad \text{and} \quad ||z_\alpha - \tilde{z}||_\infty < \delta \qquad (\alpha = 1, 2), \tag{5-45}$$

we can find sequences $(z_{1,j}, x_{1,j})$, $(z_{2,j}, x_{2,j})$ of Lipschitzian solution-sets such that as $j \to \infty$,

$$(5\text{-}46) \qquad \| z_{\alpha,j} - z_\alpha \|_\infty \to 0 \qquad \text{and} \qquad \| x_{\alpha,j} - x_\alpha \|_\infty \to 0 \qquad (\alpha = 1, 2).$$

Then the $x_{\alpha,j}(a)$ and $z_{\alpha,j}(\mathbf{t})$ will satisfy (5-45) for all large j, so (5-44) also holds. Letting $j \to \infty$ yields conclusion (ii) of Theorem (5-3), subject to the supplementary hypotheses (5-4)–(5-6).

Now we consider a system which, in addition to the hypotheses of Theorem (5-3), satisfies

(5-47) *there exists a point X of R^n and a positive δ_0 such that the cylinder*

$$C_0 = \{(t, x) \in R^{n+1} : a \leqq t \leqq b, \quad |x - X| \leqq \delta_0\}$$

is contained in G; and there is an n-vector c and a positive β such that $\sum_{i=1}^{n} c_i g^i(t, x) > \beta \; [(t, x) \in C_0]$.

We shall show that

(5-48) the conclusions of Theorem (5-3) hold for each limit set $(\bar{z}, \bar{x})$ such that

$$|\bar{x}(t) - X| \leqq \delta_0/2 \qquad (a \leqq t \leqq b).$$

There is no loss of generality in assuming $X = 0$. There exists a $\delta_1 > \delta_0$ such that the inequality in (5-47) continues to hold for all (t, x) in the cylinder

$$(5\text{-}49) \qquad C_1 = \{(t, x) \in R^{n+1} : a \leqq t \leqq b, |x| \leqq \delta_1\}.$$

There is a twice continuously differentiable function R with positive non-increasing derivative on $[0, \infty)$ such that

$$(5\text{-}50) \qquad \begin{aligned} R(r) &= r \qquad (0 \leqq r \leqq \delta_0) \\ R(r) &< \delta_1 \qquad (r \geqq 0). \end{aligned}$$

Define a map $X(\mathbf{x})$ on R^n by

$$(5\text{-}51) \qquad \begin{aligned} X^i(0) &= 0, \\ X^i(x) &= x^i[R(|x|)/|x|] \qquad (x \neq 0). \end{aligned}$$

This is a C^2 map of R^n onto C_1, and it is the identity map on C_0. Since the f^i and g^i are Lipschitzian on C_0 and $X(\mathbf{x})$ is Lipschitzian on R^n, the equations

$$(5\text{-}52) \qquad x^i(t) = x^i(a) + \int_a^t f^i(s, X(x(s)))\, ds + \int_a^t g^i(s, X(x(s)))\, dz(s)$$

satisfy the hypotheses of Theorem (5-3) and also satisfy (5-4)–(5-6). Since $\| \bar{x} \|_\infty \leqq \delta_0/2$, $(\bar{z}, \bar{x})$ is a limit set for (5-52). By the part of the proof already completed, there are positive numbers δ_2, L that serve as the δ and L of Theorem (5-3) for Eqs. (5-52). Define

(5-53) $$\delta_3 = \min\{\delta_2, \delta_0/6L\}.$$

If (z, x) is a limit set for Eqs. (5-1) and

(5-54) $$| x(a) - \bar{x}(a) | < \delta_3 \qquad \text{and} \qquad \| z - \bar{z} \|_\infty < \delta_3,$$

and c is the largest number in $[a, b]$ such that

(5-55) $$| x(t) - \bar{x}(t) | \leqq 2\delta_0/3 \qquad (a \leqq t \leqq c),$$

then (z, x) is also a limit set for Eqs. (5-52) on $[a, c]$. By conclusion (ii) of Theorem (5-3), which holds for limit sets of (5-52),

$$| x(t) - \bar{x}(t) | \leqq L(2\delta_3) \leqq \delta_0/3$$

for $a \leqq t \leqq c$. But if $c < b$, (5-55) will then continue to hold on an interval $[a, c']$ with $c < c' \leqq b$, contradicting the definition of c. So (5-55) holds on $[a, b]$, and every limit set for Eqs. (5-1) that satisfies (5-54) is also a limit set for Eqs. (5-52). This establishes statement (5-48).

Finally, let the hypotheses of Theorem (5-3) hold. For each t_0 in $[a, b]$ there is a vector $c(t_0)$ and a positive $\beta(t_0)$ such that the inequality

(5-56) $$\sum_{i=1}^{n} c_i(t_0) g^i(t, x) > \beta(t_0)$$

holds for $t = t_0$ and $x = \bar{x}(t_0)$, and therefore for all (t, x) with $a \leqq t \leqq b$ that satisfy

(5-57) $$| t - t_0 | < \delta(t_0), \qquad | x - \bar{x}(t_0) | < \delta(t_0)$$

for some positive $\delta(t_0)$. By a well-known argument, there is a positive δ_4 that serves as $\delta(t_0)$ for all t_0 in $[a, b]$; and there is a δ_5 in $(0, \delta_4)$ such that

(5-58) if t_1, t_2 are in $[a, b]$ and $| t_1 - t_2 | < \delta_5$, then $| \bar{x}(t_1) - \bar{x}(t_2) | < \delta_4/2$.

If we choose points $t_0 = a < t_1 < \cdots < t_k = b$ with

$$| t_j - t_{j-1} | < \delta_5 \qquad (j = 1, \ldots, k),$$

(5-47) is satisfied with $[t_{j-1}, t_j]$ in place of $[a, b]$. The conclusion of Theorem (5-3) holds for each $[t_{j-1}, t_j]$; we shall show by induction that it holds for each $[a, t_j]$. Suppose that the conclusions of Theorem (5-3) hold on $[a, t_j]$

with constants δ' (>0) and L' (>1); they hold on $[t_j, t_{j+1}]$ with constants δ'' and L''. Define

$$\delta = \min\{\delta', \delta''/2L'\}.$$

Then if (z, x) is a limit set on $[a, t_{j+1}]$ with

$$| x(a) - \bar{x}(a) | < \delta \qquad \text{and} \qquad \| z - \bar{z} \|_\infty < \delta,$$

we have

$$| x(t_j) - \bar{x}(t_j) | < L'(2\delta) \leqq \delta'',$$

so the initial conditions on $[t_j, t_{j+1}]$ hold. This proves that conclusion (i) holds on $[a, t_{j+1}]$. Moreover, if (z_1, x_1) and (z_2, x_2) are two limit sets such that

$$| x_\alpha(a) - \bar{x}(a) | < \delta \qquad \text{and} \qquad \| z_\alpha - \bar{z} \|_\infty < \delta \qquad (\alpha = 1, 2),$$

then for $a \leqq t \leqq t_j$ we have

$$| x_1(t) - x_2(t) | < L'\{| x_1(a) - x_2(a) | + \| z_1 - z_2 \|_\infty\},$$

and for $t_j \leqq t \leqq t_{j+1}$ we have

$$| x_1(t) - x_2(t) |$$
$$\leqq L''\{L'[| x_1(a) - x_2(a) | + \| z_1 - z_2 \|_\infty] + \| z_1 - z_2 \|_\infty\},$$

so the final inequality in Theorem (5-3) holds with

$$L = L''(L' + 1).$$

This completes the proof of Theorem (5-3).

(5-59) *COROLLARY* *Let $z(\mathbf{t})$ be a process on $[a, b]$ that satisfies* V(1-4) *and also, for all s, t such that $a \leqq s < t \leqq b$, a.s.,*

$$E([z(t) - z(s)]^6 \mid \mathfrak{F}(s)) \leqq (t - s)\psi(t - s),$$

where ψ is a nondecreasing function on $(0, \infty)$ whose limit at 0 is 0. Let f^i and g^i ($i = 1, \ldots, n$) be functions Lipschitzian and thrice continuously differentiable on the set $\{(t, x): a \leqq t \leqq b,\ x \in R^n\}$, such that the $g^i(t, x)$ do not all vanish at any point of that set. Let $x(\mathbf{t}, \boldsymbol{\omega})$ be a process with a.s. continuous sample functions that is a strict-sense solution of the canonical equaiions

(5-60) $$dx^i = f^i(t, x)\, dt + g^i(t, x)\, dz + \left(\sum_{h=1}^{n} g^i_{x^h}(t, x) g^h(t, x)/2\right) (dz)^2.$$

Then there exists a subset Ω_0 of Ω with the following properties.

(i) $P(\Omega_0) = 1$.

(ii) *If $\omega \in \Omega$ and $z(\mathbf{t}, \omega)$ is Lipschitzian, then $\omega \in \Omega_0$.*

(iii) *If $\omega_0 \in \Omega_0$ and $(z_j, x_j^1, \ldots, x_j^n)$ $(j = 1, 2, 3, \ldots)$ is a sequence of $(n + 1)$-tuples of functions on $[a, b]$ such that for each j either $z_j(\mathbf{t}) = z(\mathbf{t}, \omega_j)$ and $x_j^i(\mathbf{t}) = x^i(\mathbf{t}, \omega_j)$ for some ω_j in Ω_0 or else z_j is Lipschitzian and (z_j, x_j) satisfies (5-1), and as $j \to \infty$*

$$| x_j(a) - x(a, \omega_0) | \to 0 \quad \text{and} \quad \| z_j(\mathbf{t}) - z(\mathbf{t}, \omega_0) \|_\infty \to 0,$$

then $x_j(t)$ converges to $x(t, \omega_0)$ uniformly on $[a, b]$.

For each partition Π of $[a, b]$, we define z_Π by linear interpolation as in (3-31) $[u(y) = y]$, and as before we denote by $X_\Pi(\mathbf{t}, \boldsymbol{\omega})$ the solution of (5-1) with initial value $x(a, \boldsymbol{\omega})$ and with z_Π in place of z. By Lemma (3-36), the r.v.

$$\sup\{| X_\Pi(t, \boldsymbol{\omega}) - x(t, \boldsymbol{\omega}) | : a \leqq t \leqq b\} \tag{5-61}$$

converges in probability to 0 as mesh $\Pi \to 0$. We first choose a sequence of partitions $\Pi(1), \Pi(2), \ldots$ with mesh tending to 0. The corresponding r.v.'s (5-61) converge in probability to 0, so we can select a subsequence for which $\| X_{\Pi(n)}(\mathbf{t}, \omega) - x(\mathbf{t}, \omega) \|_\infty$ tends to 0 on a set Ω_0 with $P(\Omega_0) = 1$; this set can and shall be chosen to include all ω for which $z(\mathbf{t}, \omega)$ is Lipschitzian. Therefore all sets $(z(\mathbf{t}, \omega), x(\mathbf{t}, \omega))$ with $\omega \in \Omega_0$ are limit sets, and the conclusions of Corollary (5-59) hold by Theorem (5-3).

Theorem (5-3) has an application in communication theory, in which it is sometimes desirable to approximate a continuous function $z(\mathbf{t}, \omega)$ by a band-limited function, i.e. by a function whose Fourier transform vanishes outside of some finite interval $[a, b]$. For notational convenience, we suppose $[a, b]$ centered at 0, so that $a = -b$. Let N be a positive integer. We extend $z(\mathbf{t}, \omega)$ continuously to a larger interval; to be specific, let us extend z by setting it constantly equal to $z(b, \omega)$ on $[b, 2b]$ and to $z(-b, \omega)$ on $[-2b, -b]$. For each positive integer N, we subdivide $[-2b, 2b]$ into $4N$ congruent parts by points kb/N $(k = -2N, \ldots, 2N)$. We define

$$W = N/2b,$$

and for all integers k and all $t \neq k/2W$ we define

$$a_k(t) = [\sin \pi(2Wt - k)]/[\pi(2Wt - k)];$$

where $t = k/2W$ we define $a_k(t)$ by continuity to be 1. We now define

$$Z_N(t, \omega) = \sum_{k=-2N}^{2N} z(k/2W, \omega) a_k(t).$$

The Fourier transform of Z_N vanishes outside the interval $[-W, W]$, and as N increases, $Z_N(t, \omega)$ tends uniformly on $[-b, b]$ to $z(t, \omega)$. (See, e.g., Balakrishnan [1, Chap. I].) Now from Theorem (5-3) we conclude that if $z(\mathbf{t}, \boldsymbol{\omega})$ is any process satisfying V(1-4) and the coefficients g^i satisfy the hypotheses of Theorem (5-3), for all ω in a set of P-measure 1 the solution of (5-1), with $Z_N(\mathbf{t}, \boldsymbol{\omega})$ in place of z, will converge to the solution $x(\mathbf{t}, \omega)$ of (5-60), uniformly on $[-b, b]$, as $N \to \infty$.

It is natural to try to extend Theorem (5-3) so as to apply to at least some systems with more than one random noise. The example at the beginning of Section 3 shows that this is not always possible. We shall now show that the extension is impossible unless

(5-62)

$$g^i_{\rho\sigma}(t, x) = g^i_{\sigma\rho}(t, x) \quad (a \leqq t \leqq b; x \in R^n; i = 1, \ldots, n; \rho, \sigma = 1, \ldots, r).$$

For suppose, to be specific, that there is a j and a point $(\bar{t}, \bar{x})$ for which

(5-63) $$g^j_{12}(\bar{t}, \bar{x}) - g^j_{21}(\bar{t}, \bar{x}) > 0.$$

There is no loss of generality in supposing $\bar{t} = a$. Let $x_0(t)$ be the solution of

(5-64) $$x_0{}^i(t) = \bar{x}^i + \int_a^t f^i(s, x_0(s))\, ds.$$

If the z^ρ are continuously differentiable real-valued functions on $[a, b]$, by integration by parts in (5-1) we obtain [with $x(a) = \bar{x}$]

(5-65)
$$\begin{aligned} x^i(t) &= \bar{x}^i + \int_a^t f^i(s, x(s))\, ds \\ &\quad + \sum_{\rho=1}^{r} [g_\rho{}^i(t, x(t)) z^\rho(t) - g_\rho{}^i(a, \bar{x}) z^\rho(a)] \\ &\quad - \sum_{\rho=1}^{r} \sum_{h=1}^{n} \int_a^t g^i{}_{\rho x^h}(s, x(s)) f^h(s, x(s)) z^\rho(s)\, ds \\ &\quad - \sum_{\rho,\sigma=1}^{r} \int_a^t g^i_{\rho\sigma}(s, x(s)) z^{\sigma\prime}(s) z^\rho(s)\, ds. \end{aligned}$$

We select for $k = 1, 2, \ldots$

$$z_k^1(t) = k^{-1} \sin k^2 t, \qquad z_k^2(t) = k^{-1} \cos k^2 t, \qquad z_k^3 = \cdots = z_k^r = 0,$$

and we denote by x_k^i the solution of (5-65) corresponding to these disturbances. We assume that $x_k^i(t)$ tends to $x_0^i(t)$ uniformly on $[a, b]$ and shall arrive at a contradiction. In (5-65) the left member and the first two terms in the right member converge as $k \to \infty$ to the corresponding terms in (5-64). The integrals involving $f^h z^\rho$ all tend to 0. The remaining nonzero terms are, for $i = j$,

$$\begin{aligned}(5\text{-}66) \qquad & \int_a^t g_{11}^j(s, x_k(s)) \sin k^2 s \cos k^2 s \, ds \\ & - \int_a^t g_{12}^j(s, x_k(s)) \sin^2 k^2 s \, ds \\ & + \int_a^t g_{21}^j(s, x_k(s)) \cos^2 k^2 s \, ds \\ & - \int_a^t g_{22}^j(s, x_k(s)) \cos k^2 s \sin k^2 s \, ds.\end{aligned}$$

The first term in (5-66) is half of

$$\int_a^t [g_{11}^j(s, x_k(s)) - g_{11}^j(s, x_0(s))] \sin 2k^2 s \, ds + \int_a^t g_{11}^j(s, x_0(s)) \sin 2k^2 s \, ds;$$

the first of these integrals tends to 0 because x_k converges uniformly to x_0, the second converges to 0 by the Riemann–Lebesgue lemma, or simply by integration by parts. The last term in (5-66) converges to 0 by a similar proof. Substituting

$$\sin^2 k^2 t = (1 - \cos 2k^2 t)/2,$$
$$\cos^2 k^2 t = (1 + \cos 2k^2 t)/2$$

in the other two terms of (5-66), we find that the sum (5-66) converges to

$$(5\text{-}67) \qquad \frac{1}{2} \int_a^t [g_{21}^j(s, x_0(s)) - g_{12}^j(s, x_0(s))] \, ds.$$

But by (5-63), the integrand is positive at $s = a$, hence remains positive on a subinterval $[a, \beta]$ of $[a, b]$, and the integral (5-67) is nonzero at $t = \beta$. So as $k \to \infty$, the right member of (5-65) converges to a limit different from x_0, which is the desired contradiction.

It is important to observe that this failure of small response of the solutions of (5-1) to uniformly small change in the z^ρ has nothing to do with probability theory; it occurred when the z^ρ were deterministic and infinitely differentiable. The trouble does not lie in the use of canonical extensions or in anything else that can be done by probabilistic devices; it is present already in the model (5-1) with smooth z^ρ, which we assumed to have been given us with a certification provided by some experimental science. But of course this certification extends only to physically realizable disturbances z^ρ, which presumably all satisfy one and the same Lipschitz condition. Under this condition, discontinuous response to uniformly small changes in the z^ρ does not take place. But on the other hand such z cannot be approximated closely by Brownian motions, and the white-noise model (whether canonical or not) is open to suspicion. A careful investigation of the suitability of the white-noise model would quite possibly reveal new requirements on the mathematical structure of the models that represent classical physical systems.

As an example of a use of Theorem (5-3), we consider a problem studied by Wonham [17]. The first (and simplest) problem discussed by Wonham in that paper is an estimation problem in which x is a r.v. whose range is a finite set $a_1, \ldots, a_K$ of real numbers with respective probabilities $p_1(0), \ldots, p_K(0)$. One observes the function

(5-68) $$y(t, \omega) = x(\omega)t + \int_0^t \beta(s)\, dz(s, \omega) \qquad (t \geqq 0),$$

where β is a known continuously differentiable function with positive lower bound on $[0, \infty)$. We define $p_j(t)$ to be the conditional probability

(5-69) $$p_j(t) = P\{x = a_j \mid y(s), a \leqq s \leqq t\}.$$

Let us also, for compactness, define

(5-70) $$\phi_j(t, Z) = \left\{p_j(0) \exp\left[a_j Z - (a_j^2/2) \int_0^t \beta(s)^{-2}\, ds\right]\right\} \times \left\{\sum_{k=1}^K p_k(0) \exp\left[a_k Z - (a_k^2/2) \int_0^t \beta(s)^{-2}\, ds\right]\right\}^{-1}.$$

Wonham shows that when z is a standard Wiener process,

(5-71) $$p_j(t) = \phi_j\left(t, \int_0^t \beta(s)^{-2}\, dy(s)\right).$$

[Cf. Eqs. (75) and (5) of Wonham [17]]. The integral with respect to dy

is not an Itô integral since y does not have orthogonal increments. Equation (74) can be accepted as its definition for the purposes of Wonham [17]; for us it is defined in Chapter II. He then derives a stochastic differential equation satisfied by the $p_j(t)$ when z is a Wiener process; but as usual, the solutions of this equation are quite unstable.

For reasons of space, we shall ignore the important question of the accuracy with which (5-70) and (5-71) represent the quantities $p_j(t)$ of (5-69) when z is "nearly" a Wiener process. We shall also consider only a simple special case, a little more general than that discussed in detail in Wonham [17], in which $K = 2$, $a_1 = +1$, and $a_2 = -1$, and $p_1(0) = p_2(0) = \frac{1}{2}$.

If we define

(5-72) $$q(\mathbf{t}, \omega) = p_1(\mathbf{t}, \omega) - p_2(\mathbf{t}, \omega),$$

Eqs. (5-71) are equivalent to the single equation

(5-73) $$q(t, \omega) = \tanh \int_0^t \beta(s)^{-2}\, dy(s, \omega).$$

It is desired to construct a differential equation that will generate the value of $q(\mathbf{t}, \omega)$ as the values of $y(\mathbf{t}, \omega)$ become known. By our extension IV(2-4) of the Itô differentiation lemma, q satisfies the equation

(5-74) $$dq = (1 - q^2)\beta(t)^{-2}\, dy - q(1 - q^2)\beta(t)^{-4}\, (dy)^2.$$

When y happens to be Lipschitzian, this reduces to

(5-75) $$dq = (1 - q^2)\beta(t)^{-2}\, dy;$$

when z is a Wiener process, by (5-68), IV(4-14) and III(7-1), it reduces to

(5-76) $$dq = -q(1 - q^2)\beta(t)^{-4}\, dt + (1 - q^2)\beta(t)^{-2}\, dy,$$

which in the special case $\beta = 1$ is Eq. (14) of Wonham [17]. Neither (5-75) nor (5-76) may be applied except in the special cases specified; (5-74) applies whenever z satisfies the standing hypotheses V(1-4). However, in practice we are given, for each t, not a sample function $(y(s, \omega): 0 \leqq s \leqq t)$ of the y process, but a Lipschitzian approximation to it. We now apply Theorem (5-3), with b fixed and

$$G = \{(t, q): 0 \leqq t \leqq b, |q| < 1\}.$$

We always assume $q(0) = y(0) = 1$. Then for every Lipschitzian y, Eq. (5-75) has the solution (5-73), whose graph is in G. If y_0 is continuous on $[0, b]$ and $y_1, y_2, \ldots$ is a sequence of Lipschitzian functions converging uniformly to y_0, and all y_j vanish at 0, then for $j = 1, 2, \ldots$

$$\int_0^t \beta(s)^{-2}\, dy_j = y_j(t)\beta(t)^{-2} + 2\int_0^t \beta(s)^{-3}\beta'(s)y_j(s)\, ds.$$

So if we define q_j by (5-73), it converges uniformly to

$$q_0(t) = \tanh\left\{y_0(t)\beta(t)^{-2} + 2\int_0^t \beta(s)^{-3}\beta'(s)y_0(s)\, ds\right\}, \tag{5-77}$$

and (y_0, q_0) is a limit set. Clearly it is the only limit set (y_0, q) with $q(0) = 0$. Therefore, whenever y_0 is continuous and $y_0(0) = 0$, there is just one q_0 with $q_0(0) = 0$ such that (y_0, q_0) is a limit set, and the graph of q_0 is in G. On G the coefficient $1 - q^2$ of dy does not vanish. As in the proof of Corollary (5-59), as long as z (and therefore y) satisfies the standing hypotheses V(1-4), the pairs $(y(\mathbf{t}, \omega), q(\mathbf{t}, \omega))$ are a.s. limit sets for Eqs. (5-75). So by Theorem (5-3), for all ω in a set with probability measure 1, if $\epsilon > 0$ and if y is any Lipschitzian function with $\| y - y(\mathbf{t}, \omega) \|_\infty$ sufficiently small, and q is the solution of (5-75) corresponding to y, the difference $| q(t) - q(t, \omega) |$ will be less than ϵ for all t in $[0, b]$. If we wish to use an analog computer to generate a close approximation to the solution (5-73) of Eq. (5-74), we must program it to solve Eq. (5-75).

6 Justification of the canonical extension in stochastic modeling

In Section 4 we saw that under suitable hypotheses the approximations to solutions of equations of the canonical form (1-9) converge more rapidly than do the approximations to the solutions of other types of equations, such as (1-2). This is an obvious advantage in computation. But besides this, the rapid convergence enables us to exhibit a mathematical advantage of equations in canonical form over others; and, what is more significant, it permits us to argue that these equations are better fitted to provide stochastic models of noisy systems than are other equations.

For the sake of simplicity, we assume that (2-5) holds. By a device often used before, the general case is reducible to this if we ignore a set of sample functions $\{x(\mathbf{t}, \omega) : \omega \in \Omega_0\}$ for which $P(\Omega_0)$ is arbitrarily small.

Suppose first that the hypotheses of Theorem (4-8) are satisfied. Let Π be a Cauchy partition of $[a, b]$. If $x(\mathbf{t})$ is the solution of (1-9), and x_Π is the Cauchy–Maruyama approximation (4-7) to the solution of (1-9), then, with some constant A_1 independent of Π, we have

$$\| x_\Pi(t_j) - x(t_j) \| \leqq A_1 \text{ mesh } \Pi \tag{6-1}$$

at all division points t_j of Π. If, however, $\bar{x}$ is the solution of (1-2) and $\bar{x}_\Pi$ is the Cauchy–Maruyama approximation to $\bar{x}$, then, with some constant A_2 independent of Π we have

$$\| \bar{x}_\Pi(t_j) - \bar{x}(t_j) \| \leqq A_2(\text{mesh } \Pi)^{1/2}. \tag{6-2}$$

In general no better degree of approximation exists for solutions of (1-2), or for solutions of extensions of (1-2) that are not equivalent to the canonical extension.

The type of "nearness" of two stochastic processes x and X that seems most appropriate in many experimental situations is this. First we choose finitely many constants $s_1, \ldots, s_m$ in the time interval $[a, b]$ with which we are concerned. (In principle m can be arbitrarily large, but in practice there is usually a bound on the number of points at which we can carry out our measurements.) Then X is "near" x if the random vectors $(X(s_1), \ldots, X(s_m))$ and $(x(s_1), \ldots, x(s_m))$ are "near" in some sense of nearness of random vectors. We choose the L_p-norms; if S_0 is the finite set $\{s_1, \ldots, s_m\}$, we define

$$\rho_{p,S_0}(X, x) = \max\{\| X(s_j) - x(s_j) \|_p : s_j \in S_0\}. \tag{6-3}$$

If x is obtained by solving some stochastic differential equations with noise processes $z^1, \ldots, z^r$, the solutions of the equation depend continuously on the noises in the sense that, in the class of noise processes satisfying the standing hypotheses V(1-4) with given K, given any finite subset S_0 of T and any positive ϵ, there is a finite subset S_1 and a positive δ such that if z and Z are noise processes with $\rho_{4,S_1}(Z, z) < \delta$, the corresponding solutions X, x satisfy $\rho_{2,S_0}(X, x) < \epsilon$. For by Theorem V(4-3) we can find a Cauchy partition Π, whose set S_1 of division points includes all the points of S_0, such that the Cauchy–Maruyama approximations X_Π, x_Π to X and x, respectively, satisfy

$$\rho_{2,S_1}(X, X_\Pi) < \epsilon/4, \qquad \rho_{2,S_1}(x, x_\Pi) < \epsilon/4. \tag{6-4}$$

From the definition of x_Π we readily conclude that there is a positive δ such that

$$\text{if } \rho_{4,S_1}(Z, z) < \delta, \text{ then } \rho_{2,S_1}(X_\Pi, x_\Pi) < \epsilon/2. \tag{6-5}$$

These imply the desired conclusion. However, in general the number of division points of Π will have to be large in order to obtain (6-4), since the right member of (6-2) tends only slowly to 0. If the stochastic differential equation is of the canonical form (1-9) and the hypotheses of Theorem (4-8) are satisfied, by (6-1) we can obtain (6-4) with a partition having a larger mesh, so the number of points in S_1 can be taken much smaller in this case than for other types of equations. Thus when the hypotheses of (4-8) hold, the solutions of equations in canonical form have stronger continuity properties than do the solutions of other types of equations.

It is more important, however, to go back of the mathematics and remember the purpose of constructing stochastic models. A system is being studied which, if the noises have Lipschitzian sample functions, is governed by Eqs. (1-2). As before, to shorten notation we define $z_1(t) = t$; then $z^2, \ldots, z^r$ are the random noises. To study the statistical properties of the solutions we want to replace the Lipschitzian noises $z^2, \ldots, z^r$ by more manageable processes (usually, for sound reasons, by Brownian-motion processes), and we want this idealization to furnish solutions whose statistical properties are acceptably close approximations to those of the solutions of (1-2).

Let us suppose that, in addition to the initial values of the x^i, we are given the values of the z^ρ at the division points $\{t_1, \ldots, t_{m+1}\}$ of a Cauchy partition Π of $[a, b]$. Using the known functions $g^i(t, x)$ (which we shall suppose not to depend directly on ω) we wish to construct for each j a set of functions

$$\Phi_j^{\,i}(x(a), z_1, \ldots, z_j) \qquad (i = 1, \ldots, n) \tag{6-6}$$

such that, when $z_k^{\,1}$ is replaced by t_k and $z_k^{\,2}, \ldots, z_k^{\,r}$ are replaced either by the actual $z^\rho(t_j, \omega)$ $(k = 1, \ldots, j)$ or by some suitable Brownian-motion idealization of them, the resulting r.v. will be in some sense an acceptably close approximation $x_\Pi(t_j, \omega)$ to the r.v. $x(t_j, \omega)$ obtained by solving the differential equations (1-2).

The increment $x(t_{j+1}) - x(t_j)$ is determined by $x(t_j)$ and the functions z^ρ on $[t_j, t_{j+1}]$, as we see from (1-2). If we are to approximate the increments by functions (6-6), the difference $\Phi_{j+1} - \Phi_j$ should be independent of $z_1, \ldots, z_{j-1}$ except insofar as these affect $x_\Pi(t_j)$; and since only the differences of the values of the $z^\rho(\mathbf{t})$ $(\rho = 2, \ldots, r)$ have any significance, we should have an approximation to $x(t_{j+1}) - x(t_j)$ obtained by substituting $\Delta_j z^2, \ldots, \Delta_j z^r$ for the numbers $D_2, \ldots, D_r$ and $x_\Pi(t_j)$ for x in a function

$$\Psi(t_j, x, D_2, \ldots, D_r). \tag{6-7}$$

We are willing to use only reasonably smooth functions (6-7). For these, we expand Ψ about $D_1 = \cdots = D_r = 0$ by the theorem of the mean to, say, terms of fourth degree in the D_ρ. We also wish to follow a custom that has proved profitable in applications of analysis since the very beginnings of the calculus; we wish to approximate the finite sums

$$x_{\Pi}{}^{i}(t_j) = x^i(a) + \sum_{k=1}^{j-1} \Psi(t_k, x_{\Pi}(t_k), \Delta_k z^2, \ldots, \Delta_k z^r) \tag{6-8}$$

(which may be accurate, but are quite inconvenient for use) by their limits as mesh Π tends to 0. By Theorem III(5-1) we know that the sums of terms involving three or more factors $\Delta_k z^\rho$ will tend to 0. Hence the limit will have the form

$$\begin{aligned} x^i(t) = x^i(a) &+ \int_a^t h_1{}^i(s, x(s))\, ds + \sum_{\rho=2}^{r} \int_a^t h_\rho{}^i(s, x(s))\, dz^\rho(s) \\ &+ \sum_{\rho,\sigma=2}^{r} \int_a^t h_{\rho\sigma}^{i}(s, x(s))\, dz^\rho(s)\, dz^\sigma(s), \end{aligned} \tag{6-9}$$

where the $h_\rho{}^i$ and $h_{\rho\sigma}^{i}$ are the coefficients in the Taylor expansion of (6-7).

In particular, when all the random noises are absent, (6-9) must furnish the same solution as (1-2) does when all random noises are absent. This implies

$$h_1{}^i(t, x) = g_1{}^i(t, x). \tag{6-10}$$

Next, if substitute $z^\rho(t) = t$ for some one ρ and $z^\sigma(t) = 0$ for $\sigma \neq \rho$, because of (6-10) Eq. (6-9) reduces to

$$x^i(t) = x^i(a) + \int_a^t [g_1{}^i(s, x(s)) + h_\rho{}^i(s, x(s))]\, dt,$$

while (1-2) reduces to a similar equation but with $g_\rho{}^i$ in place of $h_\rho{}^i$. Since the two must have the same solutions, we must have

$$h_\rho{}^i(t, x) = g_\rho{}^i(t, x) \qquad (\rho = 1, \ldots, r;\quad i = 1, \ldots, n). \tag{6-11}$$

Even though the z^ρ are Lipschitzian, the increment of z^ρ over a time interval $[t, u]$ is in many applications the sum of a number of independent small inputs. If $v - u$ is large enough, and the arrival of the small inputs is homogeneous in time, the mean value of $z^\rho(u) - z^\rho(v)$ will be a multiple of $v - u$, and the distribution will be nearly normal, with variance proportional to $v - u$. Hence, after subtracting off a linear function of t, the increments of the noise will be very well approximated by the increments of a Brownian motion, provided however that $v - u$ is large enough to allow for the arrival of a great number of the individual small inputs. To save repetition, we shall suppose that each z^ρ is replaced by $c_\rho t + (z^\rho(t) - c_\rho t)$, where $c_\rho(u - t) = E(z^\rho(u) - z^\rho(t))$. Then the $c_\rho\, dt$ integral can be combined with the first integral in (6-9) and z^ρ replaced by $z^\rho - c_\rho t$; that is, we may as well suppose $E(z^\rho(u) - z^\rho(t)) = 0$. With this convention, we see that

(6-12) the $x_\Pi{}^i(t_j)$ corresponding by (6-8) to the actual noises $z^2, \ldots, z^r$ are arbitrarily close to those corresponding to some suitably chosen Brownian motion processes $z^2, \ldots, z^r$, provided that the length of the shortest interval $[t_j, t_{j+1}]$ of Π is not less than a corresponding small positive δ_1.

Since it is the differential equations that we wish to use rather than the approximations (6-8), we wish to choose coefficients in (6-9) so that the solutions with the actual noises will be closely approximated by those with the Brownian-motion idealizations. In order to estimate the closeness of the approximation we form the Cauchy–Maruyama approximations x_Π for the two systems. If these x_Π are close to the solutions of (6-9) for both systems of z^ρ, and are also close to each other for the two systems, then the solutions of (6-9) for the two systems will be near each other. If we wish $\| x_\Pi(t) - x(t) \|$ to remain under a given ϵ at each division point t_j of Π, by (6-2) we must in general require

(6-13) $$\text{mesh } \Pi \leqq (\epsilon/A_2)^2.$$

But we must also have the x_Π for the two systems close to each other, and by (6-12) this demands that all the intervals $[t_j, t_{j+1}]$ have length *at least* δ_1. This may easily be incompatible with (6-13).

On the other hand, if Eqs. (6-9) are in canonical form and the x_Π are the Cauchy–Maruyama approximations (4-7) to their solution, then if the hypotheses of Theorem (4-8) are satisfied we have by (6-1) that in order to have $\| x_\Pi(t_j) - x(t_j) \| < \epsilon$ at all division points t_j of Π, it is sufficient to require

(6-14) $$\text{mesh } \Pi < \epsilon/A_1.$$

If A_1 and A_2 are of comparable size, for small ϵ this is much less restrictive on mesh Π than (6-13) is, and may be compatible with (6-12) when (6-13) is not. Hence, when the hypotheses of Theorem (4-8) are satisfied, there is a strong reason for choosing the $h^i_{\rho\sigma}$ so that (6-9) is in canonical form. We cannot be confident that this will produce adequately accurate approximations to the statistics of the solutions corresponding to the physically plausible Lipschitzian noise functions. But we can feel very confident that if the canonical extension is not good enough, neither is any other.

All of this is subject to the assumption that the hypotheses of Theorem (4-8) are satisfied, or more specially that there is only one random noise z^2.

If this is not the case, let us investigate the solutions of Eqs. (1-2) when the z^ρ all satisfy

$$z^\rho(t, \omega) = c^\rho Z(t, \omega) \qquad (\rho = 2, \ldots, r), \tag{6-15}$$

where $c^2, \ldots, c^r$ are real numbers and $Z(\mathbf{t}, \omega)$ satisfies the standing hypotheses V(1-4). In this case (1-2), or the more traditional equation II(1-5), can be written in the form

$$x^i(t) = x^i(a) + \int_a^t f^i(s, x(s))\, ds + \int_a^t G_1{}^i(s, x(s))\, dZ^1(s), \tag{6-16}$$

where

$$G_1{}^i(t, x) = \sum_{\rho=2}^{r} c_\rho g_\rho{}^i(t, x).$$

The equation corresponding to (6-16) that has the advantages of the canonical extension is the equation obtained by adding the terms

$$\frac{1}{2}\int_a^t \sum_h G^i{}_{1,x^h}(s, x) G_1{}^h(s, x)\ (dZ^1(s))^2 \tag{6-17}$$

to the right member. But (6-17) is the same as

$$\frac{1}{2}\sum_{\rho,\sigma=2}^{r}\sum_{h=1}^{n}\left(\int_a^t g^i{}_{\rho,x^h}(s, x) g_\sigma{}^h(s, x)\ (dZ^1(s))^2\right) c_\rho c_\sigma$$

$$= \frac{1}{2}\sum_{\rho,\sigma=2}^{r}\int_a^t g^i_{\rho\sigma}(s, x)\, dz^\rho\, dz^\sigma.$$

In particular, when (6-15) holds and Z is a standard Wiener process, the estimate for $E(x^i(t))$ in which we can feel greatest confidence satisfies

$$E(x^i(t)) = E(x^i(a)) + \int_a^t E(f^i(s, x(s))\, ds + \frac{1}{2}\sum_{\rho,\sigma=2}^{r} c^\rho c^\sigma \int_a^t E(g^i_{\rho\sigma}(s, x))\, ds. \tag{6-18}$$

If the initial value of x is sure, the derivative of $E(x^i(\mathbf{t}))$ at $t = a$ is

$$f^i(a, x(a)) + \frac{1}{2}\sum_{\rho,\sigma=2}^{r} c^\rho c^\sigma g^i_{\rho\sigma}(s, x(a)). \tag{6-19}$$

Had we chosen to use, instead of the canonical extension of (1-2), some other extension with other coefficients $h^i_{\rho\sigma}$ in place of $g^i_{\rho\sigma}$, for the estimate of the derivative of $E(x^i(t))$ at $t = a$ we would have obtained an expression like (6-19) but with $h^i_{\rho,\sigma}$ in place of $g^i_{\rho\sigma}$. But unless this is to be in disagreement with the most trustworthy estimate (6-19), we must have

$$\sum_{\rho,\sigma=2}^{r} c^\rho c^\sigma h^i_{\rho\sigma}(a, x(a)) = \sum_{\rho,\sigma=2}^{r} c^\rho c^\sigma g^i_{\rho\sigma}(a, x(a)).$$

Since $c^2, \ldots, c^r$ are arbitrary this implies that at $t = a$ and $x = x(a)$ we have

$$\text{(6-20)} \qquad h^i_{\rho\sigma}(t, x) + h^i_{\sigma\rho}(t, x) = g^i_{\rho\sigma}(t, x) + g^i_{\sigma\rho}(t, x)$$

for $i = 1, \ldots, n$ and $\rho, \sigma = 1, \ldots, r$. But we could use any t for initial time a and any x for initial value $x(a)$, so (6-20) holds for all t and x, and the extension with $h^i_{\rho\sigma}$ terms is merely an equivalent form of the canonical extension.

This argument leads to the conclusion that if we wish to idealize a model by using Brownian-motion noises, no extension of the classical extension deserves any confidence except perhaps the canonical extension. But when there are several noises and the equation

$$g^i_{\rho\sigma} = g^i_{\sigma\rho}$$

does not hold identically, we should be suspicious even of the canonical extension. In such cases it seems quite possible that no white-noise idealization will be satisfactorily close to the physically realized system. As we remarked at the end of Section 5, a satisfactory resolution of this difficulty might have to wait a deep analysis of the limitations on the use of differential equations in classical models of deterministic systems.

References

[1] BALAKRISHNAN, A. V., *Communication Theory* (Inter-University Electronics Series, Vol. 6). McGraw-Hill, New York, 1968.

[2] BLANC-LAPIERRE, A., and FORTET, R., *Theory of Random Functions* (J. Gani, transl.). Gordon and Breach, London, 1967.

[3] DOOB, J. L., *Stochastic Processes*. Wiley, New York, 1953.

[4] ITÔ, K., On a formula concerning stochastic differentials, *Nagoya Math. J.* **3** (1951) 55–65.

[5] KUNITA, H., and WATANABE, S., On square integrable martingales, *Nagoya Math. J.* **30** (1967) 209–245.

[6] LAMPERTI, J., A simple construction of certain diffusion processes, *J. Math. Kyoto Univ.* **4** (1964) 161–170.

[7] LOÈVE, MICHEL, *Probability Theory*. Van Nostrand, Princeton, New Jersey, 1960.

[8] MARUYAMA, G., Continuous Markov processes and stochastic equations, *Rend. Circ. Mat. Palermo* **4** (1955) 48–90.

[9] MCKEAN, H. P., JR., *Stochastic Integrals*. Academic Press, New York, 1969.

[10] MCSHANE, E. J., A Riemann-type integral that includes Lebesgue-Stieltjes, Bochner and stochastic integrals, Memoirs of the Amer. Math. Soc. No. 88, 1969.

[11] NEVEU, J., *Mathematical Foundations of the Calculus of Probability* (A. Feinstein, transl.). Holden-Day, San Francisco, 1965.

[12] SRINIVASAN, S. K., and VASUDEVAN, R., *Introduction to Random Differential Equations and their Applications*. Amer. Elsevier, New York, 1971.

[13] WARFIELD, V. M., *A stochastic maximum principle*, doctoral dissertation, Brown University, 1971.

[14] WIENER, N., Differential-space, *J. Math. Phys.* **2** (1923) 131–174.

[15] WONG, E., and ZAKAI, M., On the relation between ordinary and stochastic differential equations, *Internat. J. Engng. Sci.* **3** (1965) 213–219.

[16] WONG, E., and ZAKAI, M., On the convergence of ordinary integrals to stochastic integrals, *Ann. Math. Statist.* **36** (1965) 1560–1564.

[17] WONHAM, W. M., Some applications of stochastic differential equations to optimal non-linear filtering, *SIAM J. Control* **2** (1965) 347–369.

Subject Index

D

E

F

H

I

L

M

N

P

A 4
B 5
C 6
D 7
E 8
F 9
G 0
H 1
I 2
J 3